Examining an Operational Approach to Teaching Probability

Alessio Drivet
Geogebra Institute of Turin, Italy

A volume in the Advances in Educational Technologies and Instructional Design (AETID) Book Series

IGI Global
PUBLISHER of TIMELY KNOWLEDGE

Published in the United States of America by
IGI Global
Information Science Reference (an imprint of IGI Global)
701 E. Chocolate Avenue
Hershey PA, USA 17033
Tel: 717-533-8845
Fax: 717-533-8661
E-mail: cust@igi-global.com
Web site: http://www.igi-global.com

Library of Congress Cataloging-in-Publication Data

Names: Drivet, Alessio, 1948- author.
Title: Examining an operational approach to teaching probability / by Alessio Drivet.
Description: Hershey, PA : Information Science Education, 2021. | Includes bibliographical references and index. | Summary: "This book elaborates on the concept of probability and is dedicated to a particular aspect of the issue: probability examined from an educational point of view informing readers how to implement a very concrete operational approach in the classroom"-- Provided by publisher.
Identifiers: LCCN 2020042219 (print) | LCCN 2020042220 (ebook) | ISBN 9781799838715 (hardcover) | ISBN 9781799872474 (paperback) | ISBN 9781799838722 (ebook)
Subjects: LCSH: Probabilities--Study and teaching (Higher) | Mathematics--Study and teaching (Higher)
Classification: LCC QA273.2 .D75 2021 (print) | LCC QA273.2 (ebook) | DDC 519.2071--dc23
LC record available at https://lccn.loc.gov/2020042219
LC ebook record available at https://lccn.loc.gov/2020042220

This book is published in the IGI Global book series Advances in Educational Technologies and Instructional Design (AETID) (ISSN: 2326-8905; eISSN: 2326-8913)

British Cataloguing in Publication Data
A Cataloguing in Publication record for this book is available from the British Library.

Advances in Educational Technologies and Instructional Design (AETID) Book Series

ISSN:2326-8905
EISSN:2326-8913

Editor-in-Chief: Lawrence A. Tomei, Robert Morris University, USA

MISSION

Education has undergone, and continues to undergo, immense changes in the way it is enacted and distributed to both child and adult learners. In modern education, the traditional classroom learning experience has evolved to include technological resources and to provide online classroom opportunities to students of all ages regardless of their geographical locations. From distance education, Massive-Open-Online-Courses (MOOCs), and electronic tablets in the classroom, technology is now an integral part of learning and is also affecting the way educators communicate information to students.

The **Advances in Educational Technologies & Instructional Design (AETID) Book Series** explores new research and theories for facilitating learning and improving educational performance utilizing technological processes and resources. The series examines technologies that can be integrated into K-12 classrooms to improve skills and learning abilities in all subjects including STEM education and language learning. Additionally, it studies the emergence of fully online classrooms for young and adult learners alike, and the communication and accountability challenges that can arise. Trending topics that are covered include adaptive learning, game-based learning, virtual school environments, and social media effects. School administrators, educators, academicians, researchers, and students will find this series to be an excellent resource for the effective design and implementation of learning technologies in their classes.

COVERAGE

- Higher Education Technologies
- Online Media in Classrooms
- Classroom Response Systems
- Digital Divide in Education
- K-12 Educational Technologies
- Curriculum Development
- Adaptive Learning
- E-Learning
- Instructional Design Models
- Web 2.0 and Education

The Advances in Educational Technologies and Instructional Design (AETID) Book Series (ISSN 2326-8905) is published by IGI Global, 701 E. Chocolate Avenue, Hershey, PA 17033-1240, USA, www.igi-global.com. This series is composed of titles available for purchase individually; each title is edited to be contextually exclusive from any other title within the series. For pricing and ordering information please visit http://www.igi-global.com/book-series/advances-educational-technologies-instructional-design/73678. Postmaster: Send all address changes to above address.

Titles in this Series

For a list of additional titles in this series, please visit: http://www.igi-global.com/book-series/advances-educational-technologies-instructional-design/73678

Teaching and Learning Practices That Promote Sustainable Development and Active Citizenship
Sandra Saúde (Polytechnic Institute of Beja, Portugal & CICS, NOVA University Lisbon, Portugal) Maria Albertina Raposo (Polytechnic Institute of Beja, Portugal & MARE, NOVA University Lisbon, Portugal) Nuno Pereira (Polytechnic Institute of Beja, Portugal) and Ana Isabel Rodrigues (Polytechnic Institute of Beja, Portugal)
Information Science Reference • © 2021 • 383pp • H/C (ISBN: 9781799844020) • US $185.00

Designing, Deploying, and Evaluating Virtual and Augmented Reality in Education
Gokce Akcayir (University of Alberta, Canada) and Carrie Demmans Epp (University of Alberta, Canada)
Information Science Reference • © 2021 • 404pp • H/C (ISBN: 9781799850434) • US $185.00

IT and the Development of Digital Skills and Competences in Education
Patricia Ordóñez de Pablos (University of Oviedo, Spain) Miltiadis D. Lytras (American College of Greece, Greece & Effat University, Saudi Arabia) and Xi Zhang (Tianjin University, China)
Information Science Reference • © 2021 • 332pp • H/C (ISBN: 9781799849728) • US $195.00

Fostering Meaningful Learning Experiences Through Student Engagement
Despo Ktoridou (University of Nicosia, Cyprus) Elli Doukanari (University of Nicosia, Cyprus) and Nikleia Eteokleous (Frederick University, Cyprus)
Information Science Reference • © 2021 • 325pp • H/C (ISBN: 9781799846581) • US $195.00

Assessment Tools for Mapping Learning Outcomes With Learning Objectives
G. R. Sinha (Myanmar Institute of Information Technology, Mandalay, Myanmar)
Information Science Reference • © 2021 • 327pp • H/C (ISBN: 9781799847847) • US $185.00

IGI Global
PUBLISHER of TIMELY KNOWLEDGE

701 East Chocolate Avenue, Hershey, PA 17033, USA
Tel: 717-533-8845 x100 • Fax: 717-533-8661
E-Mail: cust@igi-global.com • www.igi-global.com

Table of Contents

Preface

Sure, very sure, indeed probable. ~ Ennio Flaiano (2014)

The term probability derives from the Latin *probabilĭtas* which initially meant "worthy of approval" and only later began to be used in the sense in which we understand it (Ventsel, 1969). With a little forcing it is also possible to find a certain link with the word *prophet*. The term derives from the ancient Greek: πρόφημι with the meaning of foretell, predict (Waite, 2009). For prophecy to be worthy of approval it is necessary to find a basis, a valid motivation. Whoever studies and applies probability, after all, tries to formulate a prediction and is, in his own way, a prophet.

This book is, however, dedicated to a particular aspect of the problem: the probability examined from an educational point of view. Books and articles have been written on the subject in large quantities, so it is not easy to find new ideas, but there is a new aspect that perhaps can be taken into consideration: to realize a very concrete operational approach in the classroom.

It all started, several years ago, from the consideration of the inadequacy of a traditional approach to teaching mathematics. Many teachers and a not small part (perhaps a majority?) of the students often realize something is wrong and report a lack of enthusiasm in dealing with discipline.

At this point the key question must be asked: why?

The explanations are manifold. Many teachers think that certain established habits have a serious pedagogical basis and therefore it is difficult to question them. In addition, perhaps, there is also a certain fear in imagining and experimenting with new ways.

From the student's point of view, both the economic principle of minimizing costs is highly developed, so it is better to walk on a safe path rather than enter dangerously unexplored terrain, and there are hesitations based on psychological elements.

Unfortunately, the excessive use of examples and abstract formulations with exclusive reference to algebraic language distances the student from the pleasure of the discipline. Mathematics, on the other hand, requires attention and concentration, but the understanding of its meaning gives rise to interest, pleasure to discover and promotes deep learning.

These reflections and the long teaching experience explain my attempt to introduce mathematics from *objects* that students could recognize and were quite easily found. The experimentation has progressively expanded involving students of different ages and the "mathematical objects" proposed are now more than 250.

The research, initially not easy, has turned out over time to be quite simple; a trained eye is able to find at home, in a toy shop window, in pictures on the Internet, etc., an object from which to introduce the desired topics. This is not a trivial job, but it is a source of great satisfaction, especially when an unusual and unexpected link is found. Of course, all this would be impossible without the help provided, directly or indirectly, by the publications of great specialists and disseminators. The attentive reader will surely find many well-known examples, some of them presented in a different way than usual. Sometimes originality consists in collecting stimuli and proposals and arranging them in an organic framework, just as dozens of pearls can make up a necklace.

The greatest gratification consisted in the success of this approach, which, perhaps, someone could define *naïve*. Adjective that indicates behavior perhaps ingenuous, but certainly simple, unsophisticated, and therefore more acceptable in a school situation.

What is often missing in the teacher-student relationship is the ability of the former to teach "know how to see" in mathematics. The importance of this perspective is well represented by a story, a sort of fairy tale, which we could call "the hand of the princess". Here is the simple plot: a King has a beautiful daughter, a princess whose hand is disputed by a large number of suitors, but the girl refuses any approach and remains silent and sad. The King invites everyone to court and promises to give his daughter in marriage to whoever will make her smile. The suitors perform in attempts of seduction: who sings, who performs acrobatics, which does magic, who creates a shower of lights, the person making flights majestic ... Nothing. The face of the princess does not reveal any emotion. The King, desperate, no longer knows what to do until the last of the suitors takes a pair of glasses from his cloak and hands them to the princess who wears them. Finally she smiles! Who does not have the right tools to see cannot appreciate the proposed topics. The teacher's task is to supply the glasses.

One of the fields in which it is particularly urgent to provide "the right glasses" to students is the probability. In fact, in Italy, but perhaps the phenomenon is also widespread elsewhere, issues of uncertainty and data processing are often relegated to a marginal position.

This book can be considered the natural development of an elaboration that, in part, I have already had the opportunity to expose (Drivet, 2017).

Our world is full of situations that can generate interpretative and predictive doubts: earthquakes, traffic congestion, election predictions, sports results, nuclear power plant failures, errors in the transmission of information, etc. The world is also a boil of data. Probabilistic modelling is the key to analyzing the information we have and making predictions. The school (and those involved in education) is called to provide, before any other institution, the appropriate tools to move in this landscape.

If you look at the real situation, however, you will notice that probability does not have a large space in the students' curricula. It is therefore necessary to make a paradigm shift in the field of teaching. In a 2009 TED talk, Arthur Benjamin (2009) stated: "I mean, probability and statistics, it's the mathematics of games and gambling. It's analyzing trends. It's predicting the future. Look, the world has changed from analog to digital. And it's time for our mathematics curriculum to change from analog to digital, from the more classical, continuous mathematics, to the more modern, discrete mathematics – the mathematics of uncertainty, of randomness, of data -- that being probability and statistics".

However, this position is still insufficient. The main problem is well posed, but a fundamental question remains which can be summarized in a question: how to teach probability?

The answer is complex because there are no recipes, but, perhaps, it is possible to formulate some good working hypotheses. The focus of the book is right here, the guiding hypothesis rests on two pillars: concrete objects should be presented and used and IT tools used to perform simulations.

The Mathematics Framework of the OECD (2018). Pisa 2018 project introduces the theme Uncertainty and data:

In science, technology and everyday life, uncertainty is a given. Uncertainty is therefore a phenomenon at the heart of the mathematical analysis of many problem situations, and the theory of probability and statistics as well as techniques of data representation and description have been established to deal with it. The uncertainty and data content category includes recognising the place of variation in processes, having a sense of the quantification of that variation, acknowledging uncertainty and error in measurement, and knowing about chance. It also includes forming, interpreting and evaluating conclusions drawn in situations where uncertainty is central. The presentation and interpretation of data are key concepts in this category.

There is uncertainty in scientific predictions, poll results, weather forecasts and economic models. There is variation in manufacturing processes, test scores and survey findings, and chance is fundamental to many recreational activities enjoyed by individuals. The traditional curricular areas of probability and statistics provide formal means of describing, modelling and interpreting a certain class of uncertainty phenomena, and for making inferences. In addition, knowledge of number and of aspects of algebra, such as graphs and symbolic representation, contribute to facility in engaging in problems in this content category. The focus on the interpretation and presentation of data is an important aspect of the uncertainty and data category.

In commenting on the statements of OECD Pisa, it should be emphasized that a distinction should be made between uncertainty and probability. The first term refers mainly to a limited state of knowledge because information is lacking or because it is not easy to measure phenomena and, consequently, predict their evolution. The probability, however, considers events related to the case in which an estimation model must be used. Therefore, probability theory can be used to make predictions based on partial information. To explain the difference between the two concepts in a simple way, an example can be useful: we have an urn with 100 red and white balls in an unknown proportion whose composition we would like to know. We are in a situation of total *uncertainty* due to lack of information, we can only say that the probability of having k white balls is k/100, with k between 0 and 100. If we now draw a white ball we can certainly exclude k = 0 but it also changes the *probability* that at the next draw, after having put the ball back, it is still white.

Probability is therefore an important tool in many situations involving uncertainty. Two examples can give a first idea of what this means.

The first refers to Artificial Intelligence, that field of computer science dedicated to the solution of cognitive problems such as learning, problem solving and pattern recognition. Its progress has improved the ability to interpret and translate a text thanks to probabilistic analysis. For example *Word2vec* is an artificial neural network designed to process natural language: the algorithm requires a collection of selected and organized texts as input, for each word contained in this corpus a vector is built that represents the semantic distribution of words in the text. In this space words will be closer if recognized as more similar. The coding is formulated with a number between 0 and 1, i.e. a probabilistic value is given. The model has no rules or previous knowledge, it only knows the statistics and probability rules. Thanks to these new paradigms we have gone from 75-80% accuracy of speech recognition (before 2000) to today's excellent results.

A second example is related to the health field: it is known that the high presence of cholesterol in the blood can cause vascular problems, having detected a certain cholesterol value, what is the probability of thrombosis? The solution can be found by applying the theorem identified by the reverend/ mathematician Thomas Bayes (1702-1761), analyzed extensively in the first chapter.

The calculation of probabilities often involves the use of experiments but extracting a series of balls, flipping a few coins, or pulling cards from a deck is not only boring, but sometimes not so indicative for understanding the disciplinary structure. Probabilistic experiments require repeated and numerous simulations. To do this, knowledge of programming or software skills are not essential: a simple Internet search allows you to find simulators of coin tosses, dice, darts, etc.. Using these tools students approach softly non-trivial concepts such as the *Law of large numbers* (for a large number of tests the empirical frequency of an event converges to the theoretical probability) or the *Monte Carlo method* (based on an algorithm used to draw estimates through simulations). They can study phenomena, formulate conjectures, and perform experiments.

In the text, therefore, the teaching of probability will rely heavily on an operational approach.

Before continuing, it is useful to formulate some observations that could be made by any teacher:

1. In school, despite appearances, the approach is "axiomatic", i.e. rules are transmitted. Algorithmic and/or conceptual learning is often preferred. The student is assessed for his ability to memorize definitions, to know solution patterns, to acquire automatisms, to know how to perform procedures, etc.
2. Our life is largely dominated by chance. Just think of phenomena such as meeting a soul mate, misfortunes like as road accidents, hobbies such as gaming or events such as equity investments. Instead, in school little time is devoted to random situations. Perhaps we should reflect a little on this inconsistency. Recall that the term "aleatory" comes from the

Latin name of the dice, *alea*, *a* typical object whose throws produce unpredictable results.

3. Teaching about probability is often limited to the proposition of exercises based on extractions (cards, balls) or throws (coins, dice) when the world we live in is full of randomness from which more interesting concrete examples can be drawn. Why do we feel that traffic lights are always red? Why is it so difficult to win at lotteries? How reliable are the weather forecasts? How do the giants of the network understand and orient our tastes? Why, according to Eastaway and Windham (1998), do the buses always arrive three at a time? In teaching it is correct to start from simple, understandable, and easily modelled situations, but then you have to go further, taking into account the new contexts in which today's students move. It is what it says, among the others, Dave Pratt (2011), Professor of Mathematics Education: "Of course, if the modelling meaning of probability was stressed in the curriculum, it is debatable whether there is much advantage in maintaining the current emphasis on coins, spinners, dice and balls drawn from a bag. Perhaps, in days gone by when children played board games, there was some natural relevance in such contexts but, now that games take place in real time on screens, probability has much more relevance as a tool for modelling computer-based action and for simulating real-world events and phenomena".

Very often human beings tend to trust their own intuitions, but probability theory teaches us that calculation can, not infrequently, contradict what we call "common sense".

In everyday life, we often use the concept of probability intuitively and with a wide margin of inconsistency, but then we have to think again later on. As an example, consider the prediction for the output of red or black in the roulette game. As it should be known not only to gamblers, roulette is a casino game in which there are 18 red and 18 black numbers that are "drawn" from a ball that stops, after a random path, on one of these 36 numbers placed on a wheel. Suppose that red numbers have been drawn six times in a row, the player's reasoning will be: now black numbers have to come out to compensate for this series. This error is called *Gambler's fallacy*, it concerns the mistaken belief that past random events affect future events: according to many players a random event is more likely to occur because it has not occurred for a certain period of time. The reasoning can be summarized as follows: "In the long run, the number of reds and blacks will be the same, so blacks now have to go out to recover. The mistake of this reasoning is that the proportion of reds is true that it is close to half of the cases, but the number of blacks is not necessarily close to the number of reds. There are also other forms of fallacy, for example, the player might think that a random event is more likely to happen because it occurred recently and therefore could continue to invest money on the output of red. In reality roulette has no memory and therefore each result represents a unique event. To be convinced of this, all you have to do is what scientists like Einstein (1879-1955) called *Thought Experiment*, a term coined by the Danish physicist and chemist Hans Christian Ørsted (1777-1854); suppose that before the daily closing of the casino the ball stopped six times on a red number. The next day the probability that the first number generated by the roulette wheel is a red number is the same as the previous day.

In a complex society, probability applies in many fields: gaming, insurance, election forecasts, stock market trends, medical diagnostics, cryptography, weather forecasts.

Let's start with the weather forecast: when we say " I don't take an the umbrella, it won't rain today" probability is being expressed as a degree of confidence in the occurrence of a given event, by which I mean a fact that can happen. However, this is not enough to clear the field of some misunderstandings. When a meteorologist says that "tomorrow the probability of rain is 50%", what do we understand? That it will rain 50% of the time? That it will rain for half a day? That it will rain on half the territory concerned? That you have to flip a coin to decide whether or not to go out with an umbrella? More simply, it means that out of 100 simulations carried out by experts in the field based on the data collected, we can expect 50 cases of rain as a result.

It should be added that official sources also often contribute to confusion. Cognitive scientist Gerd Gigerenzer reports a press release stating that taking a highly prescribed medication for patients with high cholesterol lowered the risk of death by 22%. On a superficial reading, it was thought that out of 1000 people, 220 would benefit. A significant increase in the probability of survival! In fact, out of 1000 patients observed, there were 32 deaths among those who took the drug and 41 deaths among those who took a placebo. A benefit, but small if you consider the absolute values.

Before actually addressing the subject in concrete issue, however, it will be useful to provide some food for thought.

The term probability has many synonyms: likelihood, likeliness, prospect, expectation, chance, odds, possibility. In the text we have tried to use as little as possible this variety of expressions in order to reserve them for specific situations and therefore the term probability is the most used.

The first three chapters analyze the three main definitions of probability: classical or a priori, frequentist and subjective. By accepting the terminology of Borovcnik and Kapadia (2016) APT (a priori theory), FQT (frequentist theory) and SJT subjectivist theory, we agree with the statement: "We also firmly advocate that the three approaches of APT, FQT, and SJT have to be presented collectively when teaching ideas of probability in the curriculum from primary school". The axiomatic aspect proposed in 1933 by Andrey Nikolaevich Kolmogorov (1903-1987) is not treated, this choice derives from the need to focus more on the didactic aspects than on the formal ones. In fact, this latter approach does not give a direct definition of probability, but operationally accepts the previous three, as long as they respect the fundamental properties, assumed as axioms, from which to deduce the other properties as theorems.

Each chapter is devoted to an essential concept related to probability. It starts with an introduction, then various objects and examples are presented, which allow to deepen the theme and, in general, to open a new horizon of research. The chapters can be considered as autonomous and monothematic didactic units in which objects act as a starting point to mix theory and problems.

In the introduction we begin to mention the Law of large numbers and the Monte Carlo Method and the Gambler's fallacy and Thought Experiment keywords.

The first three chapters follow the historical development of the discipline and the timing suggested by the school curricula.

In Chapter 1 the fundamental aspects of probability are initially illustrated using objects such as Coins, Syringe, Marbles, Dice, Etruscan dice, Playing cards, but then taking into consideration less usual artifacts such as Octopus, Birthday cake, Candles, Mouse, Chess, Traffic light, Spinner, Egg surprise and Lego. In particular, starting from Coins some basic terms are introduced: Trial, Events, Random Variable, Sample space, Opposite Event, Certain Event, Impossible Event, Opposite Probability, Parameters. The fundamental laws are mentioned below: Law of Compound Probabilities, Law of Total Probabilities, Law of truly large numbers, Law of Conditional Probabilities, Empirical law of chance, Bayes' theorem, Confusion of the inverse.

In Chapter 2 the frequentist approach is treated with the help of a series of unconventional objects: Drawing pins, Spinning top, Dreidel, Token, Toast, Lu-Lu, Pass the Pigs, Astragalus, Watch, Needle. From these examples it is possible to approach the Empirical law of chance at the Expected value and the concept of Aleatory variable (also known as Random variable). The latter concept is so important as to make the famous Anglo-American mathematician David Mumford (1999) say: "I'd like to propose instead that it should be possible to put random variables into the very foundations of both logic and mathematical and arrive at a more complete and more transparent formulation of the stochastic point of view".

Chapter 3 refers to subjective probability. In this case, only six artifacts are considered: Tea, Playing card, Tap, Finger Football, Rain gauge and Tulip. The proposed objects are fewer in number than in the previous chapters because some have been used in the previous chapters. In addition, it is not always possible to find an object that serves as a stimulus to deal with this type of problem. In dealing with the subjective probability we meet some important concepts: Rule of three, Rule of four, Dutch book, Kelly criterion, Mathematical Expectation, Expected value criterion, Pay-off table, Fair game.

Chapter 4 deals with the relationship between probability and combinatorics. Here you can find many example objects: Tic-tac-toe, Smartphone, Snow White, Landscapes, Palette, Beagle Boys Mug, 15 Puzzles, Padlock, T-shirts, Poker, Stamps, Keychain. Starting from these realia are progressively introduced: Tree model, Pigeonhole principle, Simple permutations, Derangement, Permutations with repetition, Partial permutations, Combinations.

Chapter 5 opens the door to more playful aspects. Risk, Clock solitaire, Slot machine, Roulette, Blackjack, Revolver, Backgammon introduce, besides contents already seen, concepts of relevant importance such as Sunk cost bias, Player's ruin, and Expectation. The chapter offers the opportunity to concretely present the Monte Carlo method, named after the Monte Carlo Casino, symbol of gambling par excellence. It is a numerical method based on probabilistic procedures, used for solving problems with particular analytical difficulties.

Chapter 6 examines some of the main probability distributions using: Lottery, Pin, Spaghetti, Chessboard, Quincunx, Win for Life, Bottle Bulb, Toothbrush. The probability distribution is a model that associates a probability to each observable mode of a random number (or, with a more commonly used term, a random variable). The distributions examined are: Uniform, Singular, Binomial, Hypergeometric, Poisson, Rectangular, Normal, Standard Normal.

Chapter 7 shifts the focus to possible strategies that can be adopted in the presence of random events. Perudo, Elephant, Rock-paper-scissors, Matchbox, Naval battle, Darts, Monopoly, Poker says, Hexapawn, Road sign, Map are the games examined and, in this case, new terms or concepts are introduced such as Prisoner's dilemma, Tit for Tat, Travelling salesman problem, Simulated annealing.

Chapter 8 is characterized by the rediscovery of some previously used objects and the addition of new elements: Playing cards, Dice, Darts, Chessboard, Bus, Football stickers, Meterstick, Phone, Soccer ball, Planner, Polyphemus, Toucan, Snakes and Ladders. The theme of the chapter concerns the use of simulations as a fundamental tool for probability teaching. In the introduction of the chapter, a fundamental theme is mentioned, the generation by computers of random numbers that are in reality Pseudo-Random Numbers. Other concepts covered are Reed-Frost model, Median, Sample space. Another important topic is the one related to Markov chains.

Chapter 9 focuses on the relationship between probability and statistics. This theme would require further study and, certainly, a dedicated book. Several objects are presented: Book, Mini soccer, Love drops, Basketball, Table of logarithms, Chocolate, Buttons; some of them serve to introduce both Zipf's law and Gini's index. There is also room for two significant theoretical references: Simpson Paradox and Benford's law. Transitional matrix and Probabilistic sampling complete the framework.

Chapter 10 represents a digression from the theme, in fact it deals with chaos, a subject only apparently classifiable among the events related to *alea*. The objects present are: Billiard, Decision Maker, Pirate, Hexbug, Breadstick, Ants, Beer. The new concept introduced refers to Chaos, Random walk, and Logistic map. It would have been interesting to broaden the discourse to fractals, but this would have taken us too far from the objective.

REFERENCES

Benjamin, A. (2009). Ted Talk [Video]. ted.com/talks/arthur_benjamin_teach_statistics_before_calculus/transcript?language=en

Borovcnik, M., & Kapadia, R. (2016). *Reasoning with risk–A survival guide*. Academic Press.

Drivet, A. (2017). Eight tips for the theme Data and Forecasts. In *Encyclopedia of Information Science and Technology* (4th ed.). IG Global.

Eastaway, R., & Windham, J. (1998). *Why Do Buses Come in Threes?* The Hidden Maths of Everyday Life. Robson Books.

Flaiano, E. (2014). *Diario notturno* (Vol. 1). Adelphi Edizioni spa.

Mumford, D. (1999). The dawning of the age of stochasticity. *Mathematics: frontiers and perspectives*, 197-218.

OECD. (2018). Pisa 2021 mathematics framework. https://pisa2021-maths.oecd.org/

Pratt, D. (2011). Re-connecting probability and reasoning about data in secondary school teaching. Paper presented at 58th ISI World Statistics Congress, Dublin, Ireland.

Ventsel, E. S. (1969). *Probability theory*. Nauka.

Waite, M. (Ed.). (2009). *Oxford thesaurus of English*. Oxford University Press.

Introduction

The most important questions of life are indeed, for the most part, really only problems of probability. ~ Pierre Simon Laplace

PROBABILITY: A FASCINATING DISCIPLINE

According to Madhav Kaushish (2019): "The earliest known probability calculation is contained in a medieval European poem from around 1250 titled De Vetula, though the combinatorial techniques used were known in India and the Arab world much earlier (Bellhouse, 2000). De Vetula contains an account of a calculation regarding the odds of certain numbers appearing on the toss of two die".

Already in these words, we find an important starting point since ancient times humans have tried to understand phenomena difficult to control. The lightning bolt of lightning was under the control of divine entities, but the throwing of the dice could perhaps be on a human level. For a description of the interesting relationship between these two levels, we can refer to Florence Nightingale David (1962).

It is widely believed that the correspondence between Blaise Pascal (1623-1662) and Pierre de Fermat (1601-1665) gave rise to the first laws of probability and combinatorial calculus, but then the argument develops and from gambling comes to cover wider and wider fields. Like other branches of mathematics, the theory of probability presents an evident interweaving with the variety of its applications and an interesting historical development up to its axiomatization. The complexity of the problem is well summarized by Borovcnik & Kapadia (2012): "One key lesson from history is that probability has always been a pluralistic concept and has drawn its meaning from the interplay of its interpretations".

In the text the reader will find many hints at the main developments of probability calculation, but what we are interested in emphasizing is the extreme relevance of his study for the current century.

There are many examples and some of the most significant meanings are given below, here we would just like to remember how probabilistic models are the basis of the most recent and fascinating applications to artificial intelligence (self-driving cars, voice and image recognition, machine learning, social media sentiment analysis, etc.).

Fortunately now we have available computer tools (e.g.: Calc, R, KNIME among the free ones; Excel, SPSS, Minitab among the paid ones) and the possibility to obtain results both in the search for information and in forecasting open unthinkable perspectives.

Two simple examples:

1. We want to find information on Google, Google associates to the keyword search an index (quality score) on the relevance of the query, calculates the AdRank (max bid × quality score) and returns the results. When choosing keywords Google tries to predict the probability with which the user intends a certain meaning;
2. The Director of the Financial Products Unit wants to evaluate the effectiveness of employees in finding customers who will sign a bank deposit. The company database contains data for every phone call that employees make to potential customers. With data mining techniques it is possible to divide data into Training and Testing and with Decision Tree Learner and Predictor formulate predictions and evaluate the performance of the model.

PROBABILITY AS A LEVEL OF IGNORANCE AND NOT OF UNCERTAINTY

Probability is considered unanimously regarded as the science of uncertainty. More precisely we can say that it provides mathematical rules to try to overcome our ignorance. Probability does not tell us what the weather will be tomorrow or which winning football team to bet on; rather it gives us the framework in which we can move to make sufficiently sensible decisions.

According to the great mathematician Henry Poincaré (1854-1912) problems must be classified not on the basis of probability but on the degree of ignorance. Let's see what we are dealing with by distinguishing three possible levels:

First level: we have sufficient information and we know the laws that guide the phenomenon, so we can determine the outcome of the problem.
Example: We drop an object: we can predict that it will fall to the ground with a certain speed *v* after *n* seconds. The phenomenon is deterministic.

Second level: we know the laws, but we cannot determine the exact initial conditions.
Example: We flip a coin: we can write the equations of the motion of the coin. So, knowing the throw speed and the impressed rotation, we should be able to predict whether it will fall by turning the head or the tail upwards. That doesn't really happen. We can only estimate what is the probability that a given face will come out.

Third level: we do not check the initial status and the laws are unknown or undefined.
Example: We drop a drop of ink into a glass of water and we want to predict the behavior of the molecules. We can only predict the final effect. The stain gradually widens over time: gradually the solution takes on a uniform color. The time required for the solution to assume a uniform color depends on the speed of diffusion, which in turn is influenced by various factors.

The mixture of ignorance and uncertainty is a debated topic to which a much wider space should be devoted. Just look at Michael Smithson's (2012) graph (Figure 1) to realize the complexity of the problem. As the author states:" Uncertainty, as will become increasingly evident, occupies a special position as one of the most manageable kinds of ignorance. In some fields, the term is employed as a synonym for ignorance". Let's refer to the text cited in the references for those who wish to go deeper into the problem.

Figure 1. Taxonomy of ignorance

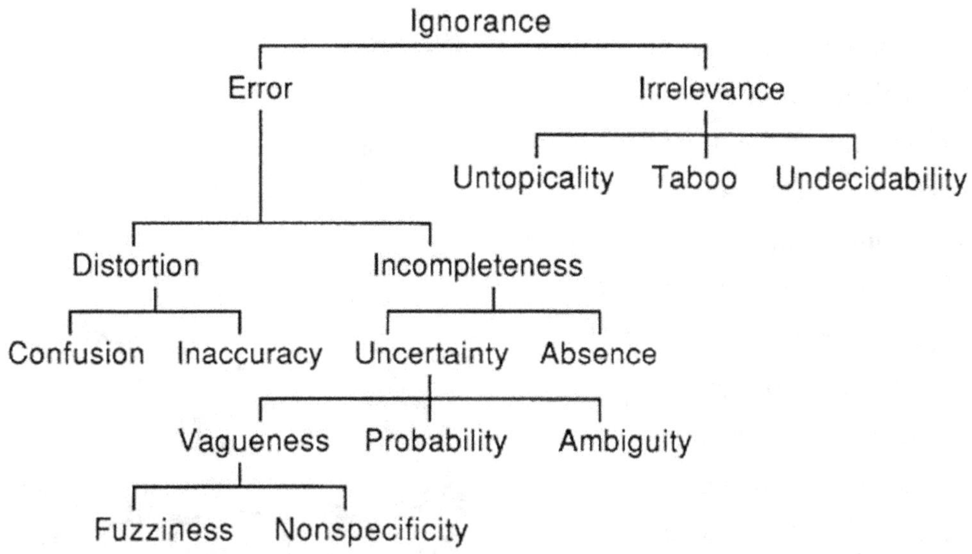

PROBABILITY AS THE ESSENCE OF THE REAL WORLD

In everyday life we make predictions about future events, as evidenced by the use of typical verbal expressions such as: "it is unlikely that", "this or that is the same for me", "I have good hopes that". The problem is that these terms are often interpreted by people in a non-univocal way. Numerous studies have shown that verbal terms of probability can have hugely different meanings. If a financial operator states in an interview that "interest rates are likely to rise in the short term" an investor could interpret the probable term as 80% and another as 60%. If a doctor says, "the operation is likely to be resolutive", can the patient feel very or slightly relieved?

Every day we are faced with choices that could be effectively guided by knowledge of probabilities. Being able to choose, for a trip, between various means of transport: which is the safest? The choice of the means of transport is normally influenced by factors such as time and cost, but also by safety. Taking a car, train or plane involves risks that can be assessed in probabilistic terms.

Theoretical knowledge helps us to evaluate the probabilities in the many situations where the result is not immediately clear and evident. In this way we can make correct decisions. In everyday life, of course, we find ourselves ignoring events whose probability is estimated to be low enough.

Figure 2. Examples of the use of probabilities

Bets: bookmakers' odds
Biological: genetics;
Computer science: treatment of big data;
Economic: insurance;
Environmental: weather forecasts;
Financial: stock performance;
Legal: assessment of the evidence;
Linguistic: cryptography;
Ludic: games of chance;
Management: queues;
Medical field: diagnostics;
Physical: statistical mechanics;
Technological: quality control.

To understand that the situations involving a probability are very numerous, it is sufficient to list some areas of use (arranged in alphabetical order) which, for the most part, will be treated in the text (Figure 2). We are already starting to present some examples in the list.

1. Betting (e.g. expected value and bookmaker shares)
   a. The expected value takes several names as the average or mathematical expectation of a random variable. It is a number that represents the idea of the average value of a random phenomenon. For example: we toss a coin and choose head, assuming a value of 1 for the victory and zero for the defeat (tail comes out). The expected value of the game is 0.5, which is the average of winnings and losses weighted according to the expected probability: $1 \times 0.5 + 0 \times 0.5 = 0.5$.
   b. The bookmaker plays the role of banker: he cashes the bettors' bets and, at the end of a sporting event (but not only), pays the winners. The betting agency makes an implicit estimate of the odds. The

fact that the dealer always wins is a consequence of the fact that the bookmaker's earnings are included in the odds.

2. Biological (e.g.: genetics and bioinformatics)
   a. Since the time of the studies on the pea crosses of the monk Mendel (1822-1884), the probability has been the basis of genetic research. For example, this simple question can be formulated: from the crossing between two heterozygous individuals, both with black hair but carrying the character of "blond" hair, what is the probability that a child with blond hair is born? The theoretical result tells us 1/4. Another example relates to a blood disease called Mediterranean anemia. Suppose two parents are both carriers of this hereditary disease: what are the chances of healthy or sick carriers?
   b. A more recent sector is bioinformatics, a rapidly expanding sector. The analysis of gene sequences allows to calculate the probability with which each amino acid undergoes a mutation.
3. Economic (e.g. insurance and econometrics)
   a. All people, at least from a certain age, have to do with insurance. Their history is ancient, and derives from a precise question: is it possible to measure and defend oneself against a risk? The primary function of insurance is that a person decides to hedge a risk in exchange for a premium. Whether it is life insurance, car insurance, insurance on the transport of goods, protection from theft, etc., the reference is to the probability that this risk becomes a reality. The calculation of probabilities gives an idea of how often that type of risk will occur within a given situation and therefore allows the insurer to predict what the average cost will be of compensating for the damage that has occurred. However, it should be noted that the calculations, in some cases, may be flawed by incomplete information. David Hand recalls that in 2011, the European Court of Justice declared it illegal to impose different insurance premiums according to gender, and from that date, gender has been in the *dark data*. The consequence was that women, who were less likely to have accidents, ended up paying more.
   b. A rather reductive definition of the term econometrics, dealing with its very etymology, is "science of measuring economic phenomena". It is not the place to discuss this definition, it is more interesting to see how many econometric models are probabilistic. Two typical questions: what is a probable range of gross domestic product

growth values next year? How much can a discount affect attracting a given number of new customers with a fixed probability?

4. Environmental (e.g. earthquakes, meteorology, climate change)
   a. A very specialized field concerns earthquakes: the seismic hazard of a site can be defined, in terms of risk, as the probability of excess variation of some parameters such as intensity and acceleration in a given time interval.
   b. Among the areas perhaps most used by the general public is that of weather forecasting, the reasons for which are so numerous that it is difficult to provide a list. They range from the individual citizen who wants to know whether he will be able to leave the city at the weekend to go to the mountains or the countryside, to the farmer who wants confirmation to decide on the best times for sowing and harvesting.
   c. Climate change is increasingly present in the political and social debate, and even scientific terminology is struggling to make its way among non-specialists: it should also be stressed that these contents are often proposed using a language in which uncertainty is probabilistically treated both qualitatively (e.g.: high evidence) and quantitatively (e.g.: likely as a probability greater than 2/3).
5. Financial (e.g.: financial product management and equity performance)
   a. The potential convenience of a financial product is assessed thanks to the probability that its value, at a certain future date, will exceed a certain level (such as, for example, the price established at the time of the agreement). The problem is that investors do not use a single way to calculate a company's probabilistic scenarios.
   b. If you look at a graph with the stock trend, you sometimes feel like you are on a roller coaster for the apparently unpredictable trend. Financial mathematics has tried to simulate the price trajectories of shares by formulating hypotheses about the law of movement of the same on the basis of probability distributions.
6. Information Technologies (e.g.: big data processing and algorithm analysis)
   ◦ Today the data sources are many and constantly increasing and therefore more and more we are faced with large masses of information for which it would seem appropriate to use the term big data; in reality what characterizes big data is not only the quantity, but also the quality of the types of data that can be recovered. This development requires both the use of old tools

and the use of new techniques: in this sense we can speak of a data science, the natural evolution of statistics and probability in the era of big data.

7. Legal (e.g.: evaluation of evidence and DNA testing)
    a. In a criminal trial the jury must assess, beyond any reasonable doubt, whether a defendant is innocent or guilty. Many theories have been developed to construct schemes for the evaluation of evidence; many, in order to address and resolve the uncertainty present in each case, use concepts and the calculation of probability (Taruffo, 2007).
    b. The discovery of a sequence of four nitrogenous bases (cytosine [C], guanine [G], adenine [A] and thymine [T]) called DNA) has made it possible to create a personalized profile with enormous potential because the order in the sequential arrangement of nucleotides constitutes the genetic information of each organism. DNA testing is certainly the most modern form of fingerprint analysis and is used both in police investigations and in cases of ascertainment of paternity. The guiding concept is that a test considered to be reliable leads to a high probability of identifying a culprit or defining paternity. Positions on this issue are not yet fully defined since, for example, distinctions can be made between random correspondence and probability of guilt. DNA testing constitutes "proof" in all respects but one fact is to say that a person with his genome was present at the scene of a crime and another is to say that he committed it.
8. Linguistic (e.g.: cryptography and text interpretation)
    a. The calculation of probabilities plays an important role in the field of cryptography, a term that has as its object of study the design of encryption methods, verification, and authentication of data integrity. To decipher a text, in the simplest case, you can determine the frequencies of the characters present in the cipher and compare them with the probability that a certain character appears in the cipher's language.
    b. Probabilistic tools are widely used in computing, particularly in the analysis of algorithms, i.e. the set of instructions that define a sequence of operations by which a class of problems is solved, and in the study of communication systems and computational models.
9. Management (e.g.: waiting files and warehouse stocks)

a. The expression "queue theory" refers to the mathematical study of a phenomenon that everyone has had the opportunity to verify concretely at the supermarket, post office, motorway toll booth, etc.. In the analysis of the phenomenon it is necessary to distinguish between deterministic processes and processes with random evolution. In the second case it is essential to know the characteristics of the distribution of arrivals and departures in order to define a parameter that represents the average time of both arrivals and service time.
b. The problem of stocks refers to the management over time of human or material resources (machinery, capital, products). In the company reality, supply needs, prices and delivery times are random variables and therefore one can speak of probabilistic models (Ullmann, 1976).

10. Medical (e.g. diagnostics and epidemiology)
    a. One of the tasks of medicine is to formulate diagnoses, studying the symptoms and identifying which are more or less likely for each disease. Another task is to identify, for each disease, how much depends on the case or on genetic and environmental factors (this is particularly important in the oncology field).
    b. Epidemiology studies the frequency and distribution of diseases or infections within a population. Probability and statistics are the basis of these studies and provide an essential aid for risk evaluation. The Environmental Protection Agency (EPA) has long since introduced the distinction between risk assessment (procedures for identifying the nature and extent of risk) and risk management (regulatory aspects). The first aspect is based on concepts such as estimation and evaluation, therefore elements typically associated with probability.
11. Physical (e.g. statistical mechanics and chaotic systems)
    a. Statistical mechanics is a branch of physics that studies the behavior of systems composed of a large number of particles. In fact, a material body is constituted by a large number of atoms but, at the macroscopic level, the thermodynamic properties can be described by few variables. To reconcile these two levels, the physicist Ludwig Boltzmann (1844-1906) suggested a statistical-probabilistic approach.
    b. There are systems that seem to depend on chance and since they show a rather irregular evolution they are treatable in terms of

probability. It would actually be more correct to talk about chaos. Who doesn't remember the famous butterfly effect of Edward Lorenz (1917-2008) that a butterfly flapping its wings in Brazil can cause a tornado in Texas? The innovative aspect of the theme is the discovery that chaos can depend on a deterministic model in which a small variation in initial conditions leads to unpredictably different results. There are two basic questions, mathematical but also philosophical: what happens in the future if I know the initial condition with a given uncertainty? Under what conditions can I formulate a probabilistic forecast in the presence of a chaotic phenomenon?

12. Recreational (e.g. games of chance and board games)
    a. Games of chance are characterized by an outcome that depends solely on chance and not on the skill of the person playing: this is the fundamental difference between, for example, bingo and poker. In the first case the player cannot influence the result, which is linked to the extraction of numbers that is carried out with electropneumatic machines, in the second case factors such as psychological intuition, financial availability, but also knowledge of the probability, come into play.
    b. Also in board games there is a random component, as in the traditional Snakes and Ladders, but strategy can play an important role, as in Monopoly.
13. Technological (e.g. quality control and risk assessment)
    a. In order to make a quality product, not only certain specifications must be respected, but also the pre-established tolerance levels. For this reason, diagrams called "control charts" are constructed. These charts make it easier to carry out tests and are based on the assumption that it is possible to determine the probability that the product meets the desired standards.
    b. The term risk can have different meanings, within the company it is the probability that a potential level of damage is achieved due to the exposure to a worker's danger. Obviously, according to the risk, assessed according to tables which indicate the different combinations of damage and probability, prevention measures should be taken.

THE CLASSIFICATION OF PROBABILISTIC PROBLEMS

Probabilistic problems may require repetitive exercises, application of formulas or schemes, operational approach to complex theories, analysis of experimental data, simulations, etc. The common thread is that in any case they are "problems". From the didactic point of view it is useful to divide this great variety of situations referring to their complexity.

You can use a model to simultaneously classify problems according to their complexity and the depth of knowledge needed to solve them.

Simple and trivial problems are solved by using sufficiently comprehensible solving formulas such as those of the opposite and compound probability. Most of the problems presented at school fall into this category. Attention, the term simple does not mean that the solution is immediately available to everyone, in the probabilistic field the possibility of errors is remarkably high, but only that the relative model is simple.

Some problems, considered more difficult, are treated in the last years of high school and can be related to well-known models such as Poisson distribution and normal distribution.

There are also difficult problems that require a deep level of knowledge, such as those related to the notion of plausibility (Ippoliti, 2007) or quantum mechanics. As for the quantum aspect, the words of professors Albrecht and Phillips (2014) are emblematic: "We argue using simple models that all successful practical uses of probabilities originated in quantum ñuctuations in the microscopic physical world around us, often propagated to macroscopic scales".

The real challenge is to define and deal with simple problems that are profound because they activate reasoning that involves the heart of the probability theme. All this can be summarized in a tetrachoric scheme (Table 1):

Table 1. Typology of problems

Deep	?	Quantum mechanics Notion of plausibility
Trivial	$P(A)=1-P(\bar{A})$ $P(A\cap B)=P(A)\cdot P(B\mid A)$	$P(n)=e^{-\lambda}\frac{\lambda^{n}}{n!}$ $f(x)=\frac{1}{\sigma\sqrt{2\pi}}e^{\frac{-(x-\mu)^2}{2\sigma^2}}$
PROBLEMS	**Simple**	**Hard**

PROBABILITY AND DIDACTICS

In conclusion, we can say that mathematics is, for the majority of students, a difficult subject: theoretical, abstract, and often without concrete applications. Probability is seen as a very demanding and often incompletely and superficially treated part of mathematics.

In educational research the fundamental problem is to find a "point of attack" to introduce topics related to the students' curriculum. The basic idea is to show how pervasive the *case* is, how close it is to the world around us and above all how it plays an important role in shaping our intuition even if we are not fully aware of it.

One of the most understandable examples comes from a 1998 film: Sliding Doors. Helen gets fired and heads for the subway. At that moment, her life is divided into two parallel dimensions:

1. Helen takes the subway and heads home where she finds her boyfriend in bed with his ex.
2. Helen loses the subway and decides to take a taxi. A man tries to steal her purse, which will make her come home later and find her boyfriend alone.

The plot indicates precisely that life is often linked to unpredictable events, to bifurcations that can lead us on different paths.

Teaching should be a priority objective to bring students to reflect on these issues and to highlight the links between the real world and probabilistic models. In this sense, several strategies can be outlined, including, for example, the study of historical development, the viewing of films and the use of computer tools to carry out simulations and examine conjectures.

For those wishing to deepen the subject both in terms of theoretical approaches and empirical research, there is an endless bibliography, initially one could examine, to get ideas from two different perspectives, the works of Kapadia & Borovcnik (2012) and Kiet & Lagrange (2017).

TEACHING PROBABILITY WITH REALIA

The term *realia* derives from the late Latin adjective reālia (real things, facts).

According to Vlahov and Florin (1986), the realia have been defined as words that denote objects or concepts typical of a certain culture and are the basis of the work of men of letters and translators. In the United States, since the late 1890s, realia has indicated everyday objects used as teaching aids for children and have been used mainly by teachers of foreign languages.

In pedagogy they generally indicate common objects that are used in instruction by a teacher to illustrate daily life or, in particular, to introduce a range of disciplinary content. For example, cinema, literature, music, comics, videogames are all tools that in language teaching are called realia and allow to understand more effectively what has already been learned through "traditional" methodologies. This didactic approach recommends, for example, after having studied the terms related to food, to consult authentic restaurant menus to facilitate understanding and remembering. In this sense, realia can be considered three-dimensional objects of real life.

If this is the meaning of the term realia, in this book we want to make a paradigmatic reversal, that is to start with "touching the menu" to introduce, from the objectivation of the context, the study of relative terms and concepts.

Lecoutre & Durand (1988) have shown that an approach disconnected from experience with a concrete dice can lead students to make mistakes when estimating the probability of getting the pair 5-6 and the probability of getting the pair 6-6 when rolling two dice simultaneously.

Important support is now offered by IT tools. This has added the "virtual realia" option. Now it is possible for the learner to see a 3D object and manipulate it: the ability to rotate, zoom and look inside this virtual reality has become an important learning tool. Without reaching this level of sophistication, however, it is possible to use, for example, dynamic geometry software (DGS) which should by now be defined as dynamic mathematics software as it includes geometry, algebra and analysis or spreadsheets that allow simulations to be performed.

This methodology can also be developed thanks to resources offered by the Internet. Several of the objects presented can be associated with digital resources: videos, simulations, games, etc.. Providing precise indications makes little sense as the addresses of the sites disappear or change with considerable frequency. With these tools it is possible to set a less conventional didactic strategy based on the introduction of the main probabilistic concepts starting from the exposition of objects that make the sense of the things studied concrete. It is important to look for a new approach, preferably very artisanal, starting from artifacts that students can recognize as part of their daily life. Identifying objects and situations that speak of "different mathematics", embodied in everyday life, offers more possibilities to deal with the mathematical illiteracy that seems to afflict a large part of our society.

This approach is not new, several authors have thought that abstract mathematical objects should be seen as concrete objects to observe, study and manipulate but in general the examples are not numerous.

In the following chapters we will introduce essential concepts related to probability starting from numerous objects (Drivet, 2017). The laws and theorems characteristic of this area will arise, hopefully clearly, from the presentation of the artifacts and not as assumptions to be uncritically assimilated.

It would be a sin of pride to think that these indications really open the way to a new teaching, but perhaps it may be useful to remember a reflection that is due to the Anglo-American mathematician David Mumford. He thinks that applied mathematics is comparable to the work of a carpenter and therefore, when faced with a problem, one must take the right tool from the toolbox.

REFERENCES

Albrecht, A., & Phillips, D. (2014). Origin of probabilities and their application to the multiverse. *Physical Review. D*, *90*(12), 123514. doi:10.1103/PhysRevD.90.123514

Borovcnik, M., & Kapadia, R. (2016). *Reasoning with risk–A survival guide. Academic Press.*

David, F. N. (1962). *Games, gods and gambling: The origins and history of probability and statistical ideas from the earliest times to the Newtonian era.* Hafner Publishing Company.

Drivet, A. (2017). *Toolbox*. Amazon.

Hand, D. J. (2020). *Dark Data: Why What You Don't Know Matters*. Princeton Univ Pr.

Ippoliti, E. (2007). *Il vero e il plausibile*. Lulu.

Kapadia, R., & Borovcnik, M. (Eds.). (2012). *Chance encounters: Probability in education* (Vol. 12). Springer Science & Business Media.

KaushishM. (2019). *Probability Education*. Retrieved from: https://www.math.arizona.edu/~kaushish/documents/probability_education.pdf

Kiet, B. A., & Lagrange, J. B. (2017). *Using Simulation by Excel and R in the Teaching of Probability and Statistics for Non-Math Major Students*. Academic Press.

Lecoutre, M. P., & Durand, J. L. (1988). Jugements probabilistes et modèles cognitifs: Étude d'une situation aléatoire. *Educational Studies in Mathematics*, *19*(3), 357–368. doi:10.1007/BF00312452

Poincaré, H. Y. (1902). *La science et l'hypothèse*. Flammarion.

Pólya, G. (1954). *Patterns ofplausible inférence* (Vol. 2). Princeton University Press.

Smithson, M. (2012). *Ignorance and uncertainty: Emerging paradigms*. Springer Science & Business Media.

Taruffo, M. (2007). Verità e probabilità nella prova dei fatti. Revista de Processo, 154, 207-222.

Ullmann, J. (1976). *Quantitative Methods in Management*. McGraw-Hill.

Vlahov, S., & Florin, S. (1986). *Neperovodimoe v perevode*. Vysšaja škola.

Chapter 1
Classical Probability

ABSTRACT

Classical probability is historically presented as the first interpretation of random phenomena. Even with its application limits, it represents the starting point for the main concepts of probability calculation. In class it is the most represented formulation, and, in some cases, it is presented as a standard definition of probability. In the chapter, after presenting the classic objects (coins, marbles, dice, playing cards) normally used to introduce the theme, the panorama is widened thanks to the introduction of less common artifacts and the use of examples that can provide useful insights for further study. The path starts from the concepts of trial and event and ends with the Bayes theorem.

INTRODUCTION

Anytime you have a 50–50 chance of getting something right, there's a 90 percent probability you'll get it wrong. ~ Andy Rooney

To understand a discipline it is essential to follow its historical developments. This is especially true for probability; it has a particularly complex development, especially in the most ancient phases. The chance was connected with mystical and religious aspects and it will take a long time before the subject is tackled rationally and takes its "mathematical" form. Suffice it to recall the case of Bishop Wibold who, just before the end of the first millennium, enumerated 56 theological virtues, each corresponding to the possible ways in which three

DOI: 10.4018/978-1-7998-3871-5.ch001

dice could be thrown and therefore chose a particular virtue to meditate in the next 24 hours. For more on the historical aspects, see Kendall.

Historically we perhaps owe to Gerolamo Cardano (1501-1576) the first writing on probability; he, in the book *Liber de ludo aleæ*, collects the observations made during his career as a fierce player and calculates the probability of the results obtained by throwing three dice.

This is no accident, in the sense that the first thoughts on the subject arise from the need to find rules in gambling. The problem of the distribution of the stakes was addressed by Luca Pacioli (1445-1517) in the *Summa de arithmetica, geometry, proportioni et proportionality* and later by Niccolò Tartaglia (1499-1557); they deal with the problem of how to distribute the stakes in case the game should be interrupted. This theme was later resolved by Blaise Pascal (1623-1662) and Pierre de Fermat (1601-1665). The story of the collaboration of the latter two great characters in solving another type of problem is extremely interesting. What is it about? A fashionable game at that time was the following: the game manager would bet at par with a player that the latter, by rolling a dice 4 times, would get the number 6 at least once. A fierce player like Antoine Gombaud, Cavalier de Méré had always thought that getting at least a 6 in 4 throws of a dice was equivalent to getting at least a double 6 in 24 rolls. However, playing according to that belief, he lost and wrote to Pascal complaining that the math was failing. Pascal and Fermat solved the problem by demonstrating that it was not mathematics that failed but the knight's erroneous reasoning. The problem is what is traditionally associated with the birth (1654) of probability calculation as a scientific discipline.

After various progresses, with J. Bernoulli (1654-1705), Abraham De Moivre (1667-1754) and Pierre-Simon Laplace (1749-1827) the classical definition of probability is crystallized as the ratio between favorable cases and possible cases supposed all equally possible.

$$P = \frac{number\ of\ favorable\ outcomes}{numer\ of\ possible\ outcomes} = \frac{f}{n}$$

The following sentence is due to Laplace: "The theory of chance consists in reducing all the events of the same kind to a certain number of cases equally possible" as reported by Dhombres.

The classical definition has the advantage of being simple and operational, but it has a negative aspect: in fact, the concept of probability is introduced by exploiting the notion of comparable events which in turn implies the knowledge of the idea of probability. The result is an evident vicious circle. Laplace himself realized this and tried to overcome the problem by introducing the "principle of sufficient reason" according to which two events are to be considered comparable when there are no valid reasons to think that one of them can occur more easily than the other.

This severely limits the scope because in reality this condition cannot always be had. As a consequence, we can say that the classical probability is applicable in all cases in which for reasons of symmetry we can think equally possible all cases: tossing coins or dice not rigged, drawing cards from a deck or balls from an urn, output of bingo numbers, etc..

COINS

Before starting the topic it is nice to start from a story, from which it is possible to develop some reasoning.

A great Japanese warrior named Nobunaga decided to attack the enemy although he had only one tenth the number of men the opposition commanded. He knew that he would win, but his soldiers were in doubt. On the way he stopped at a Shinto shrine and told his man, 'After I visit the shrine I will toss a coin. If head comes we will win; if tails we will loose. Destiny holds us in her hand.' Nobunaga entered the shrine and offered a silent prayer. He came forth and tossed a coin. Heads appeared. His soldiers were so eager to fight that they won their battle easily. 'No one can change the hand of destiny,' his attendant told him after the battle. 'Indeed not,' said Nobunaga, showing a coin, which had been doubled, with heads facing either way. (Reps & Senzaki, 1998)

Nobunaga's story can effectively introduce the argument because a coin is mentioned, the classic object that is initially presented when we talk about the calculation of probabilities for the first time.

If you ask any individual what is the probability of getting head by flipping a fair coin, you will get a unanimous answer expressed as a percentage: 50%. Alternatively, the answer could be 1/2 or less frequently 0.5, even if it would be the most appropriate response.

It is interesting to note that, in a non-formalized way and without having a clear definition of what is meant by probability, the subjects to whom this question is addressed seem to have an innate conception of this concept.

The fact that the probability is given by the ratio between the number of favorable cases and the number of possible cases (provided that these are equally possible, it should be specified) emerges almost spontaneously. These cases, obviously, can only assume positive values and the value of the numerator cannot exceed the denominator; therefore, the probability will always be characterized by a non-negative number, between 0 and 1.

Probably also the ancient Romans knew it, since one of the most practiced games was the so-called "Capita aut Navia": also in that case a coin was thrown into the air, on the faces of which were respectively the two-faced Janus' head (*Capita)* and a ship (*Navia*).

But let's go into more detail by beginning to acquire some basic terms. The starting point of a probabilistic construction is a *trial* or experiment.

The calculation of the probabilities deals with making forecasts on *events,* i.e. on facts that may occur, however characterized by the fact that the result does not appear determinable a priori.

If E indicates the event, P(E) will be the probability that such an event will occur. In practice, the simplified form P will often be used to indicate thc probability.

The result of an experiment is often an unpredictable and therefore random number or, more correctly, a *random variable*. A random variable assigns a numeric value to each possible result in the *sample space*. This expression indicates any set that lists all the possible results of some experiment or situation, for example the types of possible winds can be classified with W={constants, periodicals, variables, irregulars} and each one can be assigned a numerical probability value such as P={0.1; 0.4; 0.3; 0.2}. This number can normally be a direct expression of an experiment or the result of a processing.

Since the concept of event turns out to be central, two words must be spent on some fundamental characteristics. Firstly, for each event the *opposite event* $\bar{E}$ (complementary event) con be defined, in set terms (Figure 1):

An important distinction can be made between incompatible (the occurrence of one excludes the other) and compatible (no exclusion occurs) events. By drawing a card from a deck we can get a King or a Jack and the two events are mutually exclusive, the output of an Ace or a card of hearts instead refers to something compatible, the card of hearts could be an Ace. These distinctions are fundamental as they involve substantial changes in the calculation of probabilities.

Figure 1. Set vision of negation

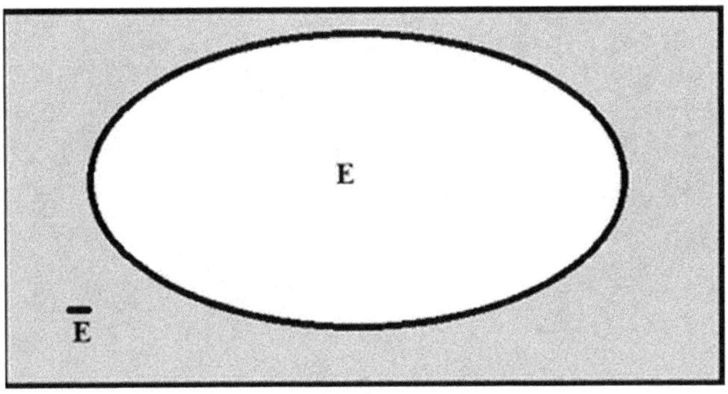

In addition, events can be independent (the occurrence of one does not affect the occurrence of the other) or dependent (the occurrence of one changes the probability of the other).

A typical example of the use of the opposite probability, extremely useful, can be applied to a situation considered important by many people: it sometimes happens to dream of people or events and believe in the realization of the dream in real life. The problem can be easily set: assuming that a premonitory dream corresponds to a real event with probability that is normally associated with an "impossible" event, i.e. 1/1000000, what is the probability that in a year there is at least one dream come true? Is this probability significant on a large scale? Assuming that the result of each dream is not conditioned by the others we will have $(1/1000000)^{365}$. However, it is convenient to use the probability of the opposite event for the calculation, therefore: $1 - (99999/1000000)^{365} = 0.000036$. On a large scale this means that the cases of premonitory dreams do not represent an irrelevant number. For example, the inhabitants of the Earth amount to about seven billion and seven hundred million, possible dreams would amount to $0.000036 \bullet 1403760000 = 277200$, definitely a significant number!

Returning to the initial anecdote, it is interesting to note that if a coin with two heads is tossed a certain number of times (Figure 2), by betting on the head, it is possible to win some bets before the other player realizes there is something strange. In fact, initially the subject thinks that whoever proposes the game is so skillful that he can always provoke the exit of the face with the head, only in a second time grows the feeling that the coin is rigged.

The use of the coin with two heads allows to define the concepts of *certain event* (the head exit) and *impossible event* (the tail exit).

Figure 2. Rigged coin

It must be emphasized that it is true that the exit of tail in the throw of a coin with two heads is impossible and therefore it is correct to attribute 0 to the probability of the event. In normal life, however, the attribution of impossibility presents interesting aspects; one must keep in mind that psychologically the individual tends to neglect extremely low probabilities. For example, if an event is considered to have a For example, if it is considered that an event has a probability of occurrence of less than one billionth, the vast majority of individuals will consider the event as impossible. This position is humanly reasonable even if it presents two difficulties, one of mathematical order and one of practical order. In the first case an exceedingly small number cannot be considered null, in the second case an event that has a probability against one billion would be considered impossible, while an event that has one against 999999999 would be possible.

About the possibility of the occurrence of a certain event, we always tend to quote the famous phrase Arthur Conan Doyle made Sherlock Holmes say: "When you have excluded the impossible, whatever remains, however improbable, must be the truth".

Reasoning about throwing a coin represents a first fundamental step in teaching because it opens the field to different considerations.

First of all the opposite probability q is defined as the probability that the event does not happen; since the event happens or does not happen the sum of its probability and the opposite one must give the unit, therefore $p + q = 1$ from which we get $q = 1 - p$.

Then it is quite legitimate to wonder why the result is unpredictable (Keller). One might think that, knowing the shape and composition of the coin, how it is thrown, forces acting on the coin, then the laws of physics should allow us to predict precisely how the coin will land. This is true in theory, but false in reality. Simplifying a bit the problem we can think that the motion of the coin is determined only by two factors: the speed of the throw and the impressed rotation. So what's stopping us from knowing which face the coin will land on? When it is thrown we give it a vertical speed v in cm/sec and a rotation speed r in rpm. These values constitute the parameters of the experiment, by this term we mean a quantity that, once defined, remains constant in a model; this quantity does not assume random values. Many probability models have one or more parameters that can be adjusted to fit the physical experiment being modelled.

The coin takes 2v/g seconds and 2rv/g rotations to return to its initial position, with g acceleration due to gravity. The boundary between head and tail is exactly at half rotation, indicating with N this whole you have N = 4vr/g. For example with v = 16 and r = 4 you will have N = 26 whose rest with divider 2 gives 0. If we modify the value of v slightly and bring it to 15.9 we will have N = 25, whose rest gives 1.

Following the indications of Ian Stewart (1997) it can be seen that by reporting the results in a spreadsheet the rectangle of possible values is divided into strips; black for head (1) and white for tail (0). Since you can only control the values of v and r within certain limits, you only need to vary the parameters a little to get the other side. Each coin toss is therefore like throwing a dart on a "zebra-skinned" target (Figure 3).

Figure 3. Coin flip results

A second aspect concerns the ease with which one is deceived by preconceptions. Let's take into consideration these possible results related to two series of twelve HTHTHTHTHTHT and HHTHTTHTTTHH launches shown in Figure 4.

Figure 4. Two sets of results

If the question arises: "which of the two sequences is more likely?", the choice often falls on the second. Choice studied from the point of view of psychological subjectivity, but not only, by Falk and Konold.

In general, individuals consider an event likely based on three ingredients: fairness, alternation, and disorder. The first characteristic occurs in both throws (six heads and six tails) and therefore the two sequences are comparable. As far as alternation is concerned, it is considered unlikely that there is a rigid alternation of heads and tails and it is common opinion that a random phenomenon is synonymous with disorder.

In fact, this opinion can be considered as a very deep-rooted preconception that can be overcome only by examining how the six events that are independent of each other are connected, i.e. the fact that the first event occurs does not alter the probability that the second event occurs (the case of dependent events will be dealt with later).

A simple tree graph allows you to view the two alternatives. In this case you can limit yourself to three throws in order not to make Figure 5 too complex.

As you can see the HHT and HTH results both occur once out of eight, so the probability is the same: 1/8. It is not difficult to see that $1/8 = 1/2 \cdot 1/2 \cdot 1/2$. We can say that for two independent events, in which the occurrence of event A is not influenced by event B, there is the rule of multiplication.

This statement corresponds to the *Law of Compound Probabilities* for independent events

$$P(A \cap B) = P(A) \cdot P(B)$$

Figure 5. Graphs related to the result of throwing two coins three times

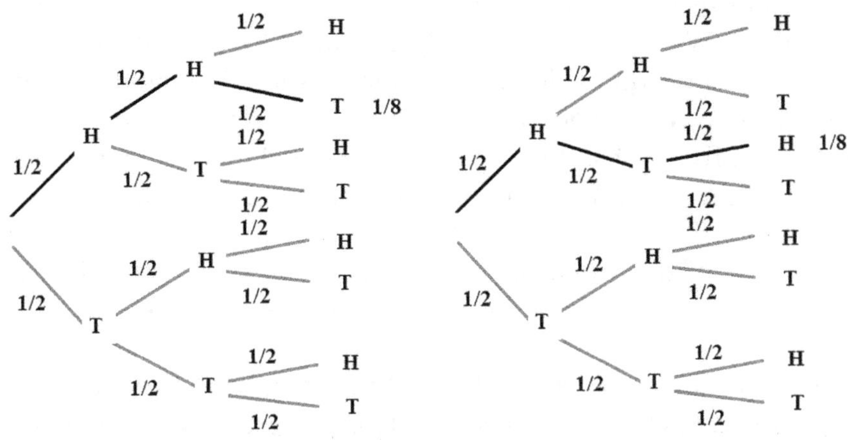

Great care must be taken when formulating the hypothesis, as in the case of coins, that events are really independent. In 2016 in New Zealand, a lady bought a pack of twelve eggs in her usual supermarket and each one contained two yolks. One might think that the event which, according to the British Information Service, occurs in one in 1000 cases, is virtually impossible. In reality it happens, especially with the larger eggs produced by young hens, that they are automatically separated from the machines and therefore can more easily contain two yolks; in this case we cannot speak of independence.

A set representation is shown in Figure 6:

Figure 6. Set vision of the logical product

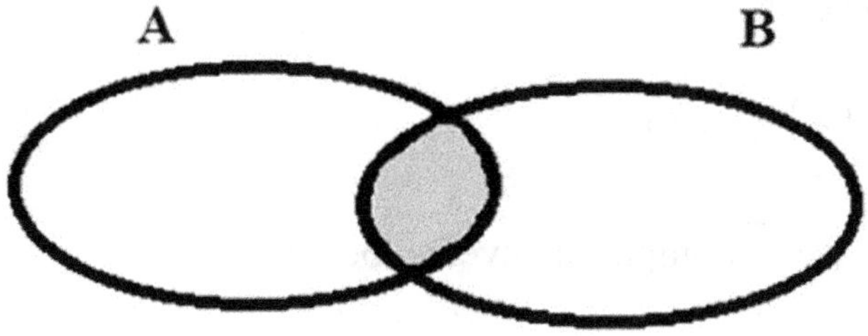

The concepts of independence and the opposite event can be examined together with a simple example: a university student thinks of going out with five different girls; the girls do not know each other, so the chances of these five dates being positive and starting a romantic story are independent. The student estimates that the probability of an appointment going well is 20% (an optimist!). What is the probability that after the five dates the student will find at least one girl suitable for a relationship? To find the probability that at least one rendezvous is successful, it is preferable to consider the product of the five probabilities that the individual dates do not go well and then subtract the result from 1 in order to obtain the probability of success:

P (at least one of the 5 independent events will take place) = $1 - (0.8)^5 = 0.67$

Some might think that the situation is a bit artificial, so let's use a naturalistic example to examine independence and dependence. At the lower end of the abdomen of fireflies (beetles belonging to the Lampyridae family) there is a flashing part. The light emitted by these insects is used for breeding: males and females attract each other in the dark to mate. Fireflies initially emit light independently but, in a small area, they tend to emit light in sync as a response to light signals. A step towards addiction according to a behavioral model analyzed, for example, by Ramírez-Ávila et al.

Returning to the example shown in Figure 6, it is important to understand the linguistic value of these considerations: HHT means that H **and** H **and** T came out, so the "**and** logic" corresponds to the multiplication along the branches of the graph. This concept is a tool to tackle, especially at the beginning, problems involving multiple events.

The case of "**or** logic" corresponds instead to the sum of two branches of the graph.

This statement corresponds to the *Law of Total Probabilities* for incompatible events.

$$P(A \cup B) = P(A) + P(B)$$

which, in the case of compatible events, becomes

$$P(A \cup B) = P(A) + P(B) - P(A \cap B)$$

In practice the law states that the probability of the logical sum of two events (A or B or both events occur) compatible with each other is given by the sum of the probabilities of the two events minus the probability of the intersection. The intersection must be subtracted in order not to count twice the events included in it. The Euler-Venn diagram in Figure 7 indicates the logical sum.

Figure 7. Set vision of the logical sum

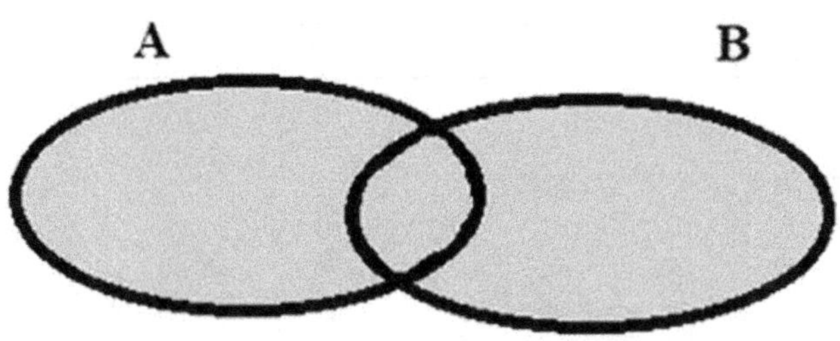

Coins are at the root of many problems related to the calculation of probabilities. Let's look at the game, from 1969, due to the mathematician Walter Penney. As we have seen, if you throw three coins you can get eight different results, all equally probable. The first player chooses one sequence, the second another. A coin is tossed and continued until one of the two chosen sequences appears; collect a point who gets his own sequence. Apparently neither of them should be favored, in reality it is possible to adopt a strategy that considerably improves the probability of victory of the second participant.

Let's imagine, for example, that the first player, A, chooses the HHH sequence and it happens once in eight. Let's suppose that the first three consecutive throws in which the head came out are n, n+1, n+2. What happened to throw n-1? Of course, the head couldn't have gone out otherwise the first player would have won first, so the tail came out. But then if the second player, B, chooses the TCC combination this will be the winner, this will happen in the other seven out of eight cases. In conclusion the probability of winning for B is P(B) = 7/8 = 0.875.

Table 1 shows the strategies and the chances of victory, in percentage form:

The mnemonic method to know which hat-trick to choose after choosing the first player may be interesting: the first element will be, as shown in Figure 8, the opposite of the second element chosen by the first player, the second and third will be respectively his first and second.

Table 1. Penney's game

Choice of A	Choice of B	Winning probability for B
HHH	THH	87.5%
HHT	THH	75.0%
HTH	HHT	66.7%
HTT	HHT	66.7%
THH	TTH	66.7%
THT	TTH	66.7%
TTH	HTT	75.0%
TTT	HTT	87.5%

Figure 8. Choice in Penney's game

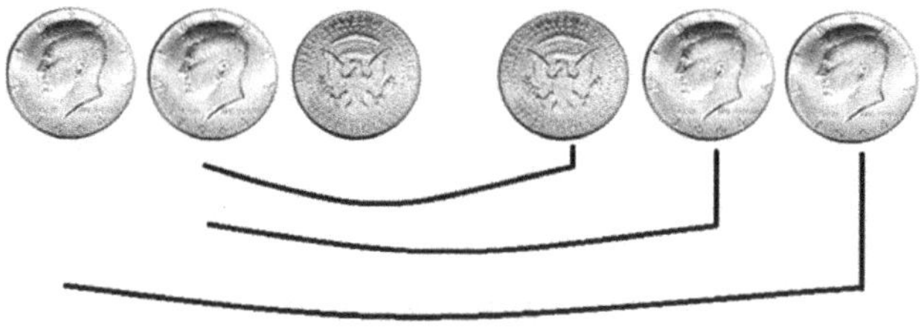

To conclude this analysis on coins we want to introduce a reasonable doubt: the probability seen from the point of view of the classics is based on the assumption that an "honest" coin has a probability to fall on the head half of the times. How sensible is this statement in a real context, even in the presence of an unrigged coin?

Persi Diaconis' team built a machine that throws coins and takes pictures in slow motion. The result? The coin falls on the same face from which it was flipped with probability 0.51and, on top of that, about once in 6000 flips it falls!

Another misconception must be clarified: it can be thought that if we toss a coin many times, say a million times, the number of heads and tails is the same. Actually, we have to expect this to be a fairly rare event. The probability that in the toss of n coins there is perfect parity in the result is with $P = \frac{1}{\sqrt{\frac{n}{2}\pi}}$

so out of a million launches the probability is a modest 2.5%.

Up to now we have dealt with problems in which coins were the protagonists, but perhaps it would be better to move the discussion towards other situations more relevant to life: we can identify remarkably close links. For example, instead of throwing two coins we can think about the birth of two children; in fact, instead of Head (H) or Tail (T) we can talk about Male (M) or Female (F). In this case instead of HH, HT, TH, TT we have MM, MF, FM, FF.

Let us try to clarify this with three examples:

1. Two women are in the park with their children, both have two. At least one of the first woman's children is female, as is the second woman's eldest daughter. The probability that both have two daughters is the same? The first woman is pregnant and only knows that she is expecting twins, what is the probability that she will have two daughters? If you were certain that at least one of them is female, does the probability change?

The probability that the first woman has two female daughters is 1/3 because of the four cases we can exclude the one with two males, while the probability that the second woman has two female daughters is 1/2 because the remaining cases are FM and FF. In the case of the pregnant woman the probability of two females is obviously 1/4, an answer that changes in the light of the information and brings the probability to 1/3 because the MM option is excluded.

2. One of the most interesting questions is the following: in the case of births of dizygotic twins (born from two different fertilized eggs) is the sex of the born independent?
   Randomly extracted the sex of the first born three situations can occur:
   a. the sex of the second one is extracted again randomly and therefore these are independent events;
   b. the sex of the second is the same as the first, there is addiction here;
   c. the sex of the second depends on a coefficient *r* which indicates the proportion of "randomly extracted" and, in this case, we can talk about related events.

In the first case ($r = 0$) the proportion must respect the theoretical one (MM = 1/4, FF = 1/4, MF or FM = 1/2), in the second case ($r=1$) the proportion must respect the theoretical one (MM = 1/2, FF = 1/2), finally in the third case the result depends on the value of a correlation coefficient normally indicated with *r*. Biologists have shown that events are related, but how much is *r*? in this case the computer comes to our rescue because it is possible to use it to simulate a large number of twin births by varying the value of *r*. Since the eighth chapter will be dedicated to this topic at this point we will simply provide the final results of a simulation. With 20000 “launches”, when $r \approx 0.1$, results are quite similar to those found in the medical literature.

3. Rosen & Krithivasan report a problem posed in 2010 by the puzzle maker Gary Foshee: Mr. Smith has two children, one of whom is a son born on a Tuesday. What is the probability that Mr. Smith has two sons?”.

The most understandable solution is due to Ivan Moscovich who presents the four possible situations shown in Table 2:

Table 2. Possible results of the Foshee problem

					Male			
		Mon	**Tue**	**Wed**	**Thu**	**Fri**	**Sat**	**Sun**
	Mon		x					
	Tue		x					
	Wed		x					
Female	**Thu**		x					
	Fri		x					
	Sat		x					
	Sun		x					
					Female			
		Mon	**Tue**	**Wed**	**Thu**	**Fri**	**Sat**	**Sun**
	Mon							
	Tue	x	x	x	x	x	x	x
	Wed							
Male	**Thu**							
	Fri							
	Sat							
	Sun							

continues on following page

Table 2. Continued

		Male 1						
		Mon	**Tue**	**Wed**	**Thu**	**Fri**	**Sat**	**Sun**
Male 2	**Mon**		x					
	Tue		x					
	Wed		x					
	Thu		x					
	Fri		x					
	Sat		x					
	Sun		x					
		Male 2						
		Mon	**Tue**	**Wed**	**Thu**	**Fri**	**Sat**	**Sun**
Male 1	**Mon**		x					
	Tue							
	Wed		x					
	Thu		x					
	Fri		x					
	Sat		x					
	Sun		x					

The possibilities are as follows: there are 7 + 7 + 7 + 6 (six because the males born on Tuesday cannot count twice) cases of children with a certain sex and day of birth and 13 of these combinations consist of two males. The answer is therefore 13/27.

SYRINGE

The syringe is a disposable medical device used both for injecting medication and for taking samples (Figure 9). Through venous blood sampling, factors indicating problems or pathologies can be assessed.

A particular type of pathology is called microcythemia, an abnormal reduction in the blood of the mean cell volume (MCV). It was the American doctor Thomas Cooley (1871-1945) who highlighted the characteristics of this particular type of hereditary anemia. Microcythemia is widespread throughout the world with different genetic characteristics in different areas. But what is the link with probability? The transmission of hereditary traits can easily

Figure 9. Syringe

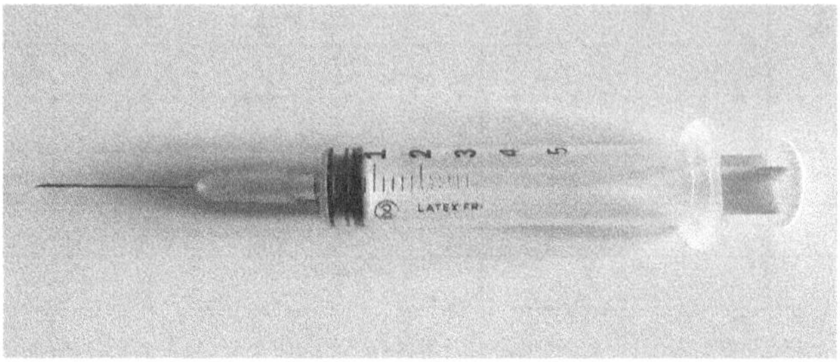

be associated with the toss of two coins. In the case of microcythemia there are three possible situations related to gametes. A gamete is a haploid cell that fuses with another haploid cell during fertilization:

1. MM: individual is healthy;
2. Mm: individual does not feel disturbed but is a carrier;
3. mm: individual is sick.

What if both parents are Mm and plan to have a child?

The possible combinations of *M* and *m* gametes lead to four possible crossings, MM (equivalent to Head-Head), mm (equivalent to Tail-Tail) and Mm or mM (equivalent to Head-Tail or Tail-Head). To the three cases seen before are therefore associated as many probabilities: 1/4, 1/4, 1/2. A couple Mm has therefore a 25% probability of generating a child suffering from Cooley's disease. This is why it is important that they undergo a blood test.

What if the parents are a healthy (MM) and a carrier (Mm)?

There are only two situations, the mm case cannot occur, but the probability of generating a carrier child (Mm) is 50%.

MARBLES

Normally, after introducing the first probabilistic concepts with coins, we move on to marbles (Figure 10). Most problems relate to extraction from bags or urns and normally represent a good method for reinforcing ideas (and for losing all enthusiasm for probability!).

Let's examine the already mentioned famous problem, called "division of the stakes" in which, instead of money, marbles are at stake.

Figure 10. Marbles

If two players have to leave a game divided into several hands before the end, how should they share the stakes? Both Blaise Pascal and Pierre de Fermat provided an elaborate solution.

Let's consider a version that has been re-proposed several times in didactic experimentations: two children, *A* and *B*, play heads and tails with a two-faced coin without rigging. *A* wins if the face shows heads, *B* wins in the opposite case. The winner of all the stakes of 24 marbles (12 of which are A's and 12 B's) is the first one to score 6 points. However, the game is interrupted on 5 to 3 for *A*. The question is how the two players should divide their stakes in such a way that the division is fair.

Apparently this is a problem of fair distribution. It is actually an important question because we need to think, in probabilistic terms, about the subsequent hypothetical bets. This problem presents a myriad of facets and pitfalls and requires a projection process: *A* wins if one of these three events H, TH, TTH occurs. For the rule of probability composed of independent events the odds will be 1/2, 1/4, 1/8. Here we insert the law of total probabilities: 1/2 + 1/4 + 1/8= 7/8 so *A* will have 21 marbles and *B* only 3.

Now let's look at an additional problem. An urn contains only 6 white marbles. You roll a regular dice and replace with black balls a number of balls equal to the result of rolling the dice. From the modified urn, a ball is drawn at random, then put it back into the urn and a new extraction is made. What is the probability that they are both black?

As we will see when we talk about the following object, but it is not a surprise, the probability that any face of the dice will come out is 1/6; assuming that the face with the one came out and that therefore there are five white marbles and a black marble in the urn, the probability that in two draws two black marbles come out is 1/6 • 1/6. In general if the marbles are x you will have x/6 • x/6. In summary:

$$P = \sum_{x=1}^{6} \frac{1}{6} \cdot \left(\frac{x}{6}\right)^2 = \frac{91}{216}$$

DICE

In addition to coins and marbles, perhaps the most commonly used object to talk about probability in the classical sense is the dice. Dice are small polyhedra, used in the context of different games to randomly generate numerical results. To obtain a random value, roll the dice on a flat surface, and consider as a "result" the value that is found on the face facing up when the dice ends its movement.

Traditional dice are cubes whose faces are marked with natural numbers from 1 to 6 (Figure 11).

Figure 11. Dice

When we say that Ω is the set Ω = {1,2,3,4,5,6} we affirm that we do not take into consideration other cases such as: the nut remains balanced on an edge, falls into a manhole and is no longer recoverable, it goes to thousand pieces, or is eaten on the fly by a pigeon. These events are not part of our hypothesis and therefore have no relevance.

Of course, the exit probability of each face is 1/6. Actually, since the numbers on the dice have small incisions, slightly more material is removed from the faces of higher value: this causes a small deviation that makes them unequal. To avoid this in casinos, transparent dice are used to show that they have not been rigged by inserting weights or drops of mercury into an internal channel and, to avoid lack of uniformity, the edges are sharp and the concavities of the marks have a minimum depth.

Together with the common cubic dice we can refer to other types of dice with faces made of regular figures. Of these dice there are only five types. Why? The demonstration is simple, just keep in mind that the angle of a regular polygon of side *n* measures (2 - 4 / n) • 90°. To form an angle you need at least three polygons, using equilateral triangles you can join three, four and five around a point; joining six of them you get a flat figure because the sum of the angles is equal to 360° and therefore it is not possible to have a dice. If we consider the square we can use only three, with pentagons it is not much better: also in this case three pentagons form an angle but there is no space for a fourth. By joining three hexagons, finally, we obtain a flat figure and therefore it is not possible to build a hexagonal nut. Figure 12 shows the five possible dice.

Obviously, the probabilities related to the results of throws with these types of dice depend on the number of faces: 4, 6, 8, 12 and 20 from which the names tetrahedron, hexahedron (or cube), octahedron, dodecahedron, and icosahedron. The theoretical probabilities give us the values 1/4, 1/6, 1/8, 1/12 and 1/20.

A simple problem: I randomly draw one of these dice from a bag and throw it; what is the probability of getting 1?

The probability of choosing any dice is 1/5 and there are five cases, they are given by the sum of the probabilities of getting the result 1 from each dice: 1/5 • 1/4 + 1/5 • 1/6 + 1/5 • 1/8 + 1/5 • 1/12 + 1/5•1/20 = 27/200 = 0.135. Of course if we had posed the problem of getting 9 the calculation would have been reduced to 1/5 • 0 + 1/5 • 0 + 1/5 • 0 + 1/5 • 1/12 + 1/5•1/20 = 4/150 = $0.02\overline{6}$ as it is impossible to get nine with the first three dice.

Figure 12. Regular dice

These solids are also called Platonic because the Greek philosopher Plato used them in his book entitled *Timaeus* to explain the natural world. Plato associated four polyhedra with what in his time were considered the four fundamental elements of nature:

- The icosahedron to the water;
- The octahedron to the air;
- The cube to the earth;
- The tetrahedron to the fire.

The dodecahedron that had served the Demiurge (God) to decorate the universe remained outside.

It should be noted that dice are not only game tools but have a naturalistic response, for example the icosahedral structure is extremely common among viruses. The capsomers (coating proteins) organize themselves to constitute a regular polyhedral structure with 20 faces consisting of equilateral triangles and 12 vertices. The tobacco virus has a protein in each vertex of each of the triangular faces of an icosahedron, for a total of sixty proteins.

The results that can be obtained by rolling a normal six-sided die raise some interesting questions, such as the use of different registers (Table 3). The following quote (Arrigo) makes it perfectly clear the key concept. "For example, if you throw an ideal dice cube and one wonders what is the probability of getting either 1 or 6, it can be answered in different ways, using different registers or different representations within the same register. In the register of native language: 'There are two possibilities on six' By conversion to the fractional register 'the probability is 2/6' or by treatment within the fractional register the probability is 1/3, by conversion to the decimal register 'the probability is 0.3', or yet by conversion to the proportion register 'the probability is 33.3%'".

If you start with the dice it is useful to show these different registers by using, for example, the different types of format available in a spreadsheet.

Table 3. Registers and representations

Registers	Representations
Linguistic	2 of 6
Fractional	1/3
Decimal	0.33
Percentage	33.3%

To conclude the dice argument, this time we can deal with an intriguing problem well told by Ian Stewart (2008). Normally, reference is made to dice with numbers between 1 and 6 on the faces. In this case two players have the same chance of winning by rolling these dice.

What happens if different numbers are on the facets?

For example, let's suppose that we have four types of dice (Figure 13) that are a bit particular as far as numbering is concerned, as shown in Table 4:

Figure 13. Strange dice

Table 4. Strange dice values

BLUE Dice	1	1	1	5	5	5
YELLOW Dice	0	0	4	4	4	4
GREEN Dice	3	3	3	3	3	3
RED Dice	2	2	2	2	6	6

In a game where the goal is to get the highest face value in a certain number of games, you can ask a few questions:

1. Is there a dice that always lets you win?
2. Is there a dice that allows you to win more?
3. Is there a dice that makes you win less?

A comparison of the results, shown in Table 5, clarifies what the answers are.

Table 5. Clashes between the dice

		BLUE							
		1	**1**	**1**	**5**	**5**	**5**		
YELLOW	**0**	BLUE	BLUE	BLUE	BLUE	BLUE	BLUE		
	0	BLUE	BLUE	BLUE	BLUE	BLUE	BLUE		
	4	YELLOW	YELLOW	YELLOW	BLUE	BLUE	BLUE	BLUE	24
	4	YELLOW	YELLOW	YELLOW	BLUE	BLUE	BLUE	YELLOW	12
	4	YELLOW	YELLOW	YELLOW	BLUE	BLUE	BLUE		
	4	YELLOW	YELLOW	YELLOW	BLUE	BLUE	BLUE		
		BLUE							
		1	**1**	**1**	**5**	**5**	**5**		
GREEN	**3**	GREEN	GREEN	GREEN	BLUE	BLUE	BLUE		
	3	GREEN	GREEN	GREEN	BLUE	BLUE	BLUE		
	3	GREEN	GREEN	GREEN	BLUE	BLUE	BLUE	BLUE	18
	3	GREEN	GREEN	GREEN	BLUE	BLUE	BLUE	GREEN	18
	3	GREEN	GREEN	GREEN	BLUE	BLUE	BLUE		
	3	GREEN	GREEN	GREEN	BLUE	BLUE	BLUE		

continues on following page

Table 5. Continued

		BLUE							
		1	1	1	5	5	5		
RED	2	RED	RED	RED	BLUE	BLUE	BLUE		
	2	RED	RED	RED	BLUE	BLUE	BLUE		
	2	RED	RED	RED	BLUE	BLUE	BLUE	BLUE	12
	2	RED	RED	RED	BLUE	BLUE	BLUE	RED	24
	6	RED	RED	RED	RED	RED	RED		
	6	RED	RED	RED	RED	RED	RED		
		YELLOW							
		0	0	4	4	4	4		
GREEN	3	GREEN	GREEN	YELLOW	YELLOW	YELLOW	YELLOW		
	3	GREEN	GREEN	YELLOW	YELLOW	YELLOW	YELLOW		
	3	GREEN	GREEN	YELLOW	YELLOW	YELLOW	YELLOW	YELLOW	24
	3	GREEN	GREEN	YELLOW	YELLOW	YELLOW	YELLOW	GREEN	12
	3	GREEN	GREEN	YELLOW	YELLOW	YELLOW	YELLOW		
	3	GREEN	GREEN	YELLOW	YELLOW	YELLOW	YELLOW		
		YELLOW							
		0	0	4	4	4	4		
RED	2	RED	RED	YELLOW	YELLOW	YELLOW	YELLOW		
	2	RED	RED	YELLOW	YELLOW	YELLOW	YELLOW		
	2	RED	RED	YELLOW	YELLOW	YELLOW	YELLOW	YELLOW	16
	2	RED	RED	YELLOW	YELLOW	YELLOW	YELLOW	RED	20
	6	RED	RED	RED	RED	RED	RED		
	6	RED	RED	RED	RED	RED	RED		
		GREEN							
		3	3	3	3	3	3		
RED	2	GREEN	GREEN	GREEN	GREEN	GREEN	GREEN		
	2	GREEN	GREEN	GREEN	GREEN	GREEN	GREEN		
	2	GREEN	GREEN	GREEN	GREEN	GREEN	GREEN	GREEN	24
	2	GREEN	GREEN	GREEN	GREEN	GREEN	GREEN	RED	12
	6	RED	RED	RED	RED	RED	RED		
	6	RED	RED	RED	RED	RED	RED		

The answers are:

1. No
2. Red

3. Yellow

We all agree that if A > B and B > C then A > C. The strange thing, with these dice, is that it is not worth the transitivity.

On a large number of throws it can be easily verify (Figure 14) that the Red dice beats the Blue which, in turn, defeats the Yellow; the latter beats the Green while Blue and Green have the same chance of winning.

Figure 14. Strange dice results

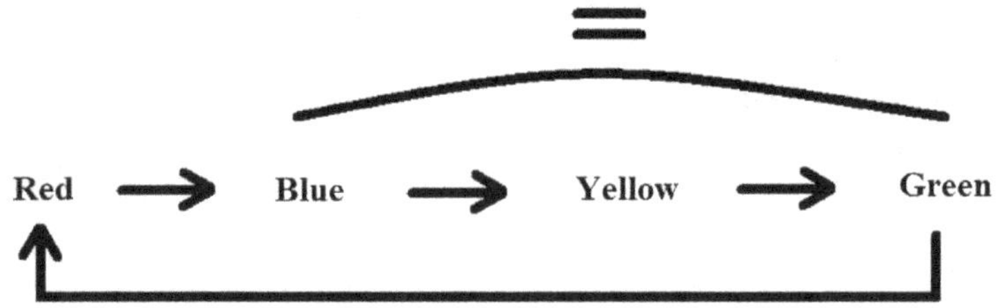

In mathematics we frequently speak of intransitivity, for example the relation "is perpendicular to ..." is not transitive: if line A is perpendicular to line B, and line B is perpendicular to line C then line A is not perpendicular to line C.

There are many contributions on the subject, on the construction of non-transitive dice we refer to the work of Angel & Davis.

ETRUSCAN DICE

To continue the discourse started with the previous object, it is interesting to capture a particular moment in the history of dice. These have been used throughout the Orient since time immemorial and, in all probability, their origin is Asian. When they arrived in the West, they emotionally involved the Romans who were passionate players. Historically, the phrase *alea iacta est* (The die is cast) pronounced by Gaius Julius Caesar when he crossed the Rubicon River with his army to march towards Rome (49 B.C.) is famous, even if not supported by certain evidence. Less well known is the fact that a previous Italic population, the Etruscans, used the dice both as entertainment and as instruments of magical ritual character to decree fate.

The Etruscan dice were marked with dots, but there is also a version that instead had the points indicated by the respective Etruscan words, from one to six (remember that the Etruscans wrote from right to left).

Most likely the five corresponded to the term makh. The six Etruscan numbers have been interpreted as follows: thu = 1; zal = 2; ci = 3; sa = 4; makh = 5; huth = 6 with a possible inversion between sa and huth. The fact remains that the Etruscans used different schemes (Artioli et al.), the best known were:

1. opposite faces show pairs of numbers in ascending order: 1-2, 3-4, 5-6 (until the 5th century BC);
2. opposite faces giving as sum seven: 1-6, 2-5, 3-4.

In reality, in the clay reproduction presented in Figure 15 and coming from the museum of Tarquinia (Italy) there is another association present in the dice: 1-2, 3-5, 4-6.

Figure 15. Etruscan dice

PLAYING CARDS

Among the objects that can be used for learning or consolidating the main probabilistic concepts are the playing cards, for example the type shown in Figure 16.

Most of the problems posed at school using this artifact are repetitive and not worth dwelling on.

Figure 16. Playing cards

It is more interesting to propose some less routine examples.

1. Give a deck of 52 cards (26 red and 26 black) well shuffled, i.e. all possible arrangements are comparable. What is the probability that the cards are arranged in such a way that there is an alternation of red and black cards? First of all we need to introduce the concept of *factorial*. In mathematics, the factorial of a natural number *n* defined as the product of positive integers less than or equal to that number. This quantity is indicated with the symbol *n!*. To have all the possible orders it is necessary to multiply 52 by 51 and then by 50 and so on. So you will have: $52! \approx 10^{68}$. The possible alternating order can be thought of as obtained by shuffling separately the 26 red cards and the 26 black cards and then alternately taking one card from each deck; taking into account that you can start with a red or a black card you have $2 \bullet 26! \bullet 26! \approx 10^{53}$. The probability sought is therefore about 10^{-15}.
2. The battle game is a game that fascinated the King of France Charles VI (1368 -1422) who spent whole days playing with his favorite Odette de Champdivers. Two decks of cards shuffled face each other card after card, whoever put the highest card could take the cards played (in case of tie two new cards were turned over). The game ended when one of the two was left without cards. The problem has been widely analyzed by Jean-Paul Delahaye and its solution is quite complex. A more affordable problem, but a typical example where a computer simulation is of great help, is the following: two players, *A* and *B*, play with their own deck by placing one card at a time on the table. If there is a coincidence (same card

same suit) A wins otherwise B wins; should we bet on A or B winning? Is there anyone favorite? The intriguing aspect of the solution is that the probabilities in favor of non-coincidence are lower, in particular the odds are worth 1/e where e, Euler's number (1707-1783), is a transcendent number whose approximate value is 2.718...

3. An interesting game is the one proposed by the French mathematician Mickael Launay in which the track consists of playing cards of the four suits between the ace (which is worth one) and the six. Arranged the 24 cards in order to form a circuit, the game consists of placing a toy car on one of the first six cards placed after the start/finish indicator and making the movements indicated by the value of the card on which the car is located. The game setting can be seen in Figure 17.

Figure 17. Circuit of cards

The interesting thing is that you can give a prediction, with p = 0.93, of where the race will end after the finish. In the example the car starts from the 4 of spades and ends its run on the 2 of diamonds, the card placed in the position indicated by the value (3) of the card placed before the finish. If you invert the 4 of spades with the 2 of diamonds the result does not change, the car will end its run on the third card placed after the start/finish point.

We used the term prediction, just as we could have used the synonyms: forecast, prognostication, prognosis, divination, augury. In this case it would have been more correct to speak of an indication of where the machine will stop. The game looks like a magic invented by a skilled magician, but in reality it is the calculation of probability that tells us what, with extremely high probability, will occur. The fact is that the journey of the toy car is only apparently independent of the value of the cards; the movement mechanism is an example of dependence.

OCTOPUS

The octopus is a cephalopod mollusk that has two interesting aspects: it has three hearts and can change color adapting to the environment. In teaching, given the difficulty of having a real octopus, you can resort to a reproduction of the type shown in Figure 18.

Figure 18. Reproduction of an octopus

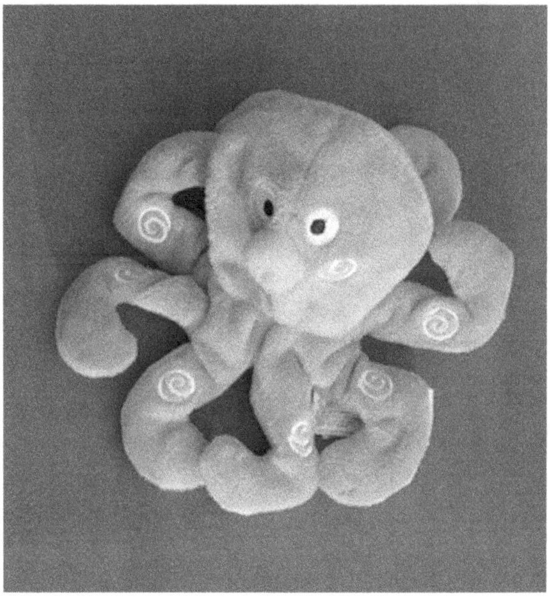

One of these animals has gained enormous popularity as an octopus predictor. The story is well known and particularly interesting. The octopus Paul (January 2008, October 2010) lived in a public aquarium at the marine life center in Oberhausen, Germany, and gained international fame at the 2010 World Cup when it was used to try to “predict” the results of football matches involving the German national football team and the final not played by Germany.

The method used to make the prediction was simple: the usual nourishment was divided into two boxes with the flags of Germany and the National challenger on top. The National team, whose flag was on the box that the octopus first opened to feed, was predicted to be the winner.

The "predictions" as shown in Table 6, were all correct.

Table 6. Table of octopus predictions and results

Opponent	Phase	Prediction	Result	Outcome
Australia	Qualifying group	Germany	4-0	Ok
Serbia	Qualifying group	Serbia	0-1	Ok
Ghana	Qualifying group	Germany	1-0	Ok
England	Eighth of final	Germany	4-1	Ok
Argentina	Quarterfinals	Germany	4-0	Ok
Spain	Semi-final	Spain	0-1	Ok
Uruguay	3rd-4th place	Germany	3-2	Ok
Netherlands - Spain	Final	Spain	0-1	Ok

By now it should be established that the probability of guessing 8 out of 8 results is 1/28, or 1/256.

This is a relatively low probability and the result is comparable to flipping a coin eight times and always getting a head (considered as a winning choice). Basically, the explanation can be traced back to a concept that has been defined by Persi Diaconis and Frederick Mosteller as *Law of truly large numbers*: given a fairly large number of opportunities, it is likely that any event can happen. What justification can be given for this seemingly surprising result from a probabilistic point of view? Let us remember that with many animals in many other countries (parrots, porcupines, crocodiles, hippos, etc.) the same experiment was attempted and failed.

With enough opportunities, even the most excessive thing is likely to happen. Suppose that the event, for people who do not live in the city, to be struck by lightning during their lifetime, has an extremely low probability of happening, say 0.0001, so the probability of the event not happening is 0.9999. For the population of a rural community of 10000 people, the probability of being struck by lightning is $1- 0.9999^{10000} = 0.63$; not a little! Of course, in the case of the octopus, we do not have a striking case of improbability, but the example is still sufficiently fitting.

One of the most shocking cases of the occurrence of an incredibly unlikely event can be traced back to 24 March 2001. In a baseball game between the San Francisco Giants and Arizona Diamondbacks, pitcher Randy Johnson hit a bird on the fly with the ball. The fact is documented by a video that still circulates on the Web today. No way? An exceedingly rare event, but one that can only find its explanation if one thinks of how many baseball games were played, since the first baseball match is dated June 19, 1846.

BIRTHDAY CAKE

One of the most important moments in a birthday is the cutting of the typical cake of this party (Figure 19). Many problems are connected with this pastry product, for example those related to the cutting of a cake. If we want to look for a witty introduction to the problem we can quote the text by Lewis Carroll (1832-1898) *Through the Looking-Glass* where we find the nonsense: "You don't know to manage Looking-glass cakes, the Unicorn remarked. Hand it round first, and cut it afterwards".

A classic question is the following: with three straight cuts you can cut a cake into a number of pieces (not necessarily equal) not exceeding 7; with 4 cuts (c) the portions (p) are 11. And with *n* cuts? The answer is in the second-degree function $p = 1/2\ c^2 + 1/2\ c + 1$.

In this case, however, given the theme, it is necessary to find a probabilistic cue.

Figure 19. Birthday cake

The best-known example is known as the birthday problem, it is a paradox (in the sense that it goes against common sense) of probability theory defined in 1939 by the American naturalized Austrian mathematician, engineer, and academic Richard von Mises (1883-1953).

The certainty that at least two people in a group have their birthday on the same day is only when there are at least 366 people, or 367 if the year is leap year. The paradoxical aspect is that already in a group of 23 people the probability is about 51% and with 50 people reaches 97%. Let's consider some football matches, where there are 22 players plus a referee, then we come to the fateful number of 23. On January 3, 2010, in Spain the match Valencia - Espanyol is played and two players (Brino & Fortin) have as date of birth January 10. On March 8, 2020, the match Juventus - Inter is played in Italy and two players (Ronaldo & de Vrij) were born on February 5. Of course there are almost as many examples to the contrary: on May 3, 2005, in England Liverpool and Chelsea clashed, but in this case there are no anniversaries. These results seem to clash against intuition, but they are correct and the explanation is quite simple using composite and opposite probabilities.

The easiest way to calculate the probability is to calculate the probability that this does not happen: given any person in the group, there are 364 cases out of 365 (in the case of a leap year, just increase the values by one) where a second person's birthday occurs on a different day; if you add a third person, there are 363 cases out of 365 where he or she has a birthday on a different day from the first two people and so on. In other words, indicating with x the persons involved and with $P(\bar{b})$ the probability that there are no coincidences of birthdays:

$$P(\bar{b}) = \frac{364}{365} \bullet \frac{363}{365} \bullet \ldots \bullet \frac{365-x+1}{365}$$

The probability that there are at least two equal birthdays is therefore:

$$P(b) = 1 - P(\bar{b})$$

Unlike the case in which you want to find at least two people celebrating the birthday on a certain day, even if in different years. In this case you have to multiply the same 364/365 fraction x times and then find the complement. In this case, 254 people are needed in order to have two people who have their birthday on a given day with a probability of 1/2 and with 365 people the probability increases to only 61%.

CANDLES

Votive candles are nothing more than candles, generally placed inside red plastic wrappers that serve to protect the small flame from the wind. A fairly common type is that of Figure 20.

Figure 20. Votive candles

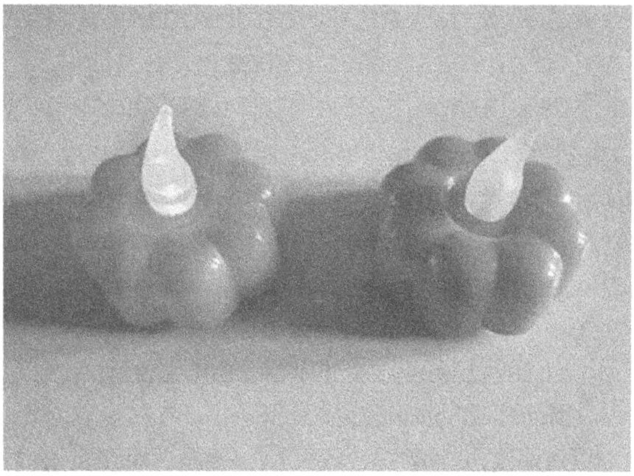

In fact, in churches, many candles are now electric and are lit for a fee. Figure 21 shows a set of candles in a church in Braga, Portugal.

Figure 21. Initial situation

The question, after the insertion of a coin, is: is the lighting of the lamp managed by an algorithm or is it by chance?

If you compare the previous image with the following one (Figure 22) you can see that the input of 10 cents allowed the lighting of the lamp located in the third place of the sixth line starting from the bottom.

Figure 22. Situation after lighting a lamp

Since there were 39 places available, it can be said that the probability of that particular ignition was 1/39 = 0.026. Based on this observation we can see that the probability, even if not to be discarded, is not high. It is quite plausible to think that perhaps there could be another cause for the ignition. In fact, it is an electronic board, therefore an algorithm, which chooses which spark plug to light and adjusts the on/off times.

The non-electric candle is a light source typically consisting of a wick immersed in a column of solid fuel, usually some kind of wax. Here, however, speaking of probability, we want to introduce a very particular type: Japanese candles.

Actually, these are not real candles but icons used in online trading, buying, and selling financial instruments over the internet. The Japanese candles are composed, for each period considered, of a body and 2 lines, one at the top and one at the bottom that represent the four prices of opening, closing, maximum and minimum. Each candle can also take a different color (red and green or black and white) depending on whether the market closing is positive or negative compared to the opening. An example is shown in Figure 23.

Figure 23. Candlestick chart

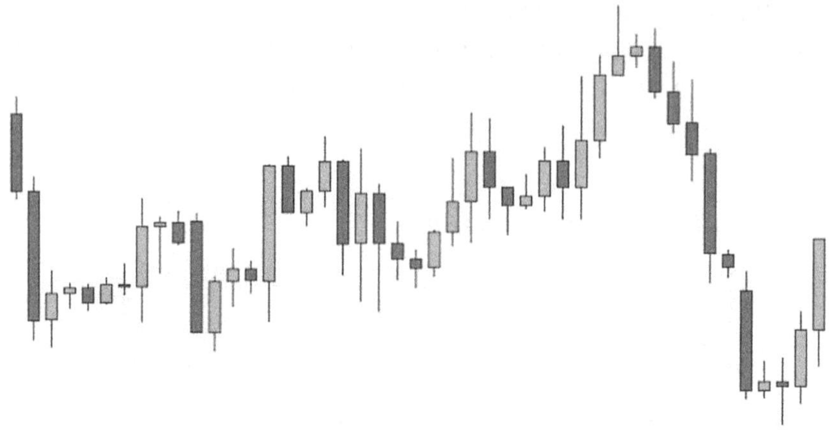

Japanese candles have a specific structure:

- Real body: indicates how much the price has changed in the trading sessions of the day; the body of the candle is proportional to the price change;
- Shadows: these are the lines that come out of the candles and indicate the maximum and minimum prices of the day;
- Open and Close: the extreme edges indicate the opening and closing prices.

The objective of each trader is to define, based on the configuration of the chart, the probability that the price will rise or fall in order to trade on the market.

For example, the larger the size of the candle body, the more likely it is that the price will continue to follow the same direction; if the two shadows are of different lengths it is likely, according to analysts, that the price will go in the direction of the shorter one.

Obviously, here, we are not interested in the examination of patterns, but only to underline how they are tools closely related to probabilistic forecasts in the presence of phenomena with a wide margin of randomness.

MOUSE

The mouse is a small rodent mammal of the Muridi family. Also in this case it is preferable to show in class a fake mouse like the one in Figure 24.

Figure 24. Toy mouse

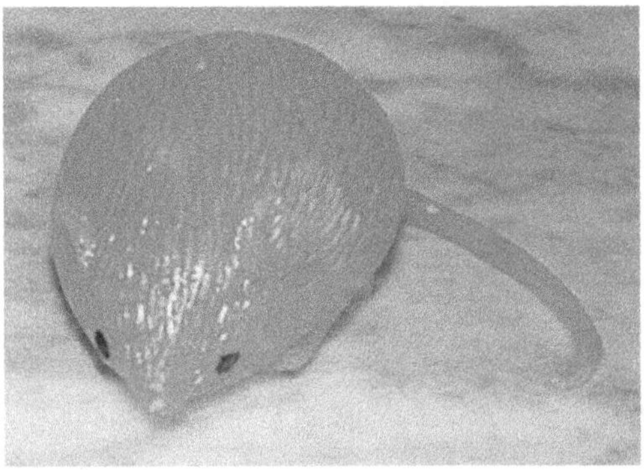

Mice are notoriously among the most widely used animals for experiments on animal behavior. Let us suppose then that a mouse is placed at the beginning of the labyrinth of Figure 25:

Figure 25. Maze

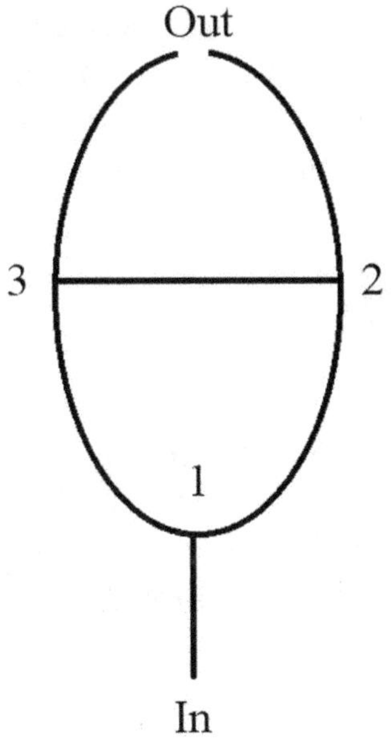

The mouse chooses at random which road to take, without going back in his steps when he meets a fork in the road, with identical probability, that is to say in the choice between two possible roads it is as if he throws a coin every time. We want to quantify the probability that the mouse returns to the starting point. In this case it can be of help to visualize with the graph of Figure 26 the possible routes.

Figure 26. Maze graph

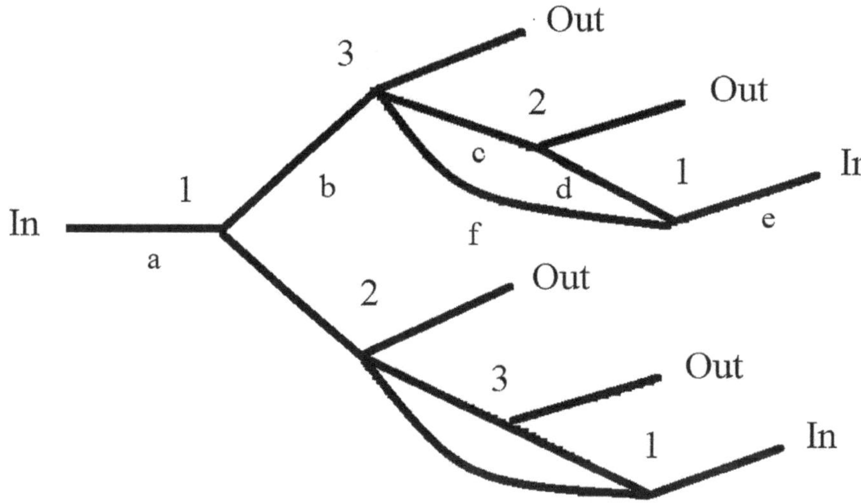

As you can see from the figure the mouse can return to the entrance if it follows the paths a-b-c-d-e or a-b-c-d-f-b-a of the upper branches; the same will happen for the lower ones so the result must be doubled.

In conclusion, the probability is double the sum of the compound probabilities:

$P = 2 \bullet (0.5^5 + 0.5^7) \approx 0{,}078.$

Let's examine another simple problem: in an experiment a mouse is placed at the beginning of the path indicated in Figure 27.

The mouse chooses the road on the left with a probability equal to one tenth of the cheeses that are in the bowls at the end of the various paths, so the left turns are 1/10, 2/10, 3/10 probability to occur. When the mouse reaches one of the bowls it stops.

Figure 27. Bowls with cheese

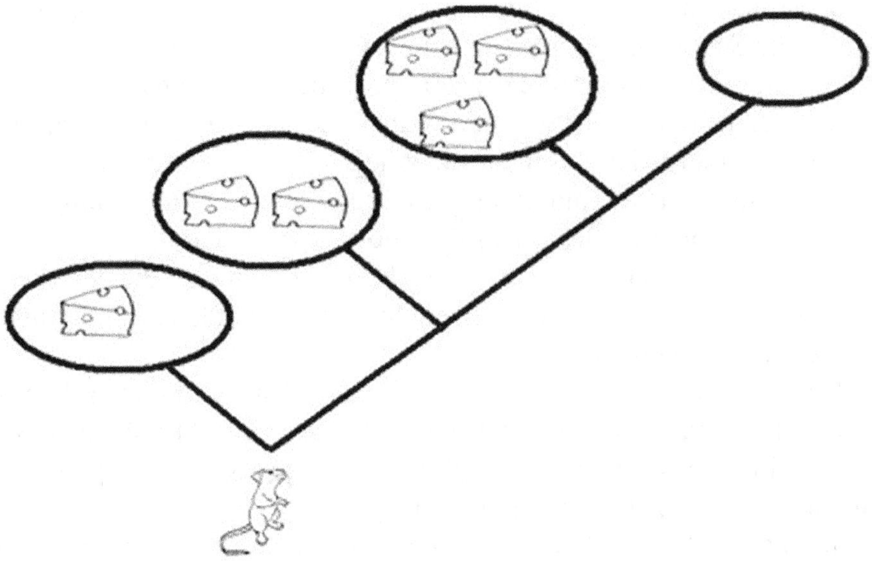

You want to know what the odds are of the rat:

1. eat a slice of cheese;
2. eat two slices of cheese;
3. eat three slices of cheese;
4. don't eat any slice of cheese.

In the first case the answer is immediate, 1/10. To answer the second question we have to think that the probability is given by the product of the single probabilities because it is a case in which the question is "don't eat in the first bowl **and** eat in the second", so P = 9/10 • 2/10 = 18/100. A similar reasoning should be made for the third question where it is "do not eat in the first bowl **and** do not eat in the second **and** eat in the third", then P = 9/10 • 8/10 • 3/10 = 216/1000. Finally, the probability that he does not eat anything is the residual probability compared to the previous cases, so P = 1 - (1/10 + 18/100 + 216/1000) = 504/1000. In conclusion:

1. 0.1
2. 0.18
3. 0.216
4. 0.504

About cheese: in the analysis of the risks associated with human error, various models have been proposed, one of particular interest is the one known as the *Swiss Cheese Model* or theory of latent errors.

What is the basic assumption in this approach? According to the professor who proposed it (Reason, 1990), accidents and errors are only the tip of the iceberg, the submerged part is made up of negative events that did not happen just because someone or something prevented them from happening.

From this vision, the idea arises that the occurrence of an accident or an error is the result of a chain of events that have overcome all the defenses put in place. Reason clarified more precisely the meaning of latent error, through the Swiss Cheese Model (Figure 28).

As he wrote in a subsequent essay: "In an ideal world each defensive layer would be intact. In reality, however, they are more like slices of Swiss cheese, having many holes — though unlike in the cheese, these holes are continually opening, shutting, and shifting their location. The presence of holes in any one "slice" does not normally cause a bad outcome. Usually, this can happen only when the holes in many layers momentarily line up to permit a trajectory of accident opportunity — bringing hazards into damaging contact with victims ". (Reason, 2000).

Figure 28. Swiss Cheese Model (Author: BenAveling)

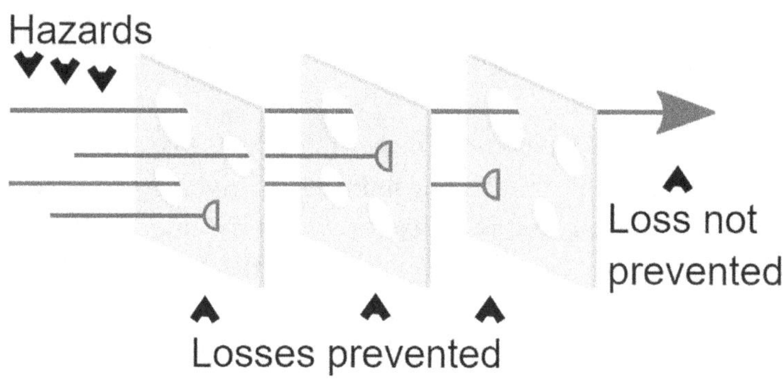

CHESS

Where the game of chess comes from is not entirely certain, the most reliable sources say that the game was probably invented in India around the sixth century. This first form of chess was played on a board called *ashtapada* (which means eight feet). The best known legend, however, traces back to

Persia the birthright of the game of chess with the story of Sussa ibn Dahir, confirmed by many texts written in ancient Persian: Sussa is the wise man who, as a prize for the invention of the game, asked "simply" grains of wheat, one on the first square of the chessboard, two on the second, four on the third and so on, always doubling until reaching the 64th square. Legend has it that Sussa put in crisis its rich sovereign, King Shiram. On the first square we have, in fact, a single grain (2^0), in the second two (2^1), in the third four (2^2) and so on until reaching 2^{63}, equal to the incredible amount of 18446744073709551615 grains. This legend was known during the Middle Ages, so much so that the poet Dante Alighieri (1265-1321) also mentions it, where it is used in the *Divine Comedy* to give the reader an idea of the number of Angels in the heavens.

Chess then spread to the West, the most famous set of this genre is the one known as Lewis's Chess built around 1200 in Scotland and currently preserved, in part, at the British Museum in London, but perhaps they are so famous because they are used in the film "Harry Potter and the Philosopher's Stone".

The most interesting aspect is that the figures have changed over time, for example there are no longer the Indian elephants or the Persian vizier and, compared to Lewis' chess, the chariot became a Rook while the elephant became the Bishop. The most modern aspect is that in the West the Queen appeared, a female figure not initially foreseen. In the analysis of the object we will obviously refer to the modern version.

For the few who didn't know, chess is a strategy game that takes place on a square board called the chessboard, formed by 64 squares on which each player has 16 pieces: a King, a Queen, two Bishops, two Knights, two Rooks and eight Pawns. Each of the pieces moves with precise rules of direction and distance. The objective of the game is to attack the opponent's King by preventing him from evading capture, this is known as checkmate (from the Persian *Shāh Māt*, the King is dead).

It is interesting to examine whether there really exist a correspondence between the force (value) attributed by the theory to each piece and the probability of attacking the King.

Imagine placing one of the pieces, at random, in one of the 64 squares and then placing the King in one of the remaining boxes. There will then be $64 \bullet 63 = 4032$ possible pairs. Now you have to consider how the various pieces move. The King moves one square in any direction. The Knight has the characteristic L-shaped movement shown in Figure 29:

Figure 29. Moves of the Knight

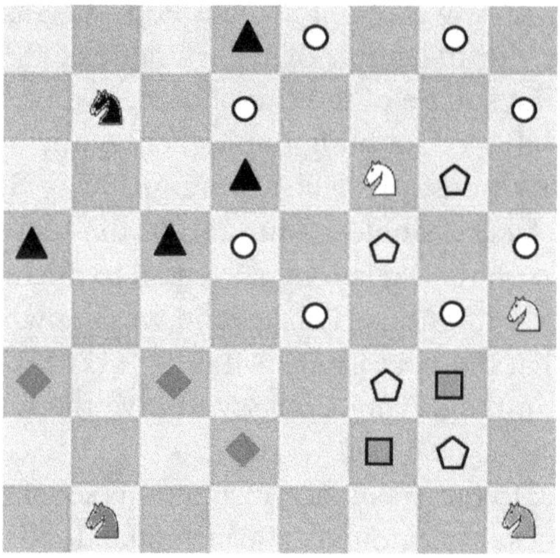

Let us examine the Knight to which strength is attributed 3. There are 4 positions where each Knight can threaten exactly 2 squares, there are 8 positions where the Knight controls 3 squares, there are 20 where he controls 4 squares, 16 where he controls 6 squares and another 16 where the Knight controls 8 squares, as shown. We can say that the offensive probability of the Knight is given by:

$$P(C) = \frac{4 \cdot 2 + 8 \cdot 3 + 20 \cdot 4 + 16 \cdot 8}{4032} = \frac{336}{4032} = \frac{1}{12}$$

The Rook, on the other hand, can move, as shown in Figure 30, on any square of the same crossbar or column with respect to the one in which it is located:

The Rook to which force 5 or 6 is attributed, depending on the chess texts, is always able to threaten 14 squares, 7 horizontally, 7 vertically and therefore there would be 64*14 situations. Very offensive piece!

However, if the King is in a square adjacent to the Rook he could eat the Rook (which cannot happen for the Knight) and this conditional fact must be taken into account.

Let's define with P(B|A) the probability of an event B conditioned by event A, with P(A) $\neq$ 0, the probability that B occurs assuming that A has occurred. In this case we have the *Law of Conditional Probabilities*:

$$P(B|A)=\frac{P(A\cap B)}{P(A)}$$

Resuming the problem of the Rook (*R*) threatened by the King (*mK*) the account of the various cases leads to:

$$P(R|mK)=\frac{36\cdot10+24\cdot11+4\cdot12}{4032}=\frac{672}{4032}=\frac{1}{6}$$

Figure 30. Rook moves

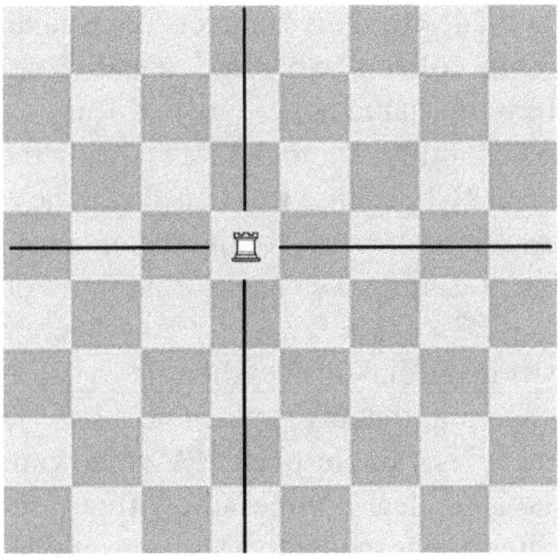

So it's correct to say that the Rook is worth almost 2 Knights. Similar reasoning can be made for the Queen moving in all directions and the Bishop moving diagonally.

Clarified some aspects related to the strength of the pieces there are a couple of interesting probabilistic problems that are worth mentioning. The first is certainly the best known but also the one that seems to belong to the class of paradoxes.

Problem 1: a medium strength player must meet two players, one strong and one weak enough. To pass the category he must win at least two consecutive matches out of three playing alternately against the other two; what is the sequence of matches that gives him the highest probability of passing the test?

A lot of people, having heard the problem, have no doubt. It is better to start playing against the weaker opponent. Actually, the only chance for our average player is to win 1 and 2 or win 2 and 3. In any case it is necessary to win the second game and therefore it is better to face the weaker player in this one. Therefore the first opponent should be the strongest competitor. To convince the skeptics we try to provide numbers. The ELO score is a system of evaluation of chess players' playing strength, let's say that the strong player (A) has a score of 1800 ELO, the weak player (C) has an ELO score of 1400, while the intermediate player (B) is halfway (1600 points). The probability of victory in case of a difference of 200 points has been estimated as 0.76. If B meets first the strongest and then the weakest, he wins if he beats A and C or if he loses the first match but then wins the next ones: probability 0.24 • 0.76 + 0.76 • 0.76 • 0.24 = 0.32. If he meets the weakest first and then the strongest wins only if he collects two wins: probability 0.76 • 0.24 = 0.18.

Problem 2: in a chess tournament there are four Russian and four American players. What is the probability that in random pairing the four Russians will meet the four Americans? Choosing a Russian player they will meet one of the remaining seven players with p = 4/7, for the second the probability becomes p = 3/5, repeating the reasoning we can say that the probability sought will be given by 4/7 • 3/5 • 2/3 • 1/1 = 8/35.

Let's try to move from the classic chessboard to a game that was extremely popular a few years ago and that takes place on a base formed by a number of boxes. "Minesweeper" is a single-player PC video game invented in 1989. The goal of the game is to clear a minefield without exploding mines. The game was later rewritten and turned into a less aggressive game with a flower instead of the mine. A number of mines are scattered on a chessboard. For example, on a field of 81 squares there are 10. Selecting a square either reveals a mine (and the game is over!) or a series of empty cells or a number. If a number appears in the square, this indicates the number of adjacent squares containing mines. If no number appears, it means that all 8 adjacent squares do not contain a mine, in which case the game automatically discovers all empty squares adjacent to the one clicked until you get to squares containing a number. The player can mark any square in which they believe a mine with a flag is present,

The aim of the game is to find all the mines on the playing field as quickly as possible, without discovering the corresponding boxes. Figure 31 shows a possible final configuration:

Figure 31. Result of a match

Now you can ask yourself a couple of questions:

1. what is the probability of hitting a mine on the first attempt?
2. what are the chances of hitting a mine on the second attempt?

The answer to the first question is trivial, the result is given by the ratio between the mines and the boxes: 10/81. The answer to the second question is more articulated, as it depends on the number n that came out: in general the probability of finding the mine in a contiguous box is n/8, in a non-contiguous box is (10 - n) / (81 - 9). That the game is not simple is demonstrated by the fact that, in 2000, Richard Kaye published a demonstration that the game is an NP-complete problem. But what does that mean? Determining whether a given set of mines can be deduced from the numbers that appear can be a difficult task, a second problem is the number of clicks needed to finish the game without exploding. This second aspect is remarkably interesting, when studying a solving algorithm one of the problems is the time needed to solve it. For example, determining which purchase cost us the most, scrolling through the receipt of a supermarket, takes a time that depends on the length of the list while sorting prices in ascending order with the *bubble sort* method requires a polynomial time, i.e. no more than the square of purchases made.

P problems, or polynomial problems, are those decisional problems whose solution requires a time that can be expressed with a polynomial $O(n^k)$ where k is a positive integer. An extension of this class of problems are those problems for which a non-polynomial time is required to verify if the solution is correct and are called *NP* (from Nondeterministic Polynomial time). For example, to factor a large number into prime numbers takes many years of work for a supercomputer. The algorithm that factors the numbers not using a trivial method like brute force takes a time that grows not in Polynomial (P) mode, but exponentially (so NP). The fastest algorithm known to date is called General Number Field Sieve, with a complexity in the order of:

$$O\left(e^{\left(\frac{64}{9}b\right)^{\frac{1}{3}}}\left(\log b\right)^{\frac{2}{3}}\right)$$

This is nothing if we consider that NP-complete problems are the most difficult problems within the NP class. One of the most famous problems is the so-called *travelling salesman problem* that we will be able to deal with later.

TRAFFIC LIGHT

The traffic light (Figure 32) has a lot to do with probability.

Figure 32. Traffic light

First of all, we must dispel the myth that the traffic light is always red for us. In reality we realize the negative event "the traffic light is red" forgetting all the cases in which the traffic light was green (or yellow).

If we indicate the probability of finding a traffic light in a certain state with p_v, p_g and p_r, the probability of finding all green lights in a street with six traffic light junctions is $(p_v)^6$ and, even in the quite unrealistic hypothesis of $p_v = 0.9$ we get a probability of finding a "green wave" of just over 50%.

An interesting phenomenon concerns the formation of queues.

Let's suppose that the traffic light has an interval of 20 seconds for green and 40 seconds for red/yellow, then a total duration of 60 seconds for each cycle. In the 20 seconds of green, let's imagine that 10 cars can pass. Let's assume that the arrival rate of the cars is relatively constant. If less than 10 cars pass every 60 seconds, traffic flows normally. But what happens when 11 cars arrive in 60 seconds? Of these, only ten will pass and one will remain blocked. The queue will continue to increase with each traffic light cycle. A critical aspect is that every time the traffic light changes color, there is a period when nobody crosses the intersection. If the signal changes frequently, the downtime takes on a relevant value compared to the total time and, therefore, tends to reduce the flow through the intersection. If, on the other hand, the signal lasts for a long time, the waiting times of the green for one of the two directions is longer and this leads to an accumulation of delays. The task of planners is to find a correct balance between these two situations. The study of traffic flows is increasingly a problem for technical and political managers and is a subject on which many experts have been confronted. Among the many works we would like to mention Novikov and al. The object of their study is the movement of vehicles on intersections controlled by signals. As the authors say: "Analysis of effectiveness of existing approaches to calculation of the traffic lights control cycle time allows development of a mathematical model for justification of algorithms for traffic lights control, taking into account widely varying characteristics of traffic flows".

SPINNER

Spinner is also known as Wheel of Fortune, a popular casino game, also played online, which later became a hit TV show. It is played with a large vertical wheel divided into several sectors in which there are numbers that correspond to the value of the prize.

Going back in time, however, we can find an origin in cards called tarot cards. In the past the Wheel of Fortune, or Rota Fortune was a symbol of the capricious nature of Fate (Figure 33). Life is a spinning wheel, sometimes you can get lucky and sometimes you can't, however there are things you cannot control, make happen or avoid.

Figure 33. Wheel of Fortune (Tarots)

Returning to Spinner we must say that it can also be a game not related to gambling, but an object that can be used as a simple pastime. In this case (Figure 34) we will examine a Spinner that is associated with the possibility of drinking from a glass.

The figure shows a Spinner divided into eight equal sectors in which four results are shown:

1. Lost turn;
2. Pass it;
3. Spin again;
4. Drink it.

Figure 34. Spinner

Of course, passing and losing the turn means you can't drink and it would seem that only a quarter of the cases are in favor, but if the Spinner stops on "Spin again" you still has a chance to turn the wheel. What are the chances of drinking? First of all you have to consider that being the sectors of equal amplitude, then the chances of stopping on a certain wedge are the same. In this case we can talk about the extension of the classic concept of probability: the favorable cases are represented by the area of the sector and the total cases by the area of the whole disc, so we can say that P = 1/8.

Returning to the initial problem the probability of winning on the first attempt is 1/4, but if it comes out "Spin again". (p = 1/4) you can try again. This means that the probability of victory is given by $1/4 + 1/4^2 + 1/4^3 + 1/4^4 + \ldots$, or, in other words, $1/4 \bullet (1+1/4+1/4^2+1/4^3+\ldots)$. To determine the value of the sum in brackets (mathematically it is a geometric series of reason q = 1/4) we must use the formula

$$S_n = \frac{1-q^{n+1}}{1-q}$$

when attempts tend to ∞ you will have, taking into account the product with the first term:

$$S_\infty = 1/4 \cdot \frac{1-1/4^\infty}{1-1/4} = \frac{1}{3}$$

Let's imagine repeating the experiment a certain number of times, the probability of drinking approaches this value: for example, if the attempts are even only 3, you will have p = 0.0332.

In the previous example the Spinner was divided into eight equal sectors and therefore the probability was the same. But what happens if the sectors are not the same size? To answer you have to determine the area of each single segment, which is possible by knowing the angle at the center of each sector and using the formula for calculating the area of the circular sector knowing the radius and angle expressed in degrees:

$$A = \pi r^2 \frac{\alpha}{360°}$$

There are online Spinners with the possibility to change the area of the sectors as you can see in Figure 35:

Figure 35. Adjustable Spinner

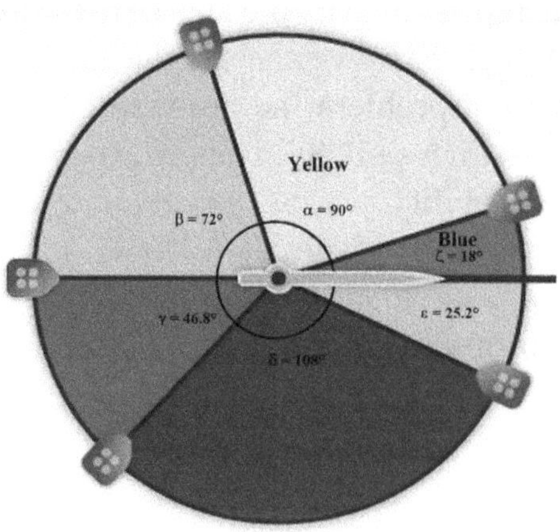

At this point the probability will be:

$$P = \frac{\pi r^2 \dfrac{\alpha}{360°}}{\pi r^2} = \frac{\alpha}{360°}$$

So, taking into consideration the yellow sector, you can easily obtain P = 0.25 while for the blue sector P = 0.05 and so on. A typical example that associates a probability with an arc of circumference.

This type of problem has been widely experienced in middle school. As an example we report the questions asked to a grade 8 class: “At the amusement park a wooden wheel, divided into colored sectors of different widths, turns quickly, while a player throws a dart, randomly hitting one of the sectors. What are the chances of winning the little doll combined with the blue sector? The operator of the carousel tells you that you win another dart if you hit the green or the red; what are your chances? Is the probability of hitting the turquoise color greater than the one that makes you win the dart?”.

It is interesting to highlight the reasons for the errors reported by the teacher: excessive haste, lack of understanding of the delivery, miscalculation. At the same time, it should be emphasized that the playful aspect and teamwork were a strong incentive.

SURPRISE EGG

Many confectionery companies sell, not only on the occasion of Easter holidays, chocolate eggs with surprises inside that children love to collect. The one shown in Figure 36 contains one of ten possible surprises, equally distributed. On average, how many eggs do you need to buy to find them all?

To facilitate the understanding of the question, let us first consider, as proposed by Frederick Mosteller, another type of problem: on average, how many rolls of a dice are needed to get a 6?

The probability with one roll is p = 1/6, with two shots is 1/6 • 5/6 where 5/6 represents the opposite probability q = 1 - p; in practice you have p • q. With three shots the probability becomes 1/6 • 5/6 • 5/6.

In general we will have the situation shown in Table 7:

Figure 36. Chocolate egg with surprise

Table 7. Table of probabilities

Attempts	Probability of success
1	p
2	$p \cdot q$
3	$p \cdot q^2$
4	$p \cdot q^3$
…	…
Σ	1

The symbol $\sum$ is a mathematical symbol that indicates, in a synthetic way, the sum of a certain set of addends. The symbol therefore indicates the sum of all probabilities, it must give 1. The average is given by the product of the probability attempts: $M = p + 2\,p \cdot q + 3\,p \cdot q^2 + 4\,p \cdot q^3 + \ldots$

Multiplying both members by *q* gives $q \cdot M = p \cdot q + 2\,p \cdot q^2 + 3\,p \cdot q^3 + 4\,p \cdot q^4 + \ldots$

Subtracting the second expression from the first one we have $M - q \cdot M = p + p \cdot q + p \cdot q^2 + p \cdot q^3 + \ldots$

Collecting M in the first member we have M (1- q) which is equivalent to the sum of the probabilities of success, i.e. 1. Being 1 - q = p we can write M • p = 1 from which we get M = 1 / p. On average we will have to do $m = \frac{1}{\frac{1}{6}} = 6$ tosses to get a six.

We now return to the egg problem. To get a surprise you obviously need to buy an egg and the probability is 10/10. To obtain a *different* surprise in the second egg the probability is 9/10 so you have to buy, on average, 1/(9/10) or 10/9 eggs. To obtain a third type of surprise, the number of eggs required rises to 10/8. Continuing you will have: 10/10 + 10/9 + 10/8 + ... + 10/1 or, picking at common factor, 10 • (1/10 + 1/9 + 1/8 +...+ 1). The sum is 29.29, so we will have to buy on average 30 eggs to get the complete set.

In general, indicating with *N* the number of surprises and with *O* the number of eggs to buy, we can write:

$$O = \sum_{i=0}^{N-1} \frac{N}{N-i}$$

There are other possible problems, for example, let's assume that each egg has a 1/5 chance of containing a surprise.

1. What is the probability that buying three eggs at least one contains a surprise?
2. How many eggs do you have to buy to have a 0.7 chance of finding at least one surprise?

The answer to the first question becomes easier if you consider the opposite probability $\overline{\mathrm{P}}$:

$$P = 1 - \left(\overline{P}_1 \bullet \overline{P}_2 \bullet \overline{P}_3\right) = 1 - \left(\frac{4}{5}\right)^3 = 0.488$$

As regards the second question, the most obvious way to determine the required value is to observe that $1 - \left(\frac{4}{5}\right)^n = 0.7$. The exponential equation must be solved and logarithms must be used to do this.

$$n = \frac{\log(0.3)}{\log\left(\frac{4}{5}\right)} = 5.4$$

Since it is not possible to buy a fraction of an egg, to have a 70% chance of finding a surprise the buyer should buy six surprise eggs.

LEGO

A completely new problem is to determine, starting from observed events, the probability of the causes that could have caused them. To clarify the concept let's examine a nice example taken from a blog that presented, in 2015, a problem called *Bayes' Theorem with Lego* and visible in Figure 37.

Even if not declared, we consider as a measure of probability the area constituted by dark grey and light grey bricks and we suppose that white bricks represent a subset of it.

Figure 37. Problem with Lego

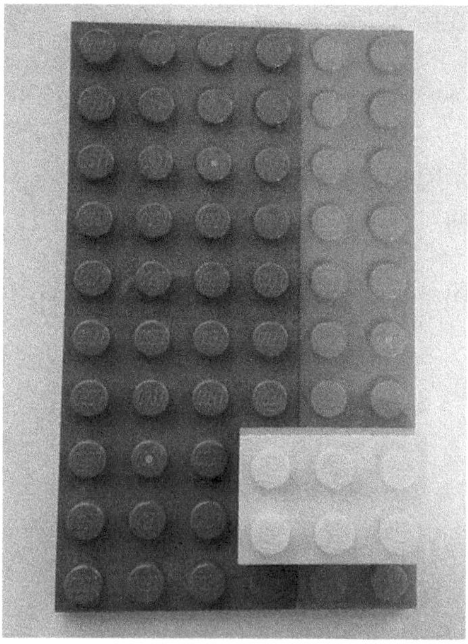

As said there are dark grey, light grey and white bricks. The whites fit on both dark gray and light gray bricks that form the base.

The dimensions, measured in "points" are Dark Grey = 40, Light Grey = 20, White = 6, the area of the base is 60. Let's now imagine to choose a point at random. We can assume that there is proportionality with respect to the area. From this information we can deduce the probability associated with each color with respect to the base:

P(DG) = 40/60 = 2/3, P(LG) = 20/60 = 1/3, P(W) = 6/60 = 1/10.

Separating the bases of different colors (Figure 38), so that four white pegs are on light gray and two on dark gray, we want to know what is the probability that the white portion is placed on the light base and on the dark base. In practice we are wondering how much P(W|GL) and P(W|GS) are worth.

Figure 38. Separate Lego

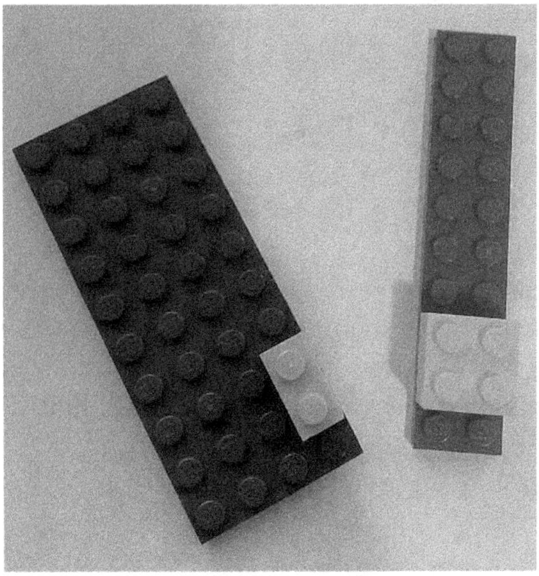

From the figure and the data provided it can be seen that 4/20 = 1/5: this is a conditional probability problem already examined.

Knowing that we are on white we now ask ourselves what is the probability that the base is light gray or dark gray, in other words how much P (GL I W) and P (GS I W) are worth. This is a more complex problem which, however, can be made clearer with the tree graph of Figure 39. In this case we can limit ourselves to the analysis of the clear basis only.

Figure 39. Lego graph

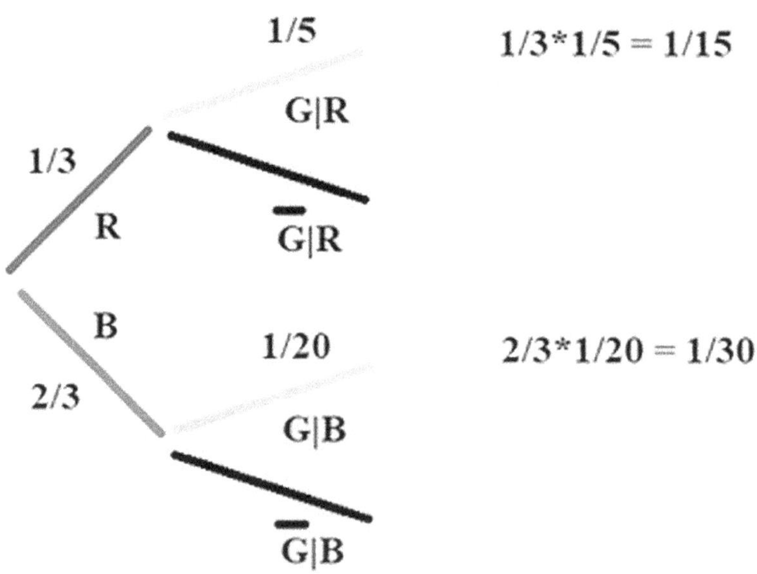

The branches that concern the white are two, but only the first one interests us because it indicates the probability that the white is on light gray while the probability of the white, in general, is given by the sum of the first and the third branch. Applying the law of conditional probabilities we can say that

$$P(GL|G)=\frac{P(GL\cap W)}{P(W)}=\frac{P(W|GL)\bullet P(GL)}{P(W)}=\frac{\frac{1}{5}\bullet\frac{1}{3}}{\frac{1}{10}}=\frac{2}{3}$$

For some examples on Bayesian probability on areas see also Impedovo.

Now it is a question of generalizing the discourse for which, considering a trial whose result E depends on H_1 events,...,H_i,...,H_n necessary and incompatible you have:

$$P\left(H_i|E\right)=\frac{P\left(H_i\right)\bullet P(E\mid H_i)}{\sum_{i=1}^{n}P\left(H_i\right)P(E\mid H_i)}$$

This formula expresses the *Bayes' theorem.*

Resuming the problem of Platonic dice seen previously the problem could become: I threw one of the five dice randomly chosen and got 1, what is the probability that the dice is the cubic one?

$$P\left(6|1\right)=\frac{\frac{1}{5}\bullet\frac{1}{6}}{\frac{1}{20}+\frac{1}{30}+\frac{1}{40}+\frac{1}{60}+\frac{1}{100}}=\frac{\frac{1}{30}}{\frac{27}{200}}=0.247$$

Bayes' theorem is also referred to as the theorem that provides the probabilities of causes, in the sense that, with it, it is possible to assign to each hypothesis, which can "explain" a certain event, the corresponding probability. The problem of determining, starting from observed events, the probability of the causes that could have caused them is of extreme importance and is used in quite different fields: medical, legal, etc.. In recent decades, a school of thought has developed, mainly in the United States, according to which the quantitative conception of probability, and in particular Bayes' theorem, would be the most appropriate conceptual tool to rationalize the judge's reasoning.

The history of this theorem is complex: the Bayesian approach was employed by Alan Turing to improve the machine used to decipher the Nazi Enigma code (McGrayne); the theorem is the basis of Big Data analysis and Data Analysis; it allows search engines to establish the importance of a site.

In the application of the theorem, based on conditional probability, one must avoid the typical error called *Confusion of the inverse* which occurs when, given two events A and B, we assume that the probability that P(A|B) is equivalent to P(B|A). In the Lego graph this means to confuse the probability that the base is red provided that the brick is yellow with the probability that the brick is yellow provided that the base is red.

This confusion can be extremely dangerous when, for example, the doctor confuses the probability that a test is positive when the patient is ill with the probability that the patient has the disease being the positive test. Given the importance of this theorem, some examples from different fields of application can be examined.

The first is an epidemiological problem which, instead of being based on hypothetical data, has fallen into a dramatically real situation. One of the strongest pandemics is the Coronavirus or Covid-19. The spread and virulence of this influenza epidemic has posed enormous human and health problems, but also has implications related to probability. The debate concerned the use of tests capable of diagnosing the positivity of individuals. On the one hand, there were those in favor of mass screening, and on the other, those in favor of a targeted use for people with a probable pathology. Screening is the examination of a larger or smaller segment of the population to detect a disease before it manifests itself with symptoms. However, it must be made clear that there is a difference between screening and diagnostic tests. Screening is a comprehensive examination of a larger or smaller segment of the population in order to detect a disease before it manifests itself with symptoms. When people who are sick (or suspected sick) are diagnosed, more precise and effective diagnostic tests are performed. Without wishing to enter into the debate, we can examine a possible Bayesian attitude to the problem. To do this we can rely on the (hopefully reliable) data that can be found from authoritative sources and that allow you to take a "photograph" of the situation in which New York found itself in 2020.

The population involved at the time amounted to 8.5 million and the percentage of Covid-19 infected in the months of March, April and May were, according to the Health Department, 28.1% of the population. A comparative analysis of different serological tests using the CLIA, CMIA and ELISA methodologies tells us that on average the sensitivity to M immunoglobulins (IgM), indicating an infectious state, is 82% and the specificity reaches 97%. The specificity of an examination is the probability that a healthy person will present a negative result; an examination is 100% specific when all healthy (H) people are negative. What is achieved by testing everyone? In other words, what is the probability that one is really sick (S) if the test is positive (P) and that one is sick if the test is negative (N)?

There are four possible cases:

1. a sick person is recognized as such: it is a True Positive (TP)
2. the patient is not diagnosed with the disease: it is a False Negative (FN)
3. a healthy person is mistakenly considered ill: it is a False Positive (FP)
4. a healthy is recognized as such: it is a True Negative (TN)

The tree graph in Figure 40 helps to highlight the results:

Figure 40. Epidemiological graph

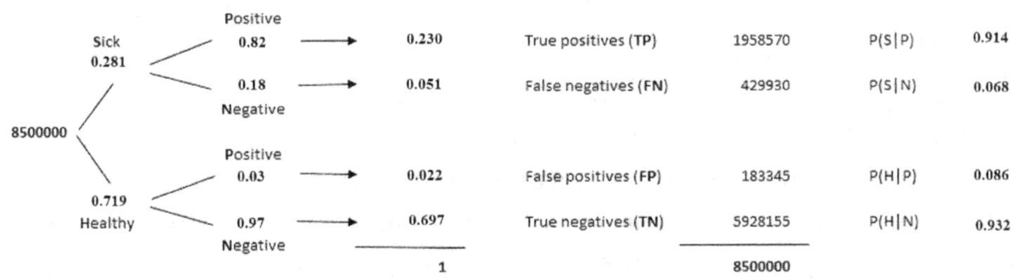

Now the question is what is the probability that an individual positive for the test actually has the disease or that an individual is infected even if the test has given a negative result. By Bayes' theorem we have:

$$P(S|P) = \frac{P(S) \cdot P(P \mid S)}{P(S) \cdot P(P|S) + P(H) \cdot P(P \mid H)} = \frac{0.0281 \cdot 0.82}{0.0281 \cdot 0.82 + 0.719 \cdot 0.03} = 0.914$$

$$P(S|N) = \frac{P(S) \cdot P(N \mid S)}{P(S) \cdot P(N|S) + P(H) \cdot P(N \mid H)} = \frac{0.0281 \cdot 0.18}{0.0281 \cdot 0.18 + 0.719 \cdot 0.97} = 0.068$$

This example shows how mass screening can lead to an incomplete assessment of reality. There are no perfect diagnostic tests: a test must combine sensitivity and specificity. Sensitivity refers to the ability to have as few false negatives as possible. A high specificity allows to reduce the cases of false positives, i.e. those who are diagnosed as carriers of infection or disease.

The second example, developed from a suggestion by blogger Raffaele Zenti, deals with a common life situation.

An SUV. in front of you throws a waste from the window. The first thought is that it's an uncivilized driver. Can we quantify that consideration?

Some data are obvious: the SUV throws a waste and its owner is uncivilized (U) or it is not $(\bar{U})$ the owner owns the SUV (S) or not $(\bar{S})$.

You want to estimate the probability P(U|S) that the driver is uncivilized knowing that he is the owner of the SUV. In order to use the formula it is essential to evaluate the probability that the owner is an uncivilized P(U); from a recent research it results that unfortunately there is a 31% of people belonging to this category P(U) = 0.31 and that 69% behave in a respectful and civilized way. $P(\bar{U}) = 0.69$. Now it remains to be defined the probability that an individual possesses an SUV being an uncivilized person. Suppose that you happened to observe 10 episodes of incorrect behavior including 2 committed by SUV owners, so we can say that P(S|U) = 0.2. Finally it remains to be established the share of SUVs among uncivilized people; a reasonable assumption is that it is proportional to the market share of this segment of cars which represents 34% of the car fleet, therefore $P(S|\bar{U}) = 0.69 \bullet 0.34 = 0.23$.

In conclusion:

$$P(U|S) = \frac{P(S\,|\,U) \bullet P(U)}{P(S|U) \bullet P(U) + P(S\,|\,\bar{U}) \bullet P(U)} = \frac{0.2 \bullet 0.31}{0.2 \bullet 0.31 + 0.23 \bullet 0.69} = 0.28$$

As you can see the reasoning is relatively simple but it is studded with many ifs. In order to apply the theorem it was necessary to estimate a series of probabilities based on the frequency of the share of uncivilized based on an opinion survey but also on certain behaviors observed personally. Moreover it was necessary to define the market share of the SUVs, in this case it was possible to refer to official data. This type of problem is related to different criteria of evaluation of the probability that will be examined in the following chapters.

To complete the framework we want to analyze another case where the theorem may be useful.

The Down Syndrome is a pathological condition which, according to the World Health Organization, has an incidence of about 1 case per 1000. The most accurate test for Down Syndrome requires a particularly invasive diagnostic test: amniocentesis. An alternative is the triple test, a prenatal screening method that can provide an estimate of the probability for a woman to have a child affected by chromosomal abnormalities. The triple test identification rate (i.e., the probability that a fetus with Down Syndrome will be correctly identified as affected by the syndrome) is a controversial figure but according to many sources it is about 0.7. The false positive rate is 0.03. What is the probability that the fetus presents this abnormality if the test is positive?

$$P(DS|positive) = \frac{0.7 \bullet 0.001}{0.7 \bullet 0.001 + 0.999 \bullet 0.03} = 0.023$$

Bayes' theorem can be difficult for a first approach, in fact there are two didactic tricks that can simplify its use. The first trick is to use the tree of alternatives to model the problem. The second trick, since the use of probabilities can cause difficulties in understanding the problem, comes from Gerd Gigerenzer. The suggestion is to use natural frequencies as shown in the tree with which the first problem was analyzed. Returning to the case of Down Syndrome the result, starting from 10000 cases, can be the following, shown in Figure 41:

Figure 41. Down Syndrome graph

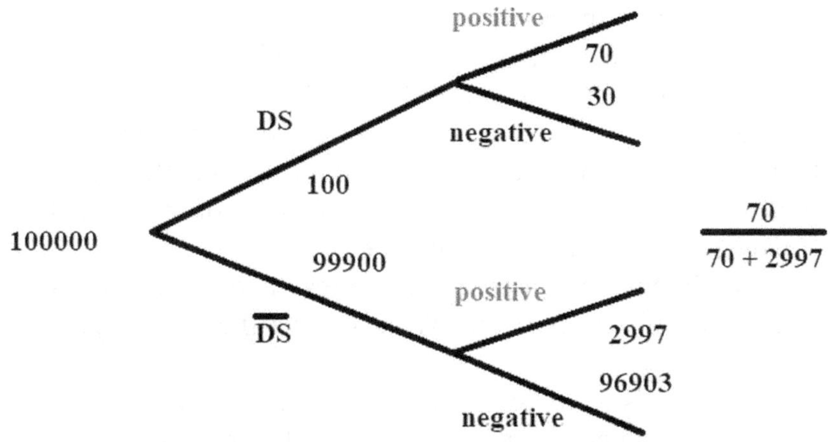

For a quick calculation of the various results and the final probability with the Bayes' Theorem you can usefully use an online calculator, within the website Online Statistics Education, developed by Rice University (Retrieved from http://onlinestatbook.com/2/probability/bayes_demo.html).

CONCLUSION

This first chapter served to introduce the focus of the book. The artifacts presented were used to introduce terminology and the fundamental laws of probability. Some are "classical" objects, others are not part of those commonly mentioned in textbooks. However, the most relevant aspect is the methodological aspect, i.e. the approach to the topic.

FUTURE RESEARCH DIRECTIONS

The objects were exploited to help understand some probability concepts. In fact, it is important that this issue be brought back into the broader cultural pathway that can link probability to other areas of mathematics.

A small example, taken from a 2017 seminar held in Florence, may serve to clarify the concept. Everyone knows the Pythagorean table, although perhaps, for someone, it is just more of a painful memory of primary school, more comfortably replaced by a pocket calculator. Another term learned in the same period is that of the remainder of a division. A deck of cards can serve to refresh the memory.

Suppose we extract the cards of two suits from the deck, discarding the figures we have two decks each with ten cards with values ranging from 1 to 10. We extract one card from each deck and multiply their value. Of this product we note only the unit number, this means determining the remainder of the division by ten. The same operation is done with the Pythagorean table, a table that presents the results of the multiplication of the digits from 0 to 9 and of which we keep only the remainder.

In this experiment we can compare the results of a random event with those of a deterministic operation. What will be the relationship between the two results?

The problem leads us to consider concepts that we will see in chapters 2, 6 and 8, but what interests us here is to underline the fact that probability lends itself very well to the introduction (or reinforcement) of mathematical contents, a strategic objective for the school.

Of course, this is only a simple example, purely indicative of a line of research.

REFERENCES

Angel, L., & Davis, M. (2017). A direct construction of nontransitive dice sets. *Journal of Combinatorial Designs*, *25*(11), 523–529. doi:10.1002/jcd.21563

Arrigo, G. (2010). Le misconcezioni degli allievi di scuola primaria relative al concetto di probabilità matematica. Rapporto di ricerca. Bollettino dei docenti di Matematica. Bellinzona.

Artioli, G., Nociti, V., & Angelini, I. (2011). Gambling with etruscan dice: A tale of numbers and letters. *Archaeometry*, *53*(5), 1031–1043. doi:10.1111/j.1475-4754.2011.00596.x

Carroll, L. (1917). *Through the looking glass: And what Alice found there*. Rand, McNally.

Dénes, J., & Keedwell, A. D. (1991). *Latin squares: New developments in the theory and applications* (Vol. 46). Elsevier.

Dhombres, J. (1992). *Leçons de mathématiques. L'École normale de l'an III, Laplace, Lagrange, Monge*. Dunod.

Diaconis, P., Holmes, S., & Montgomery, R. (2007). Dynamical bias in the coin toss. *SIAM Review*, *49*(2), 211–235. doi:10.1137/S0036144504446436

Doyle, A. C. (2004). *The Adventure of the Beryl Coronet*. Sterling Pub Co Inc.

Eastaway, R., & Windham, J. (1998). *Why Do Buses Come in Threes?* The Hidden Maths of Everyday Life. Robson Books.

Falk, R., & Konold, C. (1997). Making sense of randomness: Implicit encoding as a basis for judgment. *Psychological Review*, *104*(2), 301–318. doi:10.1037/0033-295X.104.2.301

Gigerenzer, G. (2003). *Calculated Risks: How to Know When Numbers Deceive You*. Simon & Schuster.

Impedovo, M. (2012). Chi dorme non piglia pesci, un percorso didattico sul teorema di Bayes. Progetto Alice. n. 38.pp. 229-280.

Kaye, R. (2000). *Infinite versions of minesweeper are Turing-complete*. Unpublished Manuscript.

Keller, J. B. (1986). The probability of heads. *The American Mathematical Monthly*, *93*(3), 191–197. doi:10.1080/00029890.1986.11971784

Kendall, M. G. (1969). Studies in the history of probability and statistics, XXI. The early history of index numbers. *Revue de l'Institut International de Statistique*, 1-12.

Laplace, P. S. (2007). *A Philosophical Essay on Probabilities*. Cosimo Classics.

McGrayne, S. B. (2011). *The theory that would not die: how Bayes' rule cracked the enigma code, hunted down Russian submarines, & emerged triumphant from two centuries of controversy*. Yale University Press.

Moscovich, I. (2015). *The Puzzle Universe*. Lannoo Publisher.

Mosteller, F. (1987). Fifty challenging problems in probability. Academic Press.

Novikov, A., Novikov, I., Katunin, A., & Shevtsova, A. (2017). Adaptation capacity of the traffic lights control system (TSCS) as to changing parameters of traffic flows within intellectual transport systems (ITS). *Transportation Research Procedia*, *20*, 455–462. doi:10.1016/j.trpro.2017.01.074

Penney, W. (1974). Problem 95: Penney-ante. *Journal of Recreational Mathematics*, *7*, 321.

Ramírez-Ávila, G. M., Kurths, J., Depickère, S., & Deneubourg, J. L. (2019). Modeling fireflies synchronization. In *A Mathematical Modeling Approach from Nonlinear Dynamics to Complex Systems* (pp. 131–156). Springer. doi:10.1007/978-3-319-78512-7_8

Reason, J. (1990). The contribution of latent human failures to the breakdown of complex systems. *Philosophical Transactions of the Royal Society of London. Series B, Biological Sciences*, *327*(1241), 475–484. doi:10.1098/rstb.1990.0090

Reason, J. (2000). Human error: Models and management. *BMJ (Clinical Research Ed.)*, *320*(7237), 768–770. doi:10.1136/bmj.320.7237.768

Reps, P., & Senzaki, N. (1998). *Zen flesh, Zen bones: A collection of Zen and pre-Zen writings*. Tuttle Publishing.

Rosen, K. H., & Krithivasan, K. (2012). *Discrete mathematics and its applications: with combinatorics and graph theory*. Tata McGraw-Hill Education.

Stewart, I. (1997). Does God play dice? The new mathematics of chaos. Penguin UK.

Stewart, I. (2008). *Professor Stewart's Cabinet of Mathematical Curiosities*. Jost Entreprises.

Tarry, G. (1900). *Le problème des 36 officiers*. Secrétariat de l'Association française pour l'avancement des sciences.

ADDITIONAL READING

Artigue, M., & Winsløw, C. (2010). International comparative studies on mathematics education: A viewpoint from the anthropological theory of didactics. *Recherches en didactique des mathématiques*, *30*(1), 47–82.

Borjas, G. J. (2020). *Demographic determinants of testing incidence and COVID-19 infections in New York City neighborhoods (No. w26952)*. National Bureau of Economic Research. doi:10.3386/w26952

Durrett, R. (2019). *Probability: theory and examples* (Vol. 49). Cambridge university press. doi:10.1017/9781108591034

Gnedenko, B. V. (2018). *Theory of probability*. Routledge. doi:10.1201/9780203718964

Lu, F. S., Nguyen, A., Link, N., & Santillana, M. (2020). Estimating the prevalence of COVID-19 in the United States: three complementary approaches.

Nguyen, H., & Eisenreich, H. (2018). Using School Choice to Connect to Mathematics Learning in a Statistics and Probability Course for K-8 Pre-Service Teachers. *Georgia Educational Researcher*, *15*(1), 1–15. doi:10.20429/ger.2018.15101

Shuster, J. (2006). Using a Two-Player Coin Game Paradox in the Classroom. *The American Statistician*, *60*(1), 68–70. doi:10.1198/000313006X90477

KEY TERMS AND DEFINITIONS

Bayes' Theorem: Estimates the probability of an event, based on knowledge of the conditions that could be related to the event.

Certain Event: Random event that defines the entire sample space.

Confusion of the Inverse: Is the logical fallacy when a conditional probability is exchanged for its inverse.

Events: A set of outcomes of an experiment.

Impossible Event: Random event that defines the empty set.

Law of Compound Probability: Term relating to the probability of two independent events occurring.

Law of Conditional Probabilities: A measure of the probability of an event occurring given that another event has occurred.

Law of Total Probability: It expresses the total probability of an outcome.

Law of Truly Large Numbers: With a large enough number of samples, anything is likely to be observed.

Opposite Event: Given an event, the opposite event happens when the first did not occur.

Opposite Probability: The probability of an opposite event.

Parameters: Values characterizing a particular model.

Random Variable: A variable whose values depend on outcomes of a random phenomenon.

Sample Space: All the possible outcomes of an experiment.

Swiss Cheese Model: Model used in risk analysis and risk management.

Trial: An experimental procedure.

Chapter 2
Frequentist Probability

ABSTRACT

Frequentist probability is historically presented as an attempt both to overcome the limitations of classical conception and to take into account the impressive development of experimental sciences and statistics. It is precisely because of this close link with the statistical sciences that it finds a significant place in teaching. It is also an area in which the use of IT tools is crucial. This approach also makes it possible to calculate the probability of events and from it we derive the same rules examined in the classical definition, since it is sufficient to replace the ratio between the number of favorable cases and the number of possible cases with the limit of the ratio when the repeated tests tend to infinity. One of the fundamental concepts appears in the chapter, that of a random variable capable of describing events and their distribution.

INTRODUCTION

When the facts change, I change my mind. ~ John Maynard Keynes

The classical definition of probability is based on the assumption that all cases of the event are comparable. In reality this condition occurs in relatively few cases and therefore it is necessary to overcome this limit on the basis of a different conception. In an attempt to overcome the problems inherent in the classical definition, Von Mises (1883-1953) believes that probability should be evaluated on the basis of a large mass of observations and formulates the frequentist definition: it is measured on the basis of statistical observations, i.e. the frequency with which certain phenomena are recorded.

DOI: 10.4018/978-1-7998-3871-5.ch002

If you toss a fair coin a number of times, experience says that as the number of tests made all under the same conditions increases, the frequency of the number of heads (or tails) tends to stabilize around a value that corresponds to the a priori theoretical probability. Large fluctuations are increasingly rare, in this consists the so-called *Empirical law of chance*.

In practice we can say that, called P the probability of an event, it is given by the ratio between the number s of times in which the event occurs, that is the number of successes, and the number n of tests done, to the tendency of the number to infinity:

$$P = \lim_{n \to \infty} \frac{s}{n}$$

In order to have a reliable number to give to the probability it is necessary that the frequency is calculated on a large number of cases (in theory infinite), therefore it is not possible to use this type of definition in cases where the number of repeated tests is low or where conditions have varied.

In the frequentist conception the probability is derived *a posteriori*, based on the examination of the data. It is known the case of the statistician Karl Pearson (1857-1936) who in an experiment launched 24000 times a coin obtaining 12012 heads; the frequency is equal to 0.5005 approaching the "theoretical" value. As well as the case of the mathematician prisoner of war during the Second World War who counted 10000 flips, almost equally divided between 5067 heads and 4993 tails.

This type of probability, also known as statistics, has an enormous field of application. For example, in order to construct life and death probability tables for the calculation of premiums, it is first necessary to know the probability of the event, which is estimated on the basis of the frequencies observed in the past on a large mass of cases.

By indicating with x the whole age of an individual and with l_x the number of living persons of age x, it can be noted that there is an empirical relationship $l_x = f(x)$. This relation, linked to age, constitutes a biometric function.

A table containing the l_x data, obtained with appropriate processing using the method of censuses or deaths, is called the survival demographic table. In it, starting from a fictitious group of 100000 individuals of age $x = 0$, the number of individuals who survived the various ages up to the extreme age is considered. A life table is a concise way of showing the probabilities of living or dying for a member of a particular population at a particular age.

The table is used by actuaries, demographers etc. to present the mortality experience of a population. This method is applicable to the analysis of many measurable processes and mainly in the insurance field. For detailed analysis, see Namboodiri & Suchindran (2013). From the tables it is possible to obtain information to answer questions such as: what is the probability that a 70-year-old will survive for ten years? What is the average number of years of life remaining for men who have just reached their 50th birthday?

The idea is that the probability is not calculated as the ratio between the number of heads and coin flips, but as the ratio between the survivors of a cohort.

The first survival table can be dated to the third century AD, but it is necessary to wait until the end of the 17th century to have sufficiently reliable mortality studies. We remember in particular the study of the astronomer Edmond Halley (1656-1742), who determined the probability of death on the basis of observations made on the population records of a Polish city.

Other simple examples: the probability of theft can be obtained from police records; the probability that a new medicine will give positive results in the treatment of a disease is the result of tests carried out before the medicine is put on the market; the probability of post-operative death can be inferred from the frequencies reported in the medical records of hospitals. The examples could be multiplied at will as they relate to the possibility of collecting data and processing them statistically.

Here we are content with just an example, to highlight the close relationship between statistics and frequentist probability. The question is: what is the probability of meeting an individual with a certain surname in a country like the USA? Take for example the surname Smith which refers to 2442977 individuals, or the surname Gray which occurs almost ten times less frequently (246116). Since there are about 329 million people living in the USA in 2020, the probability of meeting a Smith is 0.0074 and drops to 0.00075 in the case of the surname Gray. Another calculation considers the distribution in the various States of residence. To visualize the differences we use the frequencies in absolute value and not the probabilities. You will have two cartograms (Figures 1 and 2):

Figure 1. Distribution of the surname Smith

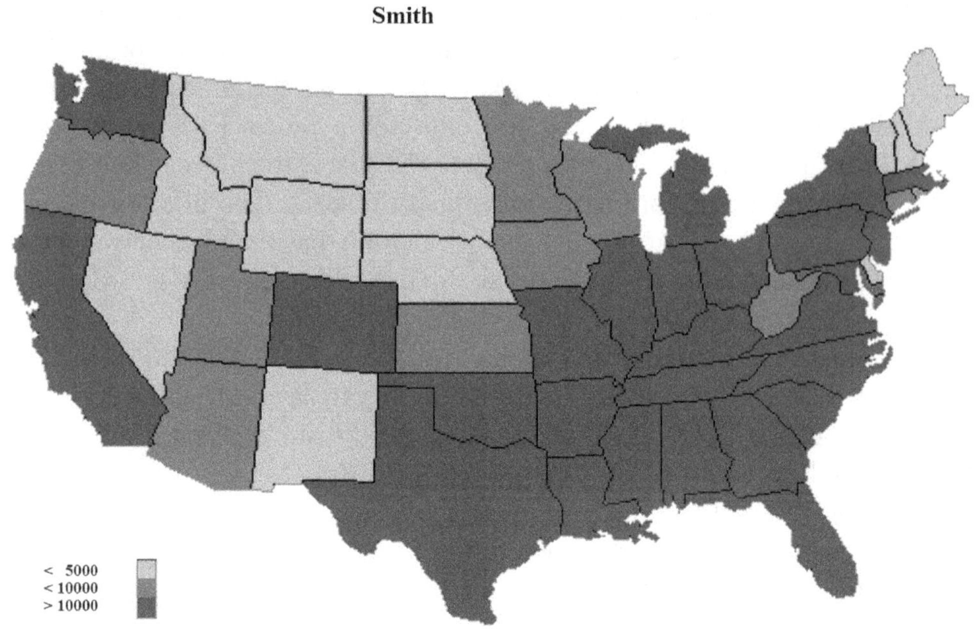

Figure 2. Distribution of the surname Smith

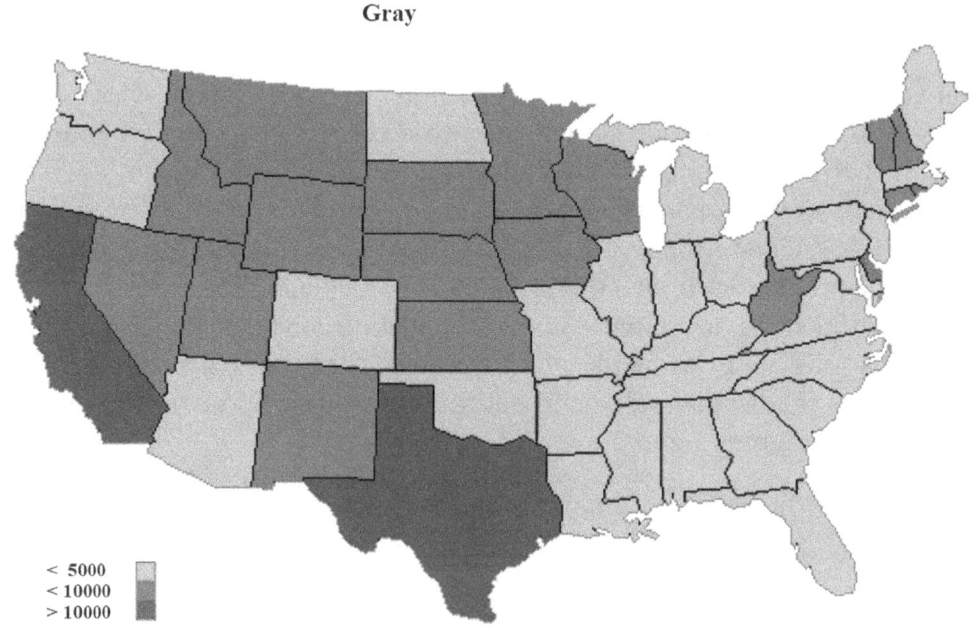

THUMBTACK

As with the coin, the throwing of the thumbtack also ends in two ways: it can fall with the tip upwards or downwards. (Figure 3). So it can be said that the probability of each outcome is 1/2?

Figure 3. Thumbtacks

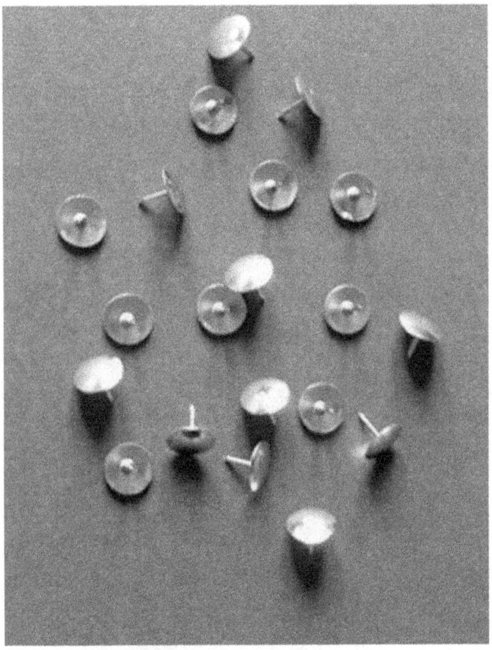

It is normal to doubt this, it is only possible to evaluate the probability of the two events only by carrying out a large number of tests and counting all the results obtained. In practice we trust that, by repeating an experiment several times, under the same conditions, the relative frequency of an event concerning that experiment is a sensible evaluation of the probability of the event itself. The advantage of this method is that the definition is operative and allows to calculate the probabilities in statistical-simulative situations. Properties and demonstrations are similar to those of the classical definition.

Let's examine the results of some tests carried out with normal thumbtacks: by throwing 200 times one thumbtack 78 times it was presented with the tip upwards; by repeating the experiment with 300 throws 120 points upwards were recorded; finally on 500 throws 205 times the thumbtack fell back with the tip upwards. What can be said on the basis of these data? Using the frequentist definition we obtain $p_1 = 78/200 = 0.39$, $p_2 = 120/300 = 0.4$, $p_3 = 205/500 = 0.41$. The probability of the event (fall with the tip up) is about 40%.

The assessment of the probability of an event occurring is all the more reliable the more evidence there is. This "law", whose validity can be detected experimentally, has also been given the name *Law of large numbers.*

Of course, the results may differ depending on the type of thumbtack used, in another experiment done at the Bangkok Patana School in 200 throws the thumbtack had fallen with the tip upwards 125 times with a relative frequency of 0.625.

It may be interesting to compare what happens with different number of attempts, assuming as probability the value 0.4 per Base (B) and 0.6 per Point (P). Of course, since the phenomenon is random, the results may vary, but the trend (Figure 4) remains.

Figure 4. Throw graph of 100, 1000, 10000, 1000000 pins with p(Base)= 0.4

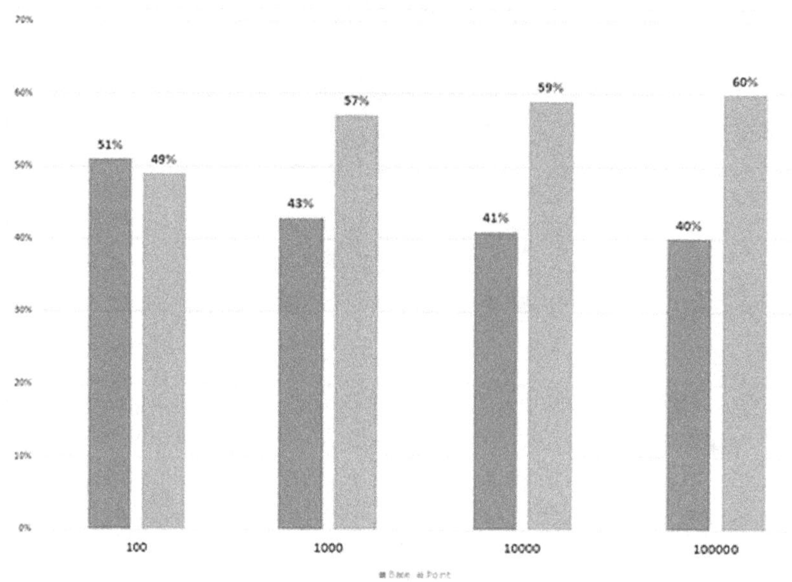

For the avoidance of misunderstandings, it must be said that the pushpin has not really been thrown a million times, but the data comes from a simulation performed with a computer.

This theme is of central importance and will be addressed in a special chapter. Does it make sense in school to make statistics with little data? There is no clear answer, but it is clear that just a few experiments ("calculate the probability by throwing the thumbtack 10 times") is limited and often misleading. It is certainly useful to work with a set of data large enough to provide materials for non-trivial considerations and inferences, when this is impossible one can resort to simulations, what is important is to use a critical mass of data to get to the heart of the problem, without neglecting the fact that the simulation, in several cases, can represent not the answer to a particular type of problem but the question. Without disturbing the concept of serendipity many empirical observations can stimulate new paths. The topic is of considerable scope and it is not easy to untangle between large and small data sets, but we can remember what Martin Lindstrom (2016) says: "Big Data is all about finding correlations, but Small Data is all about finding the causation"

With the proposed experiment we move, in a convincing way, from the classical conception to the frequentist one, which has the merit of overcoming the notion of comparable events but which has however several limits, for example it requires repeated events and a less imprecise definition of a high number of tests. However, even in this case, the frequency is a number between 0 and 1.

The understanding of the process can be strengthened by using different models of thumbtacks: flat-headed, concave-headed, with three tips as shown in Figure 5. In this way you can compare the results showing how the probability varies with the geometric properties of the object.

Figure 5. Three types of drawing pins

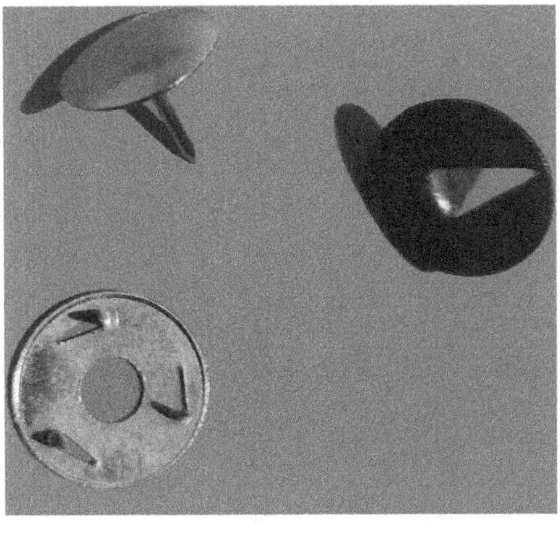

To shorten the execution times, you can think that instead of throwing a thumbtack many times, you can launch many at the same time, since the constitutive differences are not so accentuated. The incorrectness of the procedure can actually be accepted in an educational-operational key.

Let us return for a moment to the data of the Bangkok school where it was found that P (P) = 0.625 and P (B) = 0.375, we can take up the concepts of compound and total probability seen in the previous chapter. Let's ask ourselves, for example, what is the probability of obtaining two heads, two tails or one head and one tail by throwing the thumbtack twice.

You can help yourself with the tree in Figure 6:

Figure 6. Thumbtack graph

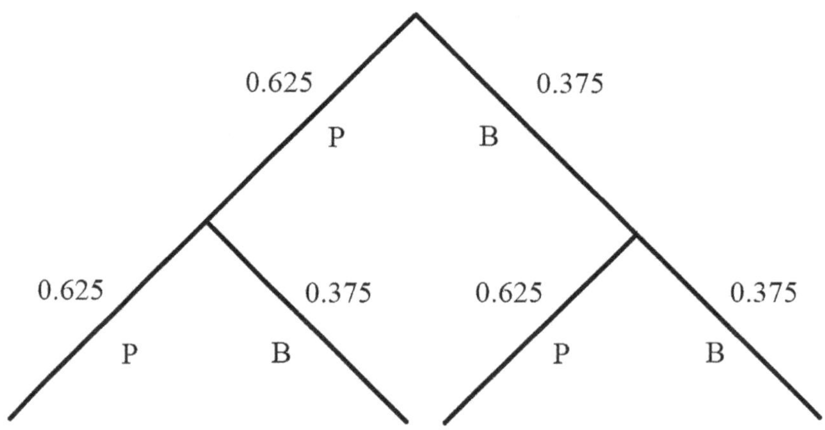

The answer to the problem is obtained by multiplying the probabilities on the various branches.

In conclusion:

$P(PP) = 0.625 \bullet 0.625 = 0.391$
$P(BB) = 0.375 \bullet 0.375 = 0.141$
$P(PB) = 0.625 \bullet 0.375 = 0.234$
$P(BP) = 0.375 \bullet 0.625 = 0.234$

In the determination of the probability of obtaining a head and a tail we have to add P(PB) with P(BP), then we will have 0.468.

Obviously, the issue is complicated if you extend the problem to a higher number of launches, in this case you have to use the binomial distribution that will be examined in Chapter 6.

SPINNING TOP

The spinning top is a toy, made of wood or other hard material, cone-shaped with a pin at one end (Figure 7). It is an object with an ancient history. The spinning top was a toy widespread among the Romans who called it "turbo". Spinning tops were also widespread among the Indians of North and South America long before the arrival of Columbus. The greatest diffusion took place, especially in England, in the 14th century, in combination with religious ceremonies and not only as a game for children.

The analysis of the operation of the spinning tops is not trivial, despite the appearance. One of the most interesting works on the motion of this artifact is due to Harold Crabtree (1914): the text provides sufficient information to understand the behavior of these devices in mathematical terms.

Figure 7. Spinning top

A traditional spinning top does not lend itself to probabilistic discourse, but with some innovations new horizons are opened up. An interesting work proposal is the one suggested by David Green (1991). In practice, a series of geometrical shapes made of cardboard that act as spinning tops are proposed. Each one has a pivot, placed in the center of gravity, so that it can rotate. A match, straw or pencil can be used as a pin.

The didactically interesting aspect is that, starting from a square shape, in order to have an equal probability, irregular shapes can be reached later. In this case, large discussion spaces can be opened in a classroom.. A few examples are enough to convince us of this. For the rectangles, some reasoning can already be done: if the sides are twice as much as each other, in what relation are the probabilities?

Is it possible to make predictions for parallelogram tops based on the angles and length of the sides?

These experiments obviously require a number of tests in order to obtain results capable of confirming or disproving some initial hypotheses. However, there are at least two questions that must be asked before the tests, the first concerns the geometric determination of the center of gravity, the second concerns the possible influence of the size of the tops.

DREIDEL

A particular type of spinning top (Figure 8) is linked to a Yiddish custom and refers to a game called *dreidel*, which is widespread in the Jewish community. It is based on a four-sided spinning top: on each of the faces there is a letter of the Hebrew alphabet: נ (Nun), ג (Gimel), ה (Hey), ש (Shin). The name is derived from the word dreyen (to turn). According to popular conjecture, the letters abbreviate the words "a great miracle happened there".

Figure 8. Four Dreidel

At the beginning of each round, each participant puts a bet on the "pot" which may consist of a unit of money, a candy, a chocolate, etc..

Every player spins the spin. Depending on the face that appears after the spinning top has stopped spinning, four cases may occur:

1. נ Nun means "nisht" (nothing). The player wins nothing.
2. ג Gimel means "gantz" (everything). The player gets everything in the pot.
3. ה Hey means "halb" (half). The player gets half of the pot, rounding up if necessary.
4. ש Shin means "shtel" (put in). The player adds the equivalent of an initial bet to the pot.

The most obvious question is whether the game favors one of the players or it is a fair game.

First of all we must assume that the exit of the faces has the same probability, this is theoretically desirable, but researchers Robert & Eva Nemiroff have verified with a large series of dreidel launches (made with different materials) that this is not true:" To explore this hypothesis, three different dreidels were each spun hundreds of times with the number of occurrences of each side recorded. It was found that all three dreidels tested -- a cheap plastic dreidel, an old wooden dreidel, and a dreidel that came embossed with a picture of Santa Claus -- were not fair".

This is a general problem: if the chosen object presents deformities such as to make the exit of a face more or less likely, we are in the presence of a significant alteration. To correctly answer the initial question, we can assume that only one spinning top is used or that the differences are small enough to justify the use of dreidels in a classroom experiment.

A different approach is the use of a random computer generator of values and therefore usable for a Monte Carlo simulation.

The first player who spins the dreidel has more chances of winning than the second, and so on, the participants who play afterwards are disadvantaged as they may find nothing in the pot. Is it possible to calculate what the expected winnings are for each player? With a simulation of ten million games (which is unthinkable in human terms) Paul Nahin (2013) has published the following table, which is sufficiently explanatory (Table 1):

Table 1. Winnings at the Dreidel

Players	1	2	3	4	5
2	1.143	0.857			
3	1.361	0.956	0.68		
4	1.617	1.102	0.757	0.514	
5	1.9	1.267	0.855	0.58	0.368

An experiment carried out in a class, divided into groups of four players, would seem to confirm these data: despite the presence of a small sample (96 games) the winnings were respectively for the first player 1.618, for the second 1.034, for the third 0.989, for the fourth 0.360.

An interesting problem would be to determine what conditions turn this game into a fair game. The subject was studied by Felicia Moss Trachtenberg, who developed an interesting proposal to which we refer.

TOKEN

The telephone token is now considered a historical find. The token was a metal disk used to pay for telephone calls made from public landlines.

In the United States telephone tokens were first used in Chicago in the late 1800s and were used until 1944. In Italy, the first specimens were minted in 1927 and were used until 2001 (Figure 9). With the advent of the devices that can be used with telephone cards, the telephone token has gone out of production and now, the use of mobile phones and smartphones has practically replaced the use of telephone booths.

The token has a special feature: it consists of a disc with three grooves, two on one side and one on the other. This makes the token an unbalanced coin and therefore, if it is thrown, there is not the same probability in the exit of the two faces. A repeated coin toss experiment can tell us how much the probability of the two faces coming out is worth. It's a simple but somewhat boring experiment.

Having thrown a token 20 times we saw that 11 times the face facing up showed two grooves. Is it fair to say that the probability is 0.55?

Launched the token 200 times the same phenomenon occurred 112 times, so with $p = 0.56$. On the basis of the law of large numbers we should be willing to change the assessment.

Figure 9. Italian telephone token

An interesting problem compares a fair coin and a token. Let's suppose that if you flip a coin, if it comes out heads you win $1, if it comes out tails you win $2 (let's pretend that the opponent is so unwise to accept). You stop playing when the amount earned reaches at least $30. How many throws do you have to make before you reach this amount?

Now let's change the game a bit: instead of a coin you use a telephone token. If the face comes out with one groove you win $1, if the face comes out with two grooves you win $2. How does this change the problem? In the first case the average value is $1.5 as $2 \bullet 0.5 + 1 \bullet 0.5 = 1.5$ and so you can expect that the result will be achieved with 20 throws (30 / 1.5). In the second experiment the average value depends on how unbalanced the token is, in this case we can set $2 \bullet p + 1 \bullet (1 - p) = p + 1$; the knowledge of p is linked to a frequent estimate, taking for good the value 0.56 obtained by the 200 throws we can say that the average value is 1.56. This means that with 19 launches we could have reached the objective.

TOAST

Toast is a slice of bread that is usually heated inside a specific appliance, called a toaster. Given the perishability of the product in class, reproductions in inorganic material can be shown (Figure 10).

Figure 10. Sponge-shaped toast

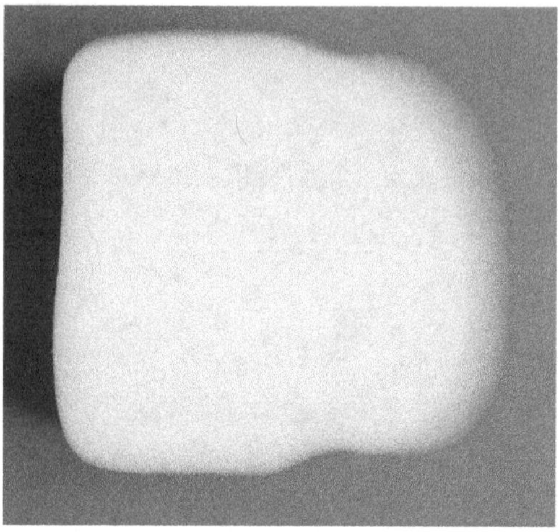

The toast is often spread with butter, margarine, or jam. This habit has created the rumor that if the toast falls on the floor, it will end its run with the buttered part on the floor.

In 1844, the satirical poet James Payn (1830-1898) wrote:

I never had a piece of toast
Particularly long and wide
But fell upon the sanded floor
And always on the buttered side.

The toast therefore seems to behave differently from a coin and this recalls the typical situation foreseen, according to Arthur Bloch (1977), by the so-called Murphy's law "If something can go wrong, it will go wrong".

Robert Matthews (1995) of the University of Aston, England, won the Ig Nobel prize in 1996 for studying this type of phenomenon. First, Matthews dismantled the first common interpretation linked to asymmetry: the mass of butter added to the toast is little compared to the mass of bread, it is spread in a thin layer and penetrates the body of the slice, so its contribution to rotational dynamics is negligible. Later it involved about 1000 students who dropped more than 9000 toasts from a table, 1.5 meters high. Half of them had to drop buttered toast and a control group operated on a butter-free toast with a large B on one side. In fact, the total number of falls with butter to the floor was 6101, so with a statistically higher number of falls on the butter side. The toast in the control group, however, also fell with the B downwards.

According to Matthews the reason for the fall with the butter down is the height from which the toast falls. From about 1.5 meters, in fact, the toast does not turn twice in time. It only turns once, and then falls with the face that at the beginning of the fall was at the top, just the one with the butter (or B).

In a test conducted for the BBC, buttered bread was thrown into the air 300 times and the results were in line with expectations of an equally likely event.

The problem of falling toast has given rise to an amusing paradox called Buttered cat paradox which would demonstrate how it is possible to achieve perpetual motion. The paradox is based on the combination of two principles considered, at least in the collective imagination, as true:

1. A cat always falls on its legs and never on its back;
2. A buttered toast always falls on the butter side

If these two "laws" are true, just take a cat and tie a slice of buttered bread on its back. If the cat is dropped, two opposing tendencies are activated: the cat would tend to land on its paws, while the slice of buttered bread would tend to fall on the butter side; a perpetual motion would then be created in which both the cat and the slice of bread would continue to rotate infinitely (Figure 11).

Figure 11. Buttered cat paradox

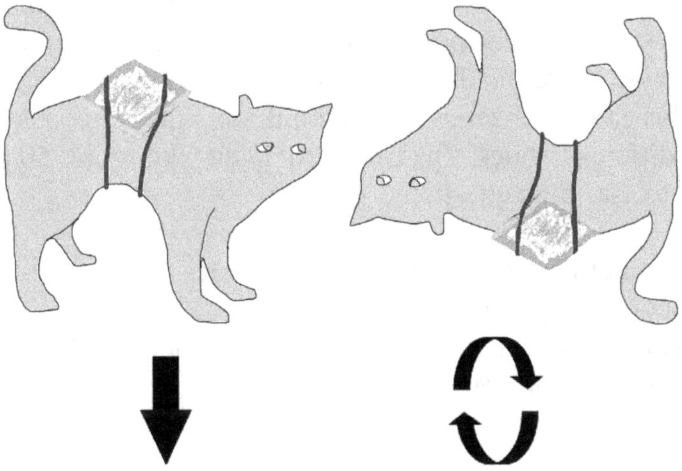

Of course, this is only a humorous consideration because, apart from any other consideration such as finding a cat so docile as to have a toast attached to its back, it is false that a cat always falls on its paws, and it is false that a slice of bread always falls on the buttered side.

Lu - Lu

Young Hawaiians play Lu - Lu (an expression that means to shake) with volcanic stone discs about 2.5 cm in diameter. On one face are indicated 1, 2, 3 or 4 points while the opposite face is empty (Figure 12).

Figure 12. Lu - Lu

The basic version of the game consists of only one throw, in turn, of the 4 stones, after which only the points on the discs that fall face up are counted. The player must shake the discs in both hands and throw them. So we can assimilate this game to the toss of four coins, only the equivalent of the head can take on different values. The first participant who scores 50 points wins.

Let's try to ask some questions:

1. What are all the possible results in a launch?
2. What scores are possible for a round?
3. How many different ways can each score be achieved?
4. What is the probability of each score?

In the first case the possible results are sixteen: 4-3-2-1; 0-4-3-2; 0-4-3-1; 0-4-2-1; 0-3-2-1; 0-0-3-4; 0-0-2-4; 0-0-2-3; 0-0-4-1; 0-0-3-1; 0-0-2-1; 0-0-0-4; 0-0-0-3; 0-0-0-2; 0-0-0-1; 0-0-0-0.

The second question can be answered that the possible scores are eleven: 0, 1, 2, 3, 4, 5, 6, 7, 8, 9, 10.

If we consider the various cases we see that 0, 1, 2, 8, 9, 10 can occur in only one way (e.g. 10 = 4 -3 - 2 - 1) while 3, 4, 5, 6, 7 can occur in two ways (e.g. 3 = 0 - 0 - 3 or 0 - 0 - 2 - 1). The probability calculation is immediate: P(0, 1, 2, 8, 9, 10) = 1/16, P(3, 4, 5, 6, 7) = 2/16 = 1/8.

A more complex version provides that the player can collect all the Lu - Lu that landed face down (no point shown) and throw them again. Also in this case it is possible to ask the same questions.

One question arises quite spontaneously: if the number of games increases, does the game score tend to stabilize? If so, around what value?

These questions involve two themes that will be dealt with more extensively in two subsequent chapters dealing with distributions and simulations. For now we can say that the average score tends to 5.

Each random experiment can be expressed by a number, e.g. each of the Lu - Lu results can be identified by the sum of the points shown by the upper face, each of these values x_1, x_2, ..., x_n is associated with a probability p_1, p_2, ..., p_n. This procedure is applicable regardless of the situation, for example you can indicate with x = 0, 1, ..., n the number of customers in line at the checkout of a supermarket, the number of accidents on a stretch of road, the value of the card pulled out of a deck, etc.. If we are able to define for each number a probability of realization we obtain a function that associates to each value that the variable can have its probability.

If we report in a table the points obtainable in each throw (x) and the probabilities (p) we can calculate the average or *Expected Value* E(x). We notice that this average value can be used as a forecast indicator in the presence of a high number of tests carried out in the same conditions (Table 2).

The concept of the statistical distribution around an average was known since ancient times, for example Plato writes in the *Fedone*: “Have you never noticed that in all things extremes are rare while intermediate aspects are frequent, indeed numerous? ”. Thesis that takes on an ethical meaning in the Latin poet Horace who states: “in medio stat virtus”, that is, virtue lies in the middle.

This Expected Value is given by:

$$E(x) = \sum_{i=1}^{n} x_i p_i$$

Table 2. Calculation of the Expected Value

points	probability	product
0	1/16	0
1	1/16	1/16
2	1/16	1/8
3	1/8	3/8
4	1/8	1/2
5	1/8	5/8
6	1/8	3/4
7	1/8	7/8
8	1/16	1/2
9	1/16	9/16
10	1/16	5/8
Sum	1	5

This table allows you to introduce two terms that have already been mentioned but have not yet been clearly defined. It shows the possible values x_i that can be assumed by *Aleatory variable* X, in this case it is a *Discrete random variable*. The whole constitutes a probability distribution in which each mode x_i is associated with the probability p_i that it occurs and in which the sum of the probabilities is 1 because all the possible cases are included. The sequence of the values p_i is called probability mass function (PMF).

Defining a random variable means finding a rule by which to associate a real number to each element of the sample space. Of course for the same sample space it will be possible to define more than one rule, among these you will choose the one that reflects the most relevant information for the problem in question. Instead of working with sets you assign numerical values to the events of an experiment. There are numerous families of distributions, each with particular characteristics, which make them more suitable than others to describe certain phenomena.

The other aspect to underline is that the theoretical result can be approached empirically through a simulation. When we compare observations with the model, we compare the *theoretical* probability distribution with the *empirical* probability distribution.

5000 tests were carried out with the aid of a spreadsheet and these are the results shown in Table 3:

Table 3. Comparison between observed and expected frequencies

x	Observed frequencies	Expected frequencies
0	0.0614	0.0625
1	0.0662	0.0625
2	0.0652	0.0625
3	0.1272	0.125
4	0.1228	0.125
5	0.1224	0.125
6	0.122	0.125
7	0.13	0.125
8	0.056	0.0625
9	0.063	0.0625
10	0.0638	0.0625

As you can see, the results are comparable.

If then you want to verify if the expected value converges (we speak of asymptotic convergence when there is a value towards which we tend to increase evidence) it is enough to represent it graphically as in Figure 13:

PASS THE PIGS

Just as we went from the coin to the thumbtack, we can switch from the regular dice to an irregular "dice".

These dice are part of a game called Pass the pigs (Porker in Italy) in which the function of randomness is given by two little pigs that can fall in six different ways: on the left side, on the right side, on the back,

standing, on the muzzle, on the left ear. Figure 14 shows a possible configuration obtained by rolling two of these "dice".

Figure 13. Simulation graph

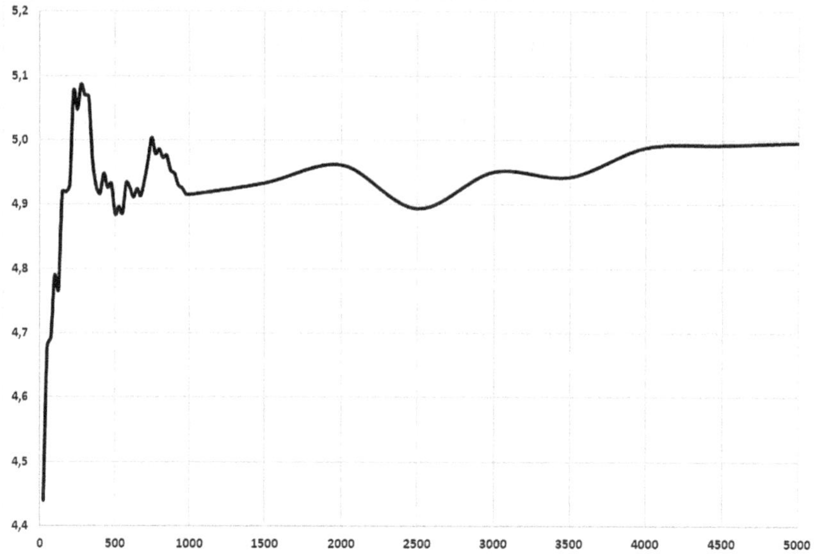

Figure 14. Pig dice

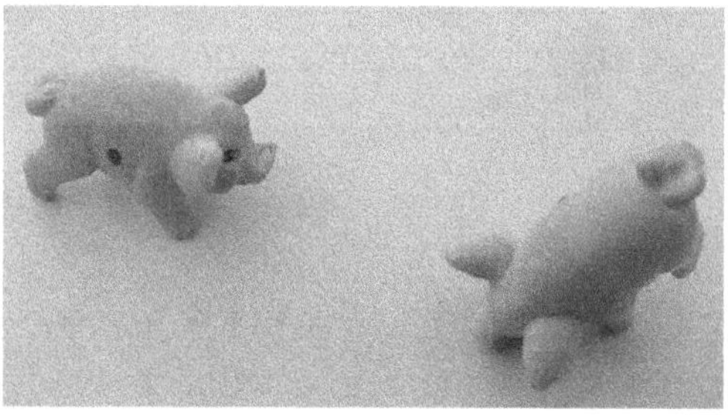

The goal is to score 100 points, with a number of throws of your choice and adding up the results obtained. Some combinations, however, lead to dangerous situations and therefore you have to decide when to stop throwing, in fact there are three negative results:

1. Pig out: the two little pigs fall on opposite sides; you lose all the points for that turn.
2. Makin' Bacon: The little pigs touch each other and all the points accumulated in the game are lost.
3. Piggy Back: The two little pigs are standing on top of each other, the player is eliminated from the game.

According to the manufacturer, different combinations can occur that lead to points, for example:

1. Sider: the two little pigs are lying on the same side, 1 point;
2. Razorback: one little piggy ends up with his back on the ground and the other on his side, 5 points; Double Razorback: both on his back 20 points;
3. Trotter: one piglet falls standing and the other on the side, 5 points; Double Trotter: both standing 20 points;
4. Snouter: one of the piglets ends with its muzzle on the ground balancing on its front legs and the other on its side, 10 points; Double Snouter: both on the muzzle 40 points;
5. Leaning Jowler: a piglet lands on the left cheek balancing on the ear and paw and the other falls on the side, 15 points, Double Leaning Jowler: both on the ear 60 points;
6. Mixed Combo: any acrobatic combination, like Trotter + Snouter = 15 points, Razorback + Leaning Jowler = 20 points.

Obviously, there must be a link between the scores attributed to the different positions and the associated probabilities, but how to determine these probabilities?

The answer should be quite obvious: by carrying out several tests. By way of example, a piglet was launched 100 times with the results shown in Table 4, these results collimate with those presented by Gorman (2012):

Table 4. Results of a piglet launch

Location	Percentage
Left Sider	35%
Right Sider	30%
Razorback	22%
Trotter	9%
Snouter	3%
Leaning Jowler	1%

In this case the data are given as a percentage but perhaps, for a simpler interpretation of the phenomenon, it would be more understandable to speak in terms of expected frequencies: instead of saying "the probability that the piglet falls on its back is 22% (or 0.22)", we can say "out of 100 throws, we would expect the event to occur 22 times".

Recall that the probability represents a number between 0 and 1. The use of the percentage, mathematically equivalent to the fraction and the decimal, is often used in common use and also by several authors (Weiss). For some probabilists it can be used only in the presence of repeated random experiments; in this example it should mean that the random experiment has been subjected to 100 independent tests and all under identical conditions.

Since there are two piglets launched, knowing that, for independent events, the compound probability (event A **and** event B occurs) is given by the product of the individual probabilities, it is possible to calculate the percentage of success. For example, for the Razorback and Trotter configuration you need to calculate $0.22 \bullet 0.09 \approx 0.02$.

We now expand the field of investigation to verify whether the attribution of the scores is correct or not. For example, the Sider configuration has a probability of realization equal to $0.35 \bullet 0.30 = 0.105$ and it is given a point while Trotter and Snouter has a probability $0.09 \bullet 0.03$ » 0.003 and allows to accumulate 15 points. Taking the Sider as a unit, we set the proportion $0.105: 1 = 0.003: x$ from which it is deduced that the score for the acrobatic combination should be 35, much higher than stated in the game instructions.

More striking the case of the Double Razorback: it has a probability of occurrence of $0.22 \bullet 0.22$ » 0.048, but, following the procedure already applied, we get that the score should be 2 and not 20!

At this point you can create Table 5 with some scores that take into account the actual probabilities, in summary:

Table 5. Rating assigned scores

Launch result	Probability	Score assigned	Score correct	Error
Double Razorback	0.048	20	2	+90%
Double Trotter	0.008	20	13	+35%
Trotter + Snouter	0.003	15	35	-133%
Razorback + Leaning Jowler	0.002	20	48	-140%
Double Snouter	0.001	40	117	-193%

It is significant to note that the game scores show exaggerated differences and that the percentage error $e = \frac{assigned\ score - correct\ score}{assigned\ score}$ is not justifiable from the point of view of the real chances of success of a given combination.

The game lends itself to different considerations from which arise as many questions: taking into account that Pig out involves 0 points, what is the probability of getting 0, 5, 10, ..., 60 points with a single throw? What is the score with the highest probability of achievement?

By summing the probabilities associated with each score you can easily construct a probability distribution. The distribution of probability allows us to answer the first question and to affirm that to the score 5 is associated a percentage of realizations of 40.6%.

Astragalus

An example of a dice, not regular, comes from antiquity. This is the Astragalus: this is a small bone (Figure 15) which is part of the foot joint in some animal species, such as the goat and the ram. At the British Museum in London there is an artefact, dated 330-340 BC, made of terracotta that depicts two young girls engaged in the game of the talus.

This bone has six faces, but if you throw it, as it is made, it can only stop on four faces and only these are the possible results.

The astragals were thrown and according to the face upwards it was possible to "read" a score: 1 (Monas), 3 (Trias), 4 (Tetras), 6 (Hexas). It should be noted that the opposite sides gave a total of 7, just as with traditional cubic dice.

Given the asymmetry the faces weigh differently and therefore the values indicated could probably be considered, as Florence Nightingale David claims, as different "weights". Perhaps it is the first attempt to recognize that more likely events must have a lower value than less frequent ones. Certainly in the players of the time were overwhelming magical and esoteric considerations compared to a rational reflection. For example, Aphrodite's shot was made up of 4 different faces and was the most coveted, while the dog's shot (the 4 lowest value faces) was the worst shot.

As in the previous example, given the conformation of the object, the exit probabilities are different and this should be taken into account for a correct evaluation.

Figure 15. Astragalus

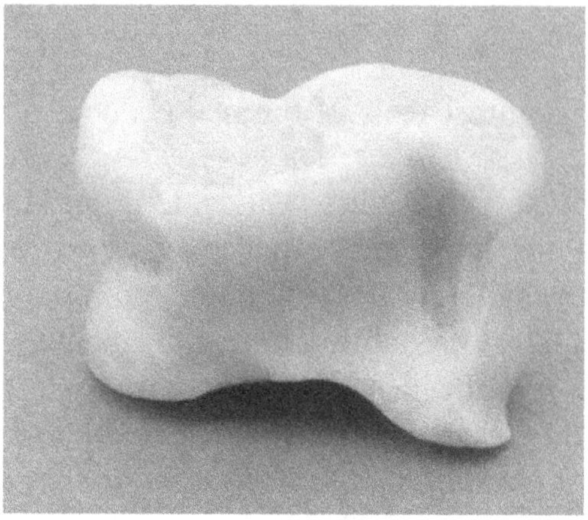

For example, by throwing a talus 100 times, the following results are shown in Table 6:

Table 6. Results in the launch of the talus

Do	Name	Score	Results	Probability
Flat with protrusion	Monas	1	7	0.07
Concave side	Trias	3	37	0.37
Convex side	Tetras	4	45	0.45
Flat	Hexas	6	11	0.11

The abhorred blow of the dog had however low probability to occur, a little more than one in 50000 throws, while the blow of Aphrodite could occur about once in a thousand.

It's not easy to find frequency searches. Robert Ineichen said: "I did more than 200 shots with the talus of a "modern" sheep [...] and came up with results $P(1) \approx P(6) \approx 10\%$ and $P(3) \approx P(4) \approx 40\%$.

The talus was used in gambling but also for the prediction of the future. In Iceland it was (and is) used to get an answer to a question. First of all the following magic phrases are read:

Segðu mér Vala mín það sem ég spyr þig að. Ég skal með gullinu gleðja þig og silfrinu seðja þig ef þú segir mér satt. En í keytunni kæfa þig og í eldinum brenna þig ef þú lýgur að mér.

Translated into English more or less sound like this:

Tell me, my Vala, the answer to my question. I shall gladden you with gold and satisfy you with silver if you tell me the truth. But I shall drown you in urine and burn you in fire if you lie to me.

Then you throw the astragalus and if the bone falls with the convex face upwards the answer will be positive, if it shows the concave face the answer is no, otherwise there is no certain answer.

This example serves to introduce an interesting theme, the link between prediction and foresight.

The first term derives from the Latin *Praedictio,* which means to announce previously the coming true of future things, the second from the late Latin *Praevisio,* which means to suppose what will happen based on conjecture. This term has always been linked to religious inspiration or clairvoyance; in the religious sphere, the prediction can indicate an oracle, a prophecy, or a possible response of divination.

It is interesting to note the attitude of humans towards the future. They see time as an oriented arrow. Crichton in his adventurous novel Congo, later translated into a film version, makes an interesting consideration attributed to the female gorilla Amy accompanying the expedition:" Amy was particularly perplexed by spatial metaphors for time, such as "that's behind us" or "that's coming up." Her trainers conceived of the past as behind them and the future ahead. But Amy's behavior seemed to indicate that she conceived of the past as in front of her—because she could see it—and the future behind her—because it was still invisible".

Making predictions is an operation that dates back to the dawn of time. For example, one form of divination was the so-called Belomancy (from the Greek *belos,* "javelin" and *mantiké*, "divination") which was achieved by using arrows without points or simple sticks. Their extraction by lot, after having previously attributed a precise meaning to each *belos*, determined the answer to a question thanks to the belief that supernatural entities intervened on the result.

Progressively we have moved from the concept of prediction to the more modern one of forecasting. In this progress, both the construction of appropriate theories and the development of increasingly powerful and efficient technologies have played a fundamental role.

For example, in meteorology we talk about weather forecasts, in economics and finance we make forecasts about the results of a company or a state (Farrar and Meyer, 1970).

WATCH

The watch is an instrument for indicating the time and has a long and interesting history. The first mechanical watches were invented as early as the Middle Ages. In more recent times, watches can be divided into analogue and digital. Digital are all those modes of time display that occur for discrete values (Figure 16).

Figure 16. Digital Clock

The use of this tool can have an application in school activities. If the teacher wants to draw lots for a student to interrogate, what tools can be used?

You can use the sum of the page number of a randomly opened book, cards with the order number, bingo numbers, a 30-sided die or a digital clock that has a stopwatch with hundredths of a second.

Which of these methods has both fairness and ease of use?

Surely the use of the book is not correct from the probabilistic point of view, just think about what happens when you add up the page numbers of a book. For example, considering a 400-page book, order number 1 can occur only in three cases (001, 010, 100) while order number 19 can occur in six modes (199, 289, 298, 379, 388, 397). The chances of extraction vary according to the order number and follow the distribution shown in Figure 17.

Figure 17. Probability of being chosen with the book

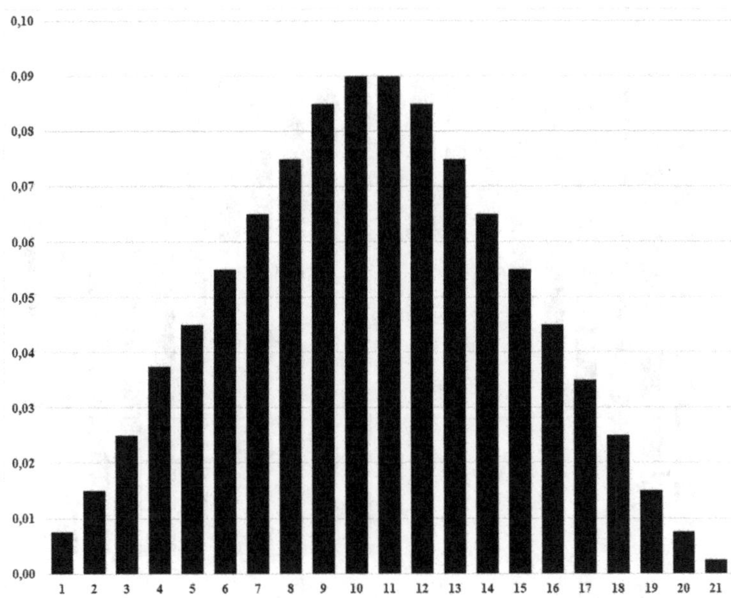

Tickets, tombola, and dice, on the other hand, represent correct methodologies. It remains to analyze the case of the clock, an instrument not particularly considered in school practice. A simple example can clarify how to use it.

Suppose we have a class with 21 students. We start the stopwatch and then stop it, then we will have a result relative to hundredths, between 00 and 99, for example 54. In this case we have to reduce the value within the twenty-one students. To do this, we must use modular arithmetic, also called clock arithmetic. Divide 54 by 21, we get 2 with the rest 12. The student chosen will be the one with number 12.

One question remains: does this method put pupils in the same situation, i.e. are they all equally likely to be extracted? The answer is positive only if the number of pupils is a divisor of one hundred, in other cases the favorite (or disadvantaged) pupils may vary. Would this practical extraction system be considered quite acceptable? To get an idea of the problem, just look at Figure 18. Even with different numbers of pupils, the probability varies by 1%.

Figure 18. Probability of being chosen with the watch

To convince students of the differences between the two modes, simply use a spreadsheet.

In the first case you have to calculate the sum that corresponds to each page (if you use a text string for the pages you must first extract each value separately, but this is possible using the library functions), the frequencies of each sum must then be represented with a bar graph. For the clock, the hundredths of a second are shown in a column and alongside the rest of the division for the number of pupils. Starting from the counting of the frequencies occurred for each pupil order number it is not complicated to build the graph. This is a remarkably simple example of how you can use software to visualize a very realistic situation in a school.

Let's now shift our attention to an analog watch, then with a hand indicating the time.

Start with a watch that marks noon and move the hand by the number of hours corresponding to the result of rolling a dice. After 20 throws, what is the probability that the hand is vertical, i.e. marks 6 or 12?

The question seems extremely complicated, but in reality even with any number of dice throws with relative movements of the hand, the result is always the same. In fact whatever the position reached before the last throw you have 1 result out of 6 to reach either 6 or 12. So the probability is 1/6. Figure 19 clarifies the result, starting from any hour between six and eleven the steps correspond to one of the possible results of throwing a dice. By symmetry, the same is obtained for an hour between twelve and five.

Figure 19. Lancet steps between 6 to 11

To conclude one last problem related to watches. Suppose a watch shop has battery-powered analog watches on its shelves. After a certain time the clocks stop because the batteries are exhausted. How many watches are required for the probability that they will show the same hour and minute, once the battery has run out, to be approximately 1/2?

First of all you have to think about the possible combinations on the dial, these are $24 \bullet 60 = 1440$. We can say that the probability that a clock stops at a certain hour and minute is given by $p = (1440\text{-}1)/1440 = 0.999$, in general we can write $p_i = (1440-i)/1440$. The probability that all N clocks stop indicating the same time is given by the product of the individual probabilities:

$$p_N = \prod_{i=1}^{N} p_i$$

We might think that it takes a large number of watches for this quite spectacular event to occur. Actually 44 watches are enough.

NEEDLE

An object as common as a needle (Figure 20) has, in fact, not indifferent mathematical references.

Buffon's needle problem is a classic random experiment dating back to 1700, it takes its name from Georges-Louis Leclerc Comte de Buffon (1707-1788), an exponent of the Enlightenment (Buffon, 1777).

Figure 20. Needle

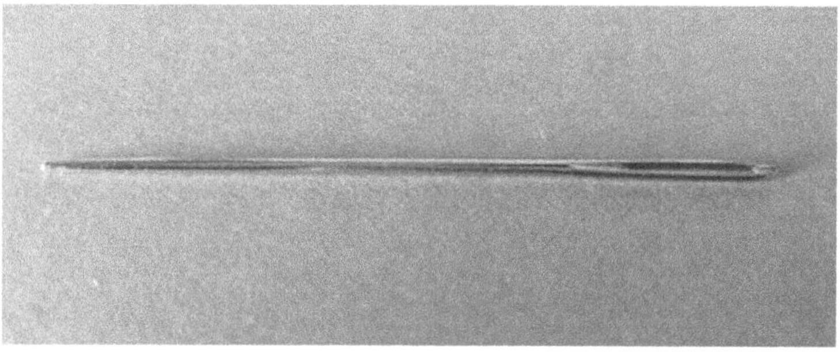

What's this about?

Suppose we have a series of parallel strips (for example a parquet floor), all of the same width, and we randomly drop a needle. What is the probability that the needle is on a dividing line between the two strips?

Before examining the demonstration, however, two mathematical premises must be formulated: the needle must be dropped at random and if the experiment is repeated the needles must be evenly distributed. Let's consider a needle of length L that is dropped on a parquet floor whose laths have width $d > L$.

It is necessary to consider the orientation of the needle to the edges of the boards and the distance of the middle stitch of the needle to the near edge. By indicating with α the angle at which the orientation of the needle in relation to the edge can be determined and with *x* the distance of the needle center of the needle from the nearest straight line you have (Figure 21):

Figure 21. Needle crossing the edge

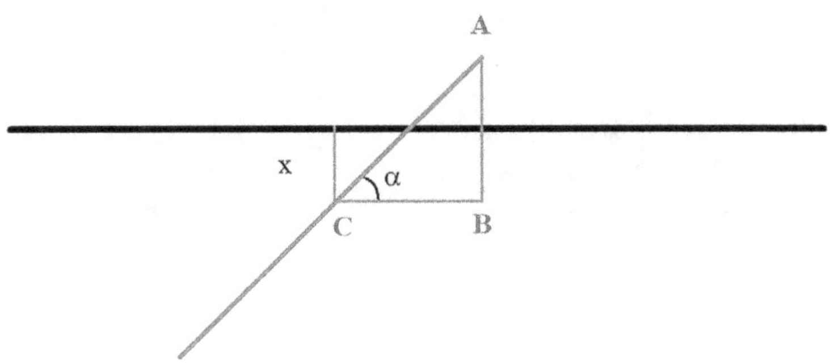

The triangle ABC in the figure is a rectangular triangle, the measure of cathetus AB is therefore equal to the hypotenuse L/2 for the sine of angle α. For the needle to intersect the straight line it must be $x < \frac{L}{2}\sin\alpha$, Buffon's needle problem will be solved if you can determine the probability $P(x<\frac{L}{2}\sin\alpha)$

To this end, the classical definition of probability as the ratio between the number of favorable cases and the number of possible cases, in a ratio between areas, is extended. In this sense it is possible to draw a rectangle of dimensions π and d/2 whose area represents all possible combinations between the orientation of the needle and the position of the center of the needle on the plane.

Favorable combinations are identified by the points below the curve (Figure 22) of equation $x = \frac{L}{2}\sin\alpha$.

At this point the difficulty is to define how much this area is worth; this is possible by using the integral calculation, that is to say the operator who associates to the function the area underlying its graph in a given interval. Using the classic definition of probability we obtain:

Figure 22. Comparison between the two areas

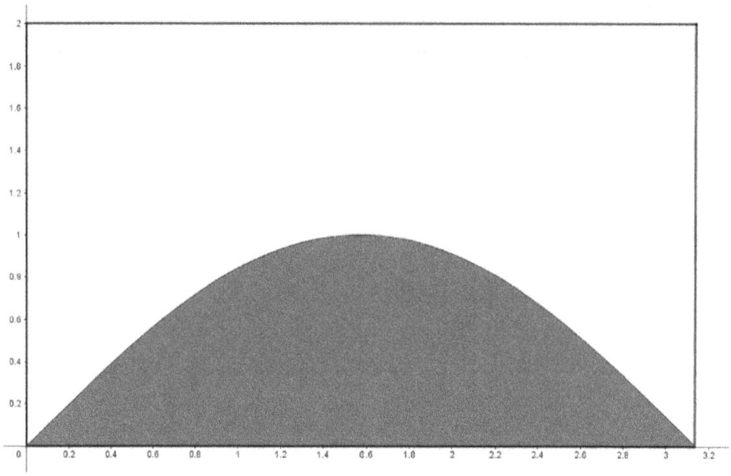

$$P(x < \frac{L}{2}\sin\alpha) = \frac{\text{area under the curve}}{\text{rectangle area}} = \frac{\frac{L}{2}\int_0^{\pi} \sin\alpha \; d\alpha}{\pi \bullet \frac{d}{2}}$$

hence

$$\int_0^{\pi} \sin\alpha \; d\alpha = 2$$

$$P = \frac{2L}{\pi \bullet d}$$

and finally

$$\pi = \frac{2L}{d \bullet P}$$

Then, from an experimental verification of the frequency of the event, an estimate of π is obtained.

A more in-depth examination of the problem can be found in a text by Herbert Solomon (1919-2004) which analyzes not only the extent of the problem due to Laplace, but also the case where the needle is long or you want to apply Buffon's (1777) problem to the third dimension. About the latter case Solomon writes: "The fourth and important modification is provided by a mathematical result. When a line of length r is placed at random in three-dimensional space containing a closed figure, for a large number of throws we may write:

$$r \bullet \frac{E(h)}{E(c)} = 4 \frac{sum\ of\ volumes}{sum\ of\ surface\ areas}$$

The experiment is physically quite problematic because it requires a high number of throws to obtain a significant result. Fortunately, many simulations can be found on Internet, it is enough to set on a search engine a request as "buffon's needle experiment simulation" (Buffon, 1777).

CONCLUSION

As we have seen, the frequentist definition is based on the empirical law of the case according to which, in a large number of tests made under the same conditions, the relative frequency of successes tends to the theoretical value of probability. This approach is extremely interesting when the classical probability is not applicable. In many cases it is possible to consider the frequency of events that have already happened to estimate the probability of future events. The scope of application is wide but above all it is significant from a school point of view. The statistical approach has a considerable didactic value: simulations can easily be carried out using the objects presented in the chapter. If we add the possibility of exploiting computer supports, the perspective is widened. Presenting the probability to pupils as the ratio between the number of successes and the number of experiments is normally very convincing.

FUTURE RESEARCH DIRECTIONS

Numerous empirical researches have verified that the majority of students have probabilistic beliefs acquired in the extra-curricular field: in fact, it is not by chance that we speak of hidden curricula. These beliefs may sometimes be correct but not occasionally wrong. Students do not always arrive at correct mathematical reasoning, expressing probability as a ratio. The use of the frequentist hypothesis can favor a positive approach to problems. Educational experimentation should focus on how to stimulate investigative trends in students. They should first try to formulate hypotheses and then try to verify them empirically by trying to verify the presence of patterns of regularity in random phenomena. Obviously, this should change the roles of the teacher and the students.

One of the most useful efforts should be directed at dealing with problems closely related to the real world.

REFERENCES

Bloch, A. (1977). *Murphy's Law and Other Reasons why Things go Wrong*. Price/Stern.

Buffon, G. L. L. Comte de (1777). *Essai d'Arithmétique Morale. Natural History. Supplément Volume Four*. Paris: De l'Imprimerie Royale.

Crabtree, H. (1914). *An elementary treatment of the theory of spinning tops and gyroscopic motion*. Longmans, Green.

David, F. N. (1955). Dicing and gaming (a note on the history of probability). *Biometrika*, *42*(1), 1–7.

Farrar, D. E., & Meyer, J. R. (1970). *Managerial Economics*. Prentice Hall.

Gorman, M. F. (2012). Analytics, pedagogy and the pass the pigs game. *INFORMS Transactions on Education*, *13*(1), 57–64. doi:10.1287/ited.1120.0088

Green, D. (1991). *A theory for spinning tops. Inductions* (Vol. 5). Gardens.

Lindstrom, M. (2016). *Small Data: The Tiny Clues that Uncover Huge Trends*. St. Martin's Press.

Matthews, R. A. (1995). Tumbling toast, Murphy's Law and the fundamental constants. *European Journal of Physics*, *16*(4), 172–176. doi:10.1088/0143-0807/16/4/005

Nahin, P. (2013). *Will You Be Alive 10 Years from Now? And Numerous Other Curious Questions in Probability*. Princeton University Press.

Namboodiri, K., & Suchindran, C. M. (2013). *Life table techniques and their applications*. Academic Press.

Nemiroff, R., & Nemiroff, E. (2016). *Dreidel Fairness Study*. arXiv preprint arXiv:1612.03730

Solomon, H. (1978). *Geometric probability* (Vol. 28). Siam. doi:10.1137/1.9781611970418

Trachtenberg, F. M. (1996). The game of dreidel made fair. *The College Mathematics Journal, 27*(4), 278–281. doi:10.1080/07468342.1996.11973792

Weiss, N. A. (2006). *Course of Probability*. Addison-Wesley.

ADDITIONAL READING

Pisarenko, V. F. (2018). The Notion of Probability and Difficulties of Interpretation. *Herald of the Russian Academy of Sciences, 88*(4), 289–293. doi:10.1134/S1019331618040056

Serradó, A., Azcárate, P., & Cardeñoso, J. M. (2006). Analyzing teacher resistance to teaching probability in compulsory education. In Proceedings of the Seventh International Conference on Teaching Statistics (pp. 1-6). Salvador (Bahia): International Association for Statistical Education and International Statistical Institute.

Spanos, A. (2019). *Probability Theory and Statistical Inference: Empirical Modeling with Observational Data*. Cambridge University Press. doi:10.1017/9781316882825

Trunkenwald, J., & Laval, D. (2019, February). Algorithms as a discovery process in frequentist approach to prediction interval. In Eleventh Congress of the European Society for Research in Mathematics Education (CERME11) (No. 35). Freudenthal Group; Freudenthal Institute; ERME.

KEY TERMS AND DEFINITIONS

Aleatory Variable: A variable whose values depend on outcomes of a random phenomenon.

Discrete Random Variable: A particular range of discrete values that the variable is permitted to take on.

Empirical Law of Chance: In a series of repeated tests each of the possible events manifests itself with a relative frequency that is approximately equal to its probability.

Expected Value: The average amount one expects to win.

Law of Large Numbers: Describes the result of performing the same experiment a large number of times.

Chapter 3 Subjective Probability

ABSTRACT

Subjective probability aims to solve a big problem: how to calculate probability when none of the definitions examined in the previous chapters is applicable. The probability that it will rain tomorrow, that a newly graduated student will find a job, or that a basketball player will pass the basket record are examples of the subjective conception of probability, as they require neither the knowledge of the mechanism that regulates the phenomenon nor the repeatability of the phenomenon. The value attributed depends on the state of information of the person making the assessment. The already mentioned Bayes' Theorem shows how to make probability evaluations and subsequently modify them in the presence of new data. Faced with the vast field of application, there is a very scarce (one would have to say nothing) attention on the part of the teachers who prefer to concentrate the little time dedicated to the subject to more standardized situations.

INTRODUCTION

Chance is the pseudonym of God when he did not want to sign. ~ Anatole France

One of the perhaps lesser-known aspects is that Greek philosophers already had a notion of probability, but in a very particular sense. In ancient times "probable" was the attribute of an opinion guaranteed by a recognized and as such credible authority. Today one would say that probability was purely

DOI: 10.4018/978-1-7998-3871-5.ch003

epistemic (a Greek word derived from the verb *epistanai*, which means to know or understand) and did not have a random aspect. This epistemic probability, according to Robert Ineichen (1999), was qualitative, not a quantitative one.

Only in more recent times have quantitative aspects been systematically taken into account, but with various theoretical-practical problems.

There are events for which none of the definitions discussed in the previous chapters is applicable because it is not possible to define the probability a priori and it is not even possible to calculate the frequency, as they are often unique events. If we ask ourselves the question "what is the probability that tomorrow it will rain?" it would be absurd to take for true the classical conception because every day the probability would be 1/2 (rain or no rain) and also it would be impossible to apply the frequentist probability because it is a unique phenomenon while this approach requires the repeatability of the test under the same conditions. At the most, we could take into account the historical series related to tomorrow's day and rely on a statistical evaluation of the event.

In this context there is, at a certain point, a change of perspective: the probability does not lie in the event, but lives in the observer: probability is subjective, probability is not a property of reality, it is a property of the subject who evaluates it according to the information in his possession; it can change from individual to individual, but it can change especially if the state of information he possesses changes.

First Bruno de Finetti (1906-1985) and then Leonard Savage (1917-1971) developed a subjective conception of probability.

Particularly strong is Bruno de Finetti's (1930) critical position towards the concepts seen previously: "The classical view is based on physical consideration of symmetry, in which one should be obliged to give the same probability to such 'symmetric' cases. But which 'symmetry'? and, in any case, why? The original sentence became meaningful is reversed: the symmetry is probabilistically significant, in someone's opinion, if it leads him to assign the same probabilities to such events. The logical view is similar, but much more superficial and irresponsible since it is based on similarities or symmetries which no longer derive from the facts and their actual properties, but merely from sentences which describe them, and their formal structure or language. The frequentistic (or statistical) view presupposes that one accepts the classical view, in that it considers an event as a class of individual events, the latter being 'trials' of the former".

For the avoidance of doubt, we make it clear that subjective does not mean arbitrary. This setting of probability can be applied, for example, if repeated tests can be performed, when tables with the results of statistical surveys on a certain phenomenon are available, as in the case of mortality tables (tables that show, for each age, what is the probability that a person of that age dies before his next birthday) or in the presence of odds defined by bookmakers based on certain algorithms; in other words, the subjective probability is not in contrast with the approaches examined above, but contains them as special cases.

The subjectivist conception of probability leads to a definition that could be formulated as follows: "the probability of an A event is the measure of the degree of belief that *coherent* individual attributes to the occurrence of A". The individual must be coherent, that is, not prey to preconceptions and able to attribute the same value of probability to similar phenomena, but above all, in a bet, must be willing to invert roles: whoever accepts to pay the sum s to receive in exchange the sum S must be willing to invert roles. In summary: the probability of an event is the price that an individual deems right to pay to receive a unit of account if the event occurs.

In general, following Bruno de Finetti's (1930) operational definition, if a certain event within a consistent and fair bet is given "x versus y", the subjective probability will be equal to:

$$P = \frac{y}{y + x}$$

This type of approach is, on the one hand, consistent, but above all applicable to most problems.

A small parenthesis: what is the substantial difference between the subjectivist conception and those that refer to other approaches, especially the frequentist one?

What changes is above all the interpretation of the Bayes' theorem: most frequent practitioners tend to apply it, while Bayesians say that every new event we need to improve the estimate and update the odds. If I have estimated to realize a penalty 75% of the times (given in line with the data in our possession) and then I start to get a series of centers, it's clear that the probability estimate will go up; in the case of the weather forecast, the percentages of rain that are given reflect precisely the historical series with a certain type of weather conditions, and these percentages will be gradually refined as the data grows.

Using survey data linked to the financial statements of Japanese listed companies, Masaaki Morikawa's (2019) study presents an estimate of companies' uncertainty about growth rates, determined subjectively, before making operational choices. It is therefore an ex-ante assessment. The results indicate that this uncertainty has a positive association with the forecasting error made; the companies with the greatest subjective uncertainty are those with the greatest forecasting errors. These results suggest that the subjective probability distribution obtained from a business survey contains valuable information to understand the size of ex-post strategy errors.

In an article in The Economist (Aug 16th, 2007) entitled *Gambling on tomorrow* there is a phrase that summarizes the concept very well:" Pascal's way of looking at the world was that of the gambler: each throw of the dice is independent of the previous one. Bayes' allows for the accumulation of experience, and its incorporation into a statistical model in the form of prior assumptions that can vary with circumstances. A good prior assumption about tomorrow's weather, for example, is that it will be similar to todays. Assumptions about the weather the day after tomorrow, though, will be modified by what actually happens tomorrow".

Having said that, you cannot hide some underlying problems. First of all, there is a problem related to the concept of the bet: many probabilists do not accept it because they believe that if the probability is subjective then there is a full interpretative possibility and the utility is zeroed while if you accept the "quibble" of the bet then you lose subjectivity and you have to resort to some more objective evaluation criteria.

Moreover, the degree of trust that a person places in the occurrence of a given event may not only differ from individual to individual, but above all is not socially neutral.

A case in point is that of the fight against smoking, many doctors and researchers have for years supported the relationship between cigarette smoking and lung cancer on the grounds that the likelihood of getting sick increased with cigarette consumption; tobacco multinationals have invested millions of dollars to create a sense of uncertainty by minimizing the risks by publishing distorted research and not mentioning unfavorable research. The debate has now subsided and it is well established that in developed countries the percentage of deaths attributable to tobacco smoke varies between 25% and 30% and that therefore the likelihood of a smoker falling ill is significant.

A second example is a no-vax controversy, i.e. the intense debate between those who argue that vaccines are useless or dangerous and those who believe it is essential to provide protection, especially for the most vulnerable. The only sensible approach should be to compare the risks in terms of probability. According to epidemiologists, the data for the MMR vaccine (a vaccine against measles, mumps, and rubella) are those in Table 1:

Table 1. Comparative risk table

Risks related to the disease	Vaccine risks
Pneumonia: 1/20	Encephalitis or severe allergic reaction: 1/1,000,000
Encephalitis:1/2000	
Death: 1/3000	

A third example, less dramatic, is related to the choice of a hotel through an online site. Not having other information is normally based on the judgment of people who have already tried that hotel; the higher the positive reviews, the greater the confidence in the goodness of the services.

Then what protocol should we follow?

The solution is that we need a continuous updating process so that non-dogmatic people can converge on the same conclusions with, of course, the same information base.

There are many examples, but they can be traced back to a single question: what is the basis for a subjective probabilistic decision? A simple answer does not exist, each of us could choose different bases: statistical data, previous experience, expert opinion, influence of propaganda, judgements of homogeneous groups, positions of influencers, etc..

The problem, as John D. Cook said in a 2010 post, resuming a work by Hanley and Lippman-Hand, is that you often have partial information or even an event that has not (yet) happened, for example in the medical field there may have been no fatal events in a number of operations. Can we therefore assume that the probability of an unfortunate event is zero? In this case the above-mentioned consultant in applied mathematics and statistics proposes a simple rule already known to statisticians. The *rule of three* says that if you have checked *N* cases so far and have not yet found what you are looking for, then a reasonable estimate is that the probability is less than 3/N. For example, if a drug is tested on 150 human subjects and no adverse event is recorded, the rule of three concludes with 95% confidence that less than 3 out of 150 people will experience an adverse event (Cook, 2010).

The justification can be frequentist: "Suppose the probability of what you're looking for is *p*. If we want a 95% confidence interval, we want to find the largest *p* so that the probability of no success out of n trials is 0.05, i.e. we want to solve $(1\text{-p})^n = 0.05$ for p. Taking logs of both sides, n log(1-p) = log(0.05) ≈ -3. Since log(1-p) is approximately –*p* for small values of *p*, we have p ≈ 3/n". But you can also propose a Bayesian interpretation:" The posterior distribution on *p* after having seen 0 successes and *N* failures has a beta(1, N+1) distribution. If you calculate the posterior probability of *p* being less than 3/N you get an expression that approaches 1 – exp(-3) as *N* gets large, and 1 – exp(-3) ≈ 0.95".

An analysis made by Alessandro Gentilini with the software R using the *prop.test* function to estimate the confidence interval of the ratio that can be inferred from the observation of 0 successes in *n* attempts would instead tend to shift the level for which there would be a *rule of four*: instead of considering three cases, four should be considered.

The meaning of this continuous adjustment process is present in a test presented by students of the Quality Class at the 2016 CIEAEM 68 conference: "A friend of yours states to be "100% sure" that he did an exceptionally good exam and that he will get the highest score. You say "Let's play a game. You have a bag with 2 red marbles and 98 white marbles. If you draw a red marble, I will give you one million dollars. Otherwise, you can wait for the exam's result and if you will get the highest score, I will give you one million dollars. What do you decide to do? Do you prefer drawing or waiting? ". If your friend decides to draw, you will conclude that he gives a value minor than 98% to the probability of getting the highest score. Now you ask him to choose between drawing from a bag with 5 red marbles and 95 white marbles. You go on asking to choose, changing the number of red and white marbles. At a certain point he will tell you it's the same. You will have measured the subjective probability, that is equal the number of red marbles over the total".

A final reflection should be made. The demand that a game be fair is sensible in the case of two players who have to agree on the stakes to be paid without anyone being favored. In reality, however, those who play do so against a managing body which, obviously, in exchange for the service offered, retains a part of the stake by way of profit. This aspect will be even clearer in chapter 5 dedicated to the Game.

TEA

Tea unpacked or in bags (Figure 1) is the favorite drink of the English but, in this context, it is only an excuse to better define the problem of subjective probability.

Figure 1. Teabag

Consider an example reported by James Berger (2013).

1. A lady, who adds milk to her tea, claims to be able to tell whether the tea or the milk was poured into the cup first. In all of ten trials conducted to test this, she correctly determines which was poured first.
2. A music expert claims to be able to distinguish a page of Haydn score from a page of Mozart score. In ten trials conducted to test this, he makes a correct determination each time.
3. A drunken friend says he can predict the outcome of a flip of a fair coin. In ten trials conducted to test this, he is correct each time.

In all three cases we are faced with experiments carried out in a controlled manner, and so we should believe all three statements equally. More reasonably we can take into account that these are three quite different situations and, based on the trust we place in the three people, we should consider the first possible event, the second certain, the third due to pure luck.

Can tea have an effect on fertility? This is a question that ordinary people would tend to answer in an absolutely subjective way; probably the majority would assign a low probability to the event.

Is there anything that could influence the answer? As we have already had occasion to say, the estimate depends on the state of information of the person making the assessment.

In this regard, the following quote attributed to Elizabeth Hatch and published in 2012 by Daily Mail: "Two cups of tea a day may help boost a woman's chances of becoming pregnant, according to a study.

It found women were 27 per cent more likely to become pregnant if they regularly drank tea compared with those who did not.".

This statement should change our estimate positively, unless we take into account the quote from an almost contemporary publication (Hatch & al., 2012): "Women who drank regular tea had moderate increases in fecundability for 2+ servings per day (FR = 1.27 [95% CI = 0.98-1.64]), with no monotonic trend according to servings per day [...] Our finding of slightly increased fecundability among tea drinkers is consistent with 2 previous prospective studies reporting an approximate 2-fold increase in fecundability among tea drinkers, but disagrees with 2 other studies that reported null associations".

Perhaps the topic is too complex to reach a clear conclusion!

Playing Cards

The ancestors of modern playing cards likely arrived in Europe at the end of the 14th century, and by that time they had already taken a similar form to the one we will refer to today. In particular, the deck contains 52 cards, divided into four suits: cups, coins, swords, and polo-sticks or hearts, diamonds, clubs, and spades. Each suit contains ten cards, numbered from 1 to 10, and three figures. In the first chapter the playing cards were used in reference to the classical probability, here instead we will need them to understand the Bayesian decision process and the connection with the subjectivist setting.

Let's formulate a simple question: what is the probability that, by randomly drawing a card from a deck of 52 cards, composed of 12 Figures (F) King, Queen, and Jack and 40 Cards (C) from the Ace to the ten, this is a C card and then the event $\overline{F}$ occurs, i.e. there is no figure (Figure 2)?

Figure 2. Ace of Spades (not Figure)

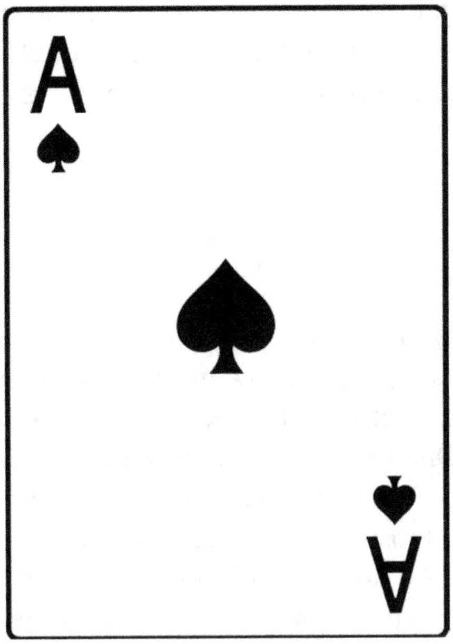

Let's ask this question to three students who do not know the probability calculation and suppose the answers are $p_a = 0.75$, $p_b = 0.50$ and $p_c = 0.25$. Since there are no elements to prefer an answer we can subjectively assign the same *a priori* probability of 1/3 for each hypothesis.

To verify what the three hypotheses compared in a particular situation would lead to, let's suppose to test them in case the CCCC sequence occurs in four extractions with re-entry. Since they are independent events, we multiply the three probabilities indicated by the students and then we obtain the following estimates: 0.316, 0.063, 0.004. At this point, by Bayes' Theorem, the a priori probability can be multiplied by these values to obtain as many values P (i) • P (CCCC l i), where *i* indicates the three possible references, which constitute the numerator of the formula; the denominator is given by their sum and is, in the given example, 0.128. The ratios give us the a *posteriori* probabilities 0.823, 0.164, 0.010. For example:

$$P(1|CCCC) = \frac{P(1) \cdot P(CCCC|1)}{\sum_{i=1}^{n} P(1)P(CCCC|1)} = \frac{0.333 \cdot 0.316}{0.128} = 0.823$$

As can be observed, the first hypothesis appears decidedly more credible and actually the first judgment is closer to the theoretical probability, equal to 40/52 = 0.769. We can draw the following conclusion: the a posteriori judgment is influenced both by the events observed and by the a priori judgment.

TAP

The tap is, notoriously, a device for regulating the flow of a liquid (or gas) installed at the end of a pipe or tank (Figure 3). Generations of students have been obsessed with problems with tanks, containers, cisterns to be filled or emptied using taps. One of the classic problems dates as far back as Chiu Chang Suan Shu (150 B.C.) and can be found in Hero (150 A.D.). To fill a cistern there are three taps, the first takes 10 hours to fill, the second takes 20 hours and the third 60 hours. If all three taps open, how long will it take the cistern to fill? This type of problem can be used, judiciously, to force the student to make logical reasoning and calculations but its abuse runs the risk of confusing the student. Assigning fictitiously realistic problems does not help, just as some examples of excessive realism can lead to questionable problems. An extreme example could be found in a book (of which we choose not to report the author) for primary school published in Pakistan in 1987: "The speed of a Kalashnikov bullet is 800 m/s. If a Russian is at a distance of 3200 m from a mujahid and the said mujahid aims at the Russian's head, calculate how many seconds will pass before the bullet hits the Russian in the forehead". In this case a realistic physics problem!

More modestly, in this context, the tap only serves to introduce a simple betting problem that well represents de Finetti's (1930) operational definition.

The bathroom faucet is leaking and the plumber was called to fix it. Unfortunately, given the precedents, the possibility that the plumber will arrive as soon as possible is not high. While waiting for the event, husband and wife discuss how likely the arrival is in the afternoon of the same day. He is willing to bet 1 against 10 on the fact that he will arrive, i.e. he gives this event a probability of 1/11. The wife decides to accept the bet. It is as if we were to draw from an urn with 1 white ball (positive event = arrives) and 10 black balls (negative event = does not arrive).

Figure 3. Tap

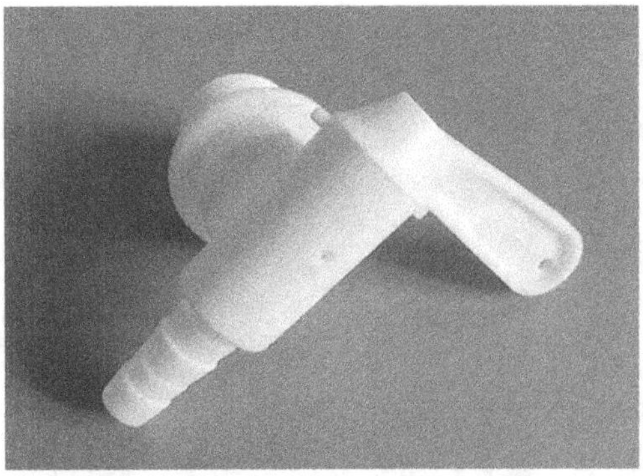

In other words, he is implicitly giving a probability assessment, and this assessment gives a 1/11 probability to the occurrence of the plumber's arrival. The fraction that expresses the probability is the one seen in the introduction to the chapter:

$$P = \frac{bet\ of\ him}{bet\ of\ him + bet\ of\ her}$$

This mechanism is the basis of betting as we will see in the following example.

FINGER FOOTBALL

Before we deal with the central theme of betting, let's discuss a small problem: a football field, due to rain, is at the limit of practicability 5 times per season (during which 19 matches are played). The probability of accidents occurring during the match is 0.6 in case of rain and 0.1 in case of good weather. Knowing that a player is injured, what is the probability that this is due to bad weather? Bayes' theorem tells us that the probability is quite high, 15/22, so playing in bad weather is a significant risk for players.

Let's move the discussion on bets related to a sporting event such as football, of course not the simulated one as in Figure 4.

One element that can guide us in evaluating and forecasting a football result, probably the most important, is the forecast of those who professionally manage the bets: bookmakers.

Figure 4. Finger Football

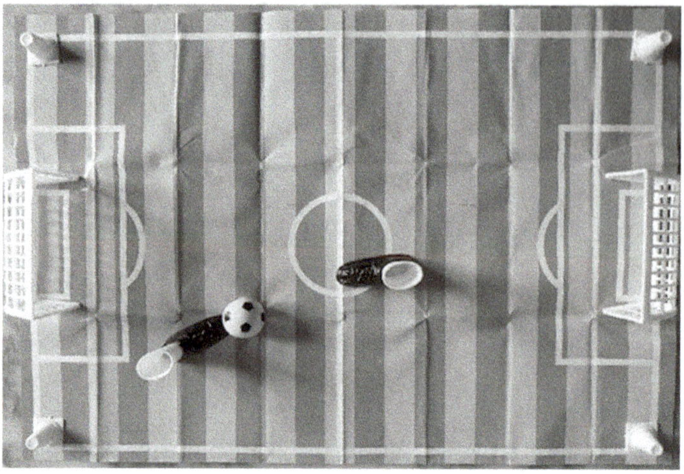

A bookmaker sets the odds for betting on a given event, such as a certain team winning, stating " I accept bets and am willing to give 5 times the amount wagered by a bettor and return the amount wagered in case that team wins ".

The odds allow you to calculate the profit linked to a bet. It represents the reciprocal of the probability of occurrence of a result. A low odds corresponds to a higher probability of winning. Let's assume that the odds associated with possible match results (1 = home team win, X = draw, 2 = visiting team win) are as shown in Table 2:

Table2. Hypothetical quotas

	1	X	2
team A vs team B	2.20	3.15	3.90

Let us calculate the reciprocal, obtaining Table 3:

Table3. Reciprocal quotas

	1	X	2	Sum
team A vs team B	0.45	0.32	0.26	1.03

Note that if you turn odds into probabilities the sum does not make 1, the difference represents the bookmaker's gain. In gambling, when a set of odds guarantees a profit, regardless of the outcome of the game, we speak of a *Dutch book*. In the language of gambling the term comes from the practice of recording the bets in a book, hence the term bookmaker to indicate the person placing the bets and then "make the book". If subjective odds do not satisfy one of the fundamental rules of probability, a "book" of bets is created that guarantees safe losses for those who accept these inconsistent beliefs. In order to obtain the correct probability, you must then divide the previous results by the sum so that the total is reduced to 1.

The correct probabilities will finally be those in Table 4:

Table 4. Probabilities associated with possible results

	1	X	2	Sum
team A vs team B	0.44	0.31	0.25	1.00

So you can, for example, bet on the host team winning with a 1/4 chance of winning.

A more precise definition of subjective probability can be given at this point.

This P(A) is the probability of an event A, it is the price that an individual believes it is correct to pay to receive 1 if the event occurs and 0 if the event does not occur; the probability must be attributed so that it is not possible to obtain a certain win or loss with a set of bets.

It is a generalizable and operational definition thanks to the fairness condition, even if the fair betting scheme is not mechanically applicable to all risk situations (a lottery is not insurance) and is too much based on the subjective (but not arbitrary!) opinion of the evaluator of the event.

If a bettor's goal is to increase his chances of winning, is there a criterion for how much he should bet on each individual bet? In order to get the maximum profit, it is necessary to estimate the real probability of success of the bet. In 1956 John Larry Kelly Jr. (1923-1965) proposed a criterion for maximizing long-term bankroll growth by determining the optimal stake for each bet. By indicating with Q the odds proposed on the event by the bookmaker and with P the odds that the player assigns to the event you get the *Kelly criterion*:

$$Bankroll = \frac{Q \bullet P - 1}{Q - 1}$$

Let's take the example of a football match where the available odds on a draw are 3.15 (with an implied probability of 31.7%), however the player has estimated a probability of 40%. The bankroll is therefore 0.121. Assuming you have $100 available you would have to invest 12% in the draw. Obviously, everything depends on the ability to determine with some precision the real probability of success. With an overestimation the stake found will be too high and will lead to lose money in the long term; if the estimates are lower than the reality the gain will be lower than expected. With the previous data a simulation of 100000 bets allowed us, starting from a capital of $100, to get 161. However, before we go into reckless betting it must be said that a new attempt has shown less comforting values; the $100 has been reduced to 18!

This is all genuinely nice but it is right to move from theory to practice. If you look at the results of the last five years of the Italian Serie A football championship, you can see that the bookmakers have got a bit more than half of the results right (55% in the period considered) as they correctly predicted 46, 54, 62, 59 and 54% of the results.

Since we can assume that bookmakers provide quite reliable odds (although it must be said that the odds change over time according to the bettors' bets) based on a mixture of statistical odds and public opinion, a possible trivial game strategy could be to bet on the team considered to be stronger, therefore with a lower odds.

Who, hypothetically, had bet 1€ on all the results with the lowest odds, for all the games of the league championship in a year, how much would have won or lost?

The answer varies from year to year, for example in 2014-15 there would have been a loss of €47.32, in 2015-16 a loss of €2.73, in 2016-17 a win of €41.89, in 2017-18 a win of €5.67, in 2018-19 a loss of €8. These data seem to contradict the predictive quality of the bookmakers seen above; in fact, it is necessary to take into consideration, economically speaking, the concept of Expected Value or *Mathematical Expectation*.

To explore the theme we can take our cue from *Newsies*, a musical film, produced by Disney in 1992, about young people selling newspapers on the street.

Each guy buys newspapers for $1.5 and sells them for $3. Each unsold copy can be returned to the distributor who will pay $0.50 for it. On the basis of information collected from some comrades, who have already done the newsboys for some time, a young estimate that the newspapers have the following odds of being sold (Table 5):

Table 5. Probability of sales

Newspapers	*20*	*30*	*40*	*50*
Probability	0.4	0.3	0.2	0.1

How many newspapers does he have to buy? We need to define a decision criterion that allows for an optimal choice. It is also known as the *Expected value criterion*: this method is applicable when there are two or more alternatives, to which are associated gains (or losses) in the face of two or more incompatible events of which the probability is known or can be estimated. The decisions, events and results that depend on random phenomena are placed in a table called *Pay-off table*.

In our example, the gain is given by:

gain = revenue from sale + revenue from return - purchase cost.

For example, if the student buys 50 newspapers and sells 40 you will have:

$$G = 3 \bullet 40 + 0.50 \bullet (50\text{-}40) - 1.5 \bullet 50 = 50$$

If all cases are taken into account, Table 6 is obtained:

Table 6. Gains from sales

		Demand			
		20	*30*	*40*	*50*
Purchased	*20*	30	30	30	30
	30	20	45	45	45
	40	10	35	60	60
	50	0	25	50	75
	p	0.4	0.3	0.2	0.1

Now let's transform the table by putting in each box the products between the economic results and the probabilities. Then we calculate the average value of sales taking into account all possible options. In Table 7 you can see all the results.

Table 7. Average sales value

	20	*30*	*40*	*50*	Average
20	12	9	6	3	30
30	8	13.5	9	4.5	35
40	4	10.5	12	6	32.5
50	0	7.5	10	7.5	25
p	0.4	0.3	0.2	0.1	

At this point, we can say that the young person will be better off buying 30 newspapers, for an average daily income of $35.

In general, the expected value is the sum of the possible values assumed by the variable in question multiplied by the probability of realization. It is thus possible to define the Expected value as the value that I would win (or lose) on average in each bet if the game continued indefinitely.

A simple example: if we bet 1 € on the favorite team that has a chance of winning 0.44 we have to expect a loss because the Expected value is 1 • 0.44 + (-1) • 0.56 = - 0.2.

We can only talk about *Fair game* if the total Expected value is worth zero.

For example in a lottery a person plays with probability p to win a sum s and with probability q = 1-p to lose a sum r (obviously won by the opponent).

The game is fair if:

$$M = s \bullet p + r \bullet q = 0$$

We can find a literary quote in Shakespeare. In *History of Henry IV*, Part II, Act I we find written:

We all that are engaged to this loss
Knew that we ventured on such dangerous seas
That if we wrought out life 'twas ten to one;
And yet we ventur'd, for the gain propos'd

We could define these words as a reference to Expected value.

RAIN GAUGE

The rain gauge (or pluviograph) shown in Figure 5 is the instrument used to measure the amount of rain falling over a given period. A simple manual rain gauge consists of a conical or cylindrical vessel equipped with a graduated scale. The height of the water filling the container is equivalent to the fallen rain. A double scale rain gauge allows you to compare the measurement in mm with inches.

Figure 5. Pluviometer

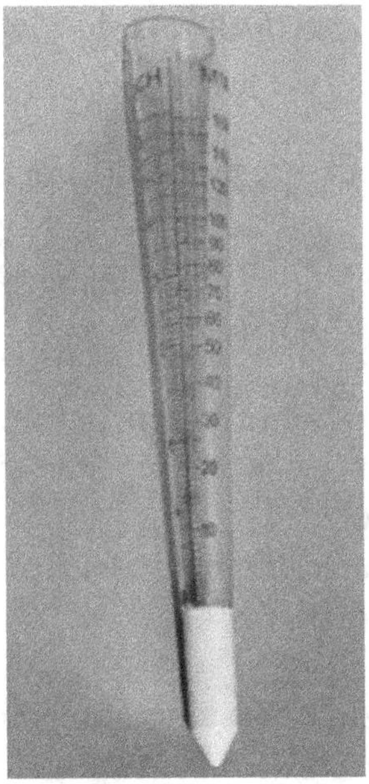

This object introduces an important aspect of everyday life: the weather forecast.

Almost everyone watches on television or specialized websites what will happen, from a meteorological point of view, the next day or the following days. But how do these forecasts work?

Forecasters analyze the results of weather model processing. Initially, the analysis covers the meteorological situation on a large scale, and then focuses on limited-area models covering a smaller area. It should be noted that the results of the simulations are not, in themselves, forecasts. The outputs of meteorological models become forecasts of the most likely evolutionary scenario. It should also be stressed that the weather factor plays an essential role: a short-term forecast is more reliable than a long term one.

So it is a question of estimating the frequency of the rain event conditioned to the rain forecast and the frequency of the non-rain event conditioned to the non-rain forecast, respectively.

Verifying the weather forecast is essentially a matter of examining its reliability by comparing the observations made with the forecast. Various methods and techniques exist in the literature to verify the reliability of forecasts, whose guidelines are dictated by the Joint Working Group on Forecast Verification Research of the World Meteorological Organization.

The indexes used are:

1. Accuracy: allows you to determine the fraction of adjusted forecasts in relation to total forecasts.
2. Probability of Detection: allows you to assess the percentage of times that rain has been correctly predicted compared to all times that it has rained.
3. False Alarm Ratio: allows to evaluate the percentage of times when rain was expected but did not occur compared to the total number of times when rain was expected.
4. Bias: is the ratio between the number of expected rain events and the number of rain events actually observed.

Let us dwell on two questions, one complementary to the other. The first is "if it is expected to rain what is the probability that it will rain?", the second can be formulated as "if it is expected that it will not rain what is the probability that it will not rain? In other words, it is about verifying Probability of Detection and False Alarm Ratio. As an example in the Italian Air Force site you can find two interesting graphs. In the first one you can observe the frequency of the events in case it was expected rain with its Trend line and the climatology of the period. As can be seen (Figure 6) this probability varies from a value of about 60% for the first 12 hours and decreases up to a value of 50% for the last 12.

Figure 6. Probability of Detection

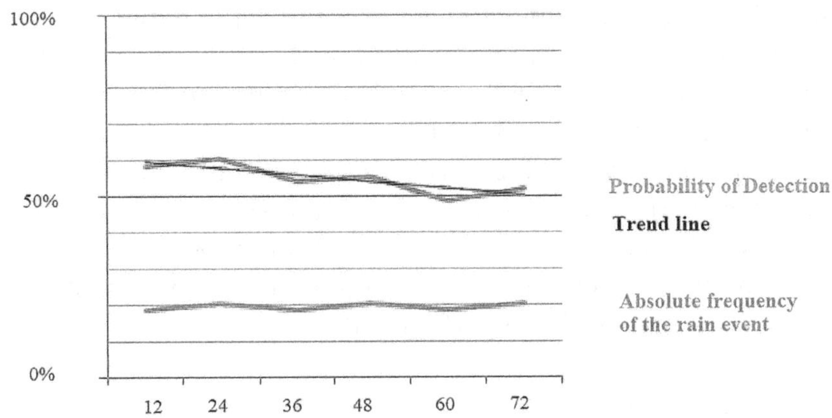

In the second graph (Figure 7) you can observe the relative frequency of the "non-rain" event conditioned to the forecast (it will not rain), i.e. the probability of success of the forecast. The probability is 94% in the first 12 hours, then it goes down slowly.

Figure 7. False Alarm Ratio

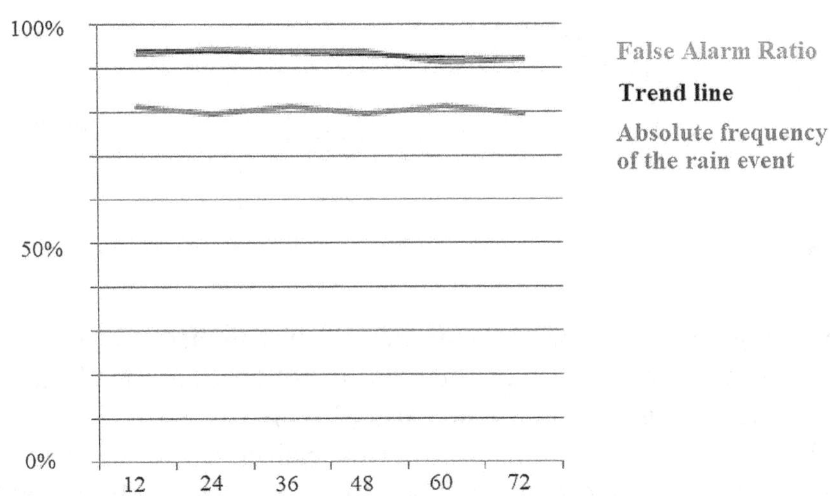

To go into the subject in more detail, we can deal with rainfall probability (or possibility) curves. They express the relationship between maximum heights and rainfall durations that may occur, in a given period, in a given area. These curves are obtained from a sample of data considered as randomly extracted. Using statistical inference techniques, the properties are estimated from those of the available sample. The problem is to determine what is the best form taken by the associated probability distribution and therefore what are the parameters of the distribution.

Different probability distributions are used in hydrology, the most used is the one named after the statistician Emil Julius Gumbel (1891-1966), but his analysis would take us beyond the limits of this book. To make the work of interpreting the data easier, special maps are used. By way of example, we can observe a map by Gumbel (Figure 8).

Figure 8. Gumbel Map

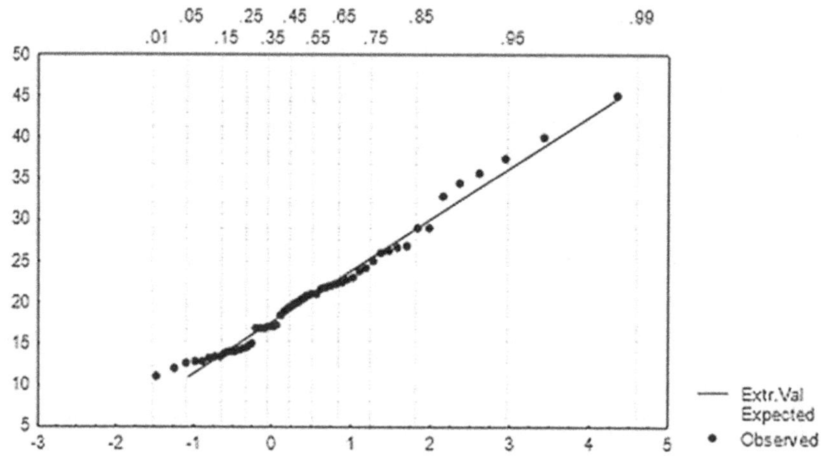

TULIP

The tulip (Figure 9) is not only a beautiful flower with different colors, but it reminds us of a historical fact of considerable importance described effectively by British writer and historian Mike Dash (2011).

In the Netherlands, the first major financial crisis occurred in 1637, triggered by the trade and sale of the bulbs of these flowers. Imported from Turkey, they spread mainly in Holland and were quickly considered as luxury goods, highly desired by the bourgeoisie and rich merchants. The demand for new flowers soon outstripped their supply, so that prices rose so high that tulip bulbs became a solid investment. Prices soon developed completely unrelated to their real value, giving rise to what we would now call a speculative bubble. It even went so far as to sell real estate to buy the rights to the most valuable bulbs. They negotiated the "rights on the bulb", i.e. the tulip futures, immediately paying only a deposit and paying the balance at the delivery of the flower. The ceiling was reached by Alkmaar's auction on 5 February 1637. However, it was enough for an auction to go deserted to cause panic and to plunge market prices throughout the country. We can consider this event as the first futures market in history, the explosion of the bubble plunged speculators and investors into a spiral of economic ruin.

Figure 9. Tulip

But what does this speculation have to do with probability?

Individuals often make mistakes both because they allow themselves to make predictions based solely on past trends and because they tend to react non-rationally to new information. Because information on various probabilities is often incomplete, individuals can conceive of a subjective probability much more favorable than the real one. In the case of tulips the subjective probability that the price increase would continue was universally accepted and therefore the expected value of this "bet" was not evaluated correctly.

Economist John Maynard Keynes (1883-1946) wrote: "A conventional valuation which is established as the outcome of the mass psychology of a large number of ignorant individuals is liable to change violently as the result of a sudden fluctuation of opinion due to factors which do not really make much difference to the prospective yield; since there will be no strong roots of conviction to hold it steady" (Keynes, 1921).

To remain within the realm of subjective predictions, we can examine a fictitious case that has very realistic aspects.

A computer store has to decide on the purchase of a batch of integrated circuits for memory expansion. It can choose between three UK importers offering different unit prices. Prices depend on the value of the pound sterling at the time of delivery, but the contract must be made now. In summary, we have Table 8:

Table 8. Prices of integrated circuits

	Price	
Importer	≤ on the rise	≤ in flexion
A	2	4
B	3	3
C	5	2

By indicating with *p* the probability that the pound is growing and therefore with 1-p the opposite probability, we will have that the three cost functions will be expressed by:

$$z_A = 2p + 4\ (1\text{-}p) = -\ 2p + 4$$
$$z_B = 3p + 3\ (1\text{-}p) = 3$$
$$z_C = 5p + 2\ (1\text{-}p) = 3p + 2$$

subject to the obvious condition $0 \leq p \leq 1$.

In this case *p* is a variable which is beyond the control of the decision-maker but which can be estimated using, for example, expert evaluations. Figure 10 shows the expected trend of the ≤ - € exchange rate made by a site of economic forecasts for the year 2021. Obviously, these forecasts have a wide range of uncertainties, as currency exchange rates are influenced by many factors: change in GDP value, trade balance data, inflation, interest rates, political events, etc.

Figure 10. Expected exchange rate trend €-≤

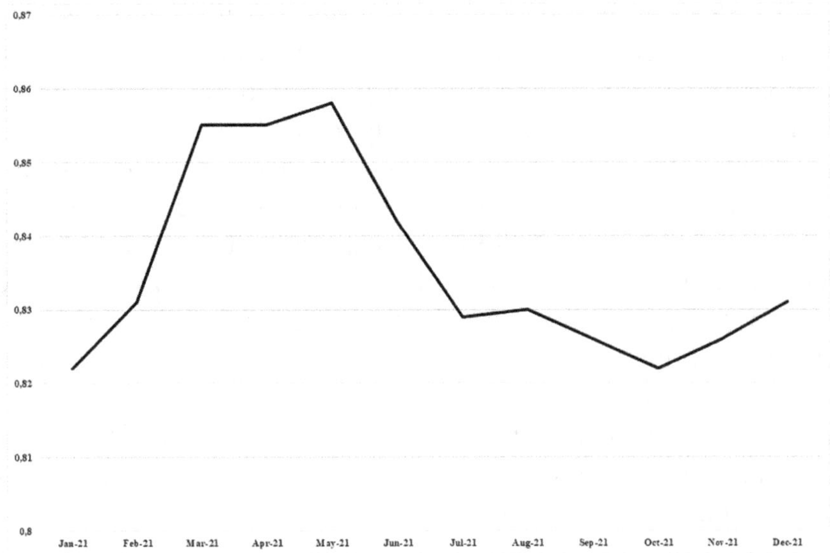

Unless he wants to try to formulate his own hypothesis based on past experience, or to accept the advice of a seer who has read the future in the crystal ball, the shopkeeper should trust (always with caution) the judgment of experts. Experts who, in this case, are essentially based on several elements: technical factors related to the Forex (Foreign Exchange Market) chart, the announcement of interest rates made by monetary authorities or the development of gross domestic product.

CONCLUSION

At this point the reader interested in the didactic aspects could be disoriented by the presence of three different models (classic, frequentist and subjective) which have lights and shadows, strengths, and weaknesses. This is not surprising, however, a great intellectual like John Maynard Keynes wrote in his General Theory: “there is no scientific basis on which to form any calculable probability whatever. We simply do not know. Nevertheless, the necessity for action and for decision compels us as practical men to do our best to overlook this awkward fact and to behave exactly as we should ”. Perhaps it would be more honest to declare that the probability of an event is closely related not to our degree of knowledge but to our degree of ignorance!

As we saw in the chapter, the subjectivist approach solves a number of problems and leaves others unresolved. If the betting method set by de Finetti (1930) as the cornerstone of its approach seems to be convincing in the formulation of numerical evaluation of the probabilities, from a practical point of view it often happens that subjective judgments are not sufficiently reliable, then the choices may prove to be suboptimal. Indeed, in the assessment, two individuals may use the same information elements differently, or they may have different information. All of this is especially true when you are faced with unique events, such as a football match, a person's life, the value of a bargaining asset, etc. Despite some problems, the confidence-based approach remains, in the opinion of many, the most convincing. Any approach you want to consider, however, must accept that subjective probability judgments represent a fundamental aspect in the decision-making process in conditions of uncertainty. For this reason, it is important to develop, from childhood, a conscious mental approach.

FUTURE RESEARCH DIRECTIONS

One of the most interesting aspects on which to focus efforts is the analysis of the gap between the probability subjectively perceived by students and the real frequency of the event. This comparative analysis, initially limited to simple and easily controllable situations, can be extended to wider areas. Think, for example, of one of the problems that has most affected the younger generations: climate change. The IPCC (Intergovernmental Panel on Climate Change) presents various hypotheses on possible climate change and its effects on humanity. Obviously, it is important to understand the degree of change and the estimated probability of the various scenarios that are proposed. What is interesting is that some researchers have underlined the fact that there is a "preference" for extreme low probability outcomes especially if the information is provided verbally.

Another terrain in which objective and subjective confrontation can be seen in the workplace. For example, according to Sun & Philips (2020): "Industry workplace injuries and illness's underreporting cannot be inferred through the form of the questionnaire. Because it is a subjective probability problem when the workers judge whether they are injured or not. The subjective probability value will be affected by many factors, some factors will increase the subjective probability value, and some factors will reduce the subjective probability value. However, when the workers fill out the questionnaire, the factors that reduce the subjective probability value do not work. So, using the questionnaire method to estimate industry workplace injuries and illness' rate naturally enlarges occupational injury and illness's rate".

REFERENCES

Berger, J. O. (2013). *Statistical decision theory and Bayesian analysis.* Springer Science & Business Media.

Cook, J. D. (2010). Estimating the chances of something that hasn't happened yet. *Endeavour.*

Dash, M. (2011). Tulipomania: The story of the world's most coveted flower and the extraordinary passions it aroused. Hachette UK.

de Finetti, B. (1930). Fondamenti logici del ragionamento probabilistico. *Bollettino UMI*, *9*, 258–261.

de Finetti, B. (2017). *Theory of Probability: A critical introductory treatment.* Wiley. doi:10.1002/9781119286387

Hanley, J. A., & Lippman-Hand, A. (1983). If nothing goes wrong, is everything all right?: Interpreting zero numerators. *Journal of the American Medical Association*, *249*(13), 1743–1745. doi:10.1001/jama.1983.03330370053031 PMID:6827763

Hatch, E. E., Wise, L. A., Mikkelsen, E. M., Christensen, T., Riis, A. H., Sørensen, H. T., & Rothman, K. J. (2012). Caffeinated beverage and soda consumption and time to pregnancy. *Epidemiology (Cambridge, Mass.)*, *23*(3), 393–401. doi:10.1097/EDE.0b013e31824cbaac PMID:22407137

Ineichen, R. (1999). *Nuts, astragales and the beginnings of probability calculation.* University of Southern Switzerland.

Kelly, J. L. Jr. (1956). A new interpretation of information rate. *The Bell System Technical Journal*, *35*(4), 917–926. doi:10.1002/j.1538-7305.1956.tb03809.x

Keynes, J. M. (1921). *A treatise on probability.* Mac Millan.

Morikawa, M. (2019). *Firms' Subjective Uncertainty and Forecast Errors. Research Institute of Economy, Trade and Industry.* RIETI.

Sun, K., & Philips, P. (2020). *Is Construction Industry Workplace Injuries and Illness's Underreporting a Severe Problem: An Answer Based on Subjective Probability Analysis?* Academic Press.

ADDITIONAL READING

Baker, J. D. (2019). Assessing Credibility in Subjective Probability Judgment (Doctoral dissertation, University of Pennsylvania).

Bassett, W., & Lumsdaine, R. (2001). Probability Limits: Are Subjective Assessments Adequately Accurate? *The Journal of Human Resources*, *36*(2), 327–363. doi:10.2307/3069662

Juanchich, M., & Sirota, M. (2017). How much will the sea level rise? Outcome selection and subjective probability in climate change predictions. *Journal of Experimental Psychology. Applied*, *23*(4), 386–402. doi:10.1037/xap0000137 PMID:28816471

Nickerson, R. S. (2002). The production and perception of randomness. *Psychological Review*, *109*(2), 330–357. doi:10.1037/0033-295X.109.2.330 PMID:11990321

KEY TERMS AND DEFINITIONS

Dutch Book: A set of odds and bets which guarantees a profit.

Expected Value Criterion: One of the criteria in the decision theory.

Fair Game: A game in which there is an equal chance of winning or losing.

Kelly Criterion: A formula for bet sizing.

Mathematical Expectation: A probability-weighted average of a random variable.

Pay-Off Table: A useful way to represent and analyze a scenario.

Rule of Three: Establishes the occurrence rate when a given event is not found in a sample. Rule of four: an alternative hypothesis to the previous one.

Chapter 4
Probability and Combinatorics

ABSTRACT

Combinatorial calculus is a branch of mathematics oriented to determine the number of distinct cases that can occur in an experiment or the number of elements that make up a set. One question often remains open in the debate about the role of combinatorics: Does the introduction of combinatorics in school teaching have an autonomous mathematical value or is it a simple tool for teaching probability? Problems of a combinatory nature can be found in many areas of mathematics, but in this chapter, emphasis will be placed on a limited circle of topics with a close link to probability. The objects presented will serve as an introduction to situations that technically relate to permutations and combinations.

INTRODUCTION

Combinatorics is not a science, it's an attitude. ~ Mark Haiman

Combinatorial analysis has an ancient history. For example, magic squares are an interesting object of recreational mathematics and represent a fun way to approach combinatorial mathematics. Magic squares are arrays of numbers with the property that the sum of the numbers placed on the rows, columns and diagonals give the same result. Examples can be found in a Chinese book dating back to the 12th century BC. The binomial coefficients, generated by the expansion of the binomial $(a + b)^n$, were known to the 12th century Bhāskara Indian mathematician.

DOI: 10.4018/978-1-7998-3871-5.ch004

In reality, only in the late 1800s did the combinatorics become an independent branch of mathematics, losing its mystical and recreational aspect. A simple definition is owed to Lorenz Halbeisen (2011): "Combinatorics is the branch of Mathematics which studies collections of objects that satisfy certain criteria, and is in particular concerned with deciding how large or how small such collections might be".

The two main types of combinatorics problems are related to "existence" and "enumeration" problems. The term combinatorics therefore denotes that branch of mathematics that studies the ways to group and/or order the elements of a finite set of objects according to given rules. Combinatorics methods are successfully used in different fields such as algebra, set theory, graphs, but especially with reference to probability. For a more in-depth examination of the topic, we refer to the texts of Loehr and Roberts & Tesman (2017).

Given the use in the field we are interested in, a question arises: does the introduction of combinatorics in school teaching have an autonomous mathematical value or is it simply an operational tool for solving probabilistic problems? The discussion would take too long and it would probably be difficult to arrive at a shared answer; it is perhaps better to focus on the utilitarian aspect by emphasizing the fundamental principle of calculus, which is based on the principle of multiplication.

When a set A is finished it is possible to associate to each element of the set a natural number distinct and successive between 1 and n. This number n expresses the numerosity of A and is called the cardinality of the set. From the combinatorial point of view, there are several cases that can be classified in two main categories: permutations and combinations. These in turn can be simple or with repetition. The symbology to represent these cases is quite variable and not uniform and therefore the one adopted may differ from that present in other texts.

Starting from two sets A and B, if you know the cardinality of each set you can determine the card A×B that is their Cartesian product, simply by making the product of the cardinalities of the two sets. A typical example is represented by the "wardrobe" problem: having three trousers and two jackets at your disposal, how many ways can you dress? To each trousers you can associate two jackets, but there are three pants so you can dress in $3 \bullet 2 = 6$ possible ways.

If you want to list the possible results you can use, at least for the simplest problems, two models. The first one is called *tree model* and is nothing else than the well-known diagram, constituted by an initial node from which successive branches depart until we reach the terminal leaves. For example, the problem of the "wardrobe" can be seen as an experiment in two phases leading to the six modes of dressing.

The second model is that of boxes, also called the *pigeonhole principle*, it states that if *m* objects are put in *n* drawers, with m > n then at least one drawer must contain more than one object. The principle was first made explicit by Johann Peter Gustav Lejeune Dirichlet (1805-1859) in 1834. Apparently, this is an obvious statement, but let's think about how this problem could be solved: in Paris there are at least two people with the same number of hair?

Some considerations based on the scalp area and average hair density lead us to think that the maximum number of hair of an individual does not exceed 200000, since there are more than 2 million inhabitants in Paris it is obvious that there will be many Parisians, and in any case at least two, with the same number of hair.

A geometric example could be the following: demonstrate that if 10 points are selected at random in an equilateral triangle on the L side there are at least 2 points closer than L/3. In this case, just think that the triangle can be divided into nine equal equilateral triangles on the L/3 side, so at least one point must end together with another and the distance will obviously be less than that of the side of these small triangles.

That the fields of application of combinatorics are various can be understood by examining some examples, the second place by several authors, including Eastaway:

1. How much genetic information is carried by a four-letter nucleic acid code taken at three at a time?
2. What is the minimum number of socks you have to take out of the drawer containing two black socks, two blue and two green, without looking, to make sure you have a pair of the same color?
3. How many ways can you change $1 for 50 ¢, 25 ¢, 10 ¢, 5 ¢ ?
4. A number of people meet and each shake hands with the others. How many handshakes occur?

If we consider even the fourth question, we can see how a seemingly banal example opens the way to hugely different considerations and approaches.

The problem itself is not complicated. If we examine the simplest case consisting of two people it is evident that there will be only one handshake; if there are three people (say A, B and C) then A shakes hands with B and C and B shakes hands with C for a total of three handshakes.

Three possible interpretations / solutions of the problem can be identified:

1. Graphical representation (Figure 1): each individual is indicated with a dot while the handshake is represented by a segment. Mathematically it is a graph. Obviously between two subjects there will be only one handshake and therefore we can speak of an *undirected graph*.

Figure 1. Handshake graph

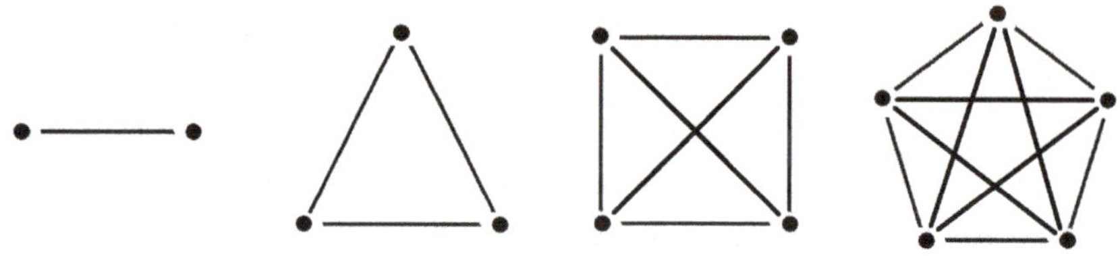

Handshakes will be: 1, 3, 6, 10, ...

This solution is of great help for understanding the phenomenon but, obviously, unmanageable if the number of people present grows.

2. Iterative solution: Iteration is the operation in which a process is repeated with the aim of getting closer to a desired result. In this case we start from the situation in which there are two people and there is only one handshake, at the next step n is added to the result, so 1 + 2 = 3, 3 + 3 = 6, 6+ 4 = 10, … Shortly:

$$\begin{cases} x_{n=2} = 1 \\ x_{n+1} = x_n + n \end{cases}$$

This solution lends itself to the use of IT tools, it is a simple cycle in which the second relationship is repeated n times.

3. Formula: to find a general solution, a logical reasoning must be made, for example if there are four individuals each shaking hands with the remaining 3; so we can think of 4 • 3 handshakes. But in this way we have counted each handshake 2 times and therefore we have to divide by two. Generally speaking if we denote by n the persons involved we will have:

$$x(n) = \frac{n(n-1)}{2}$$

Using the formula allows us to immediately answer the initial question without resorting to drawings or incremental procedures.

Combination experiments can also be found in special fields. A classic example is due to the great composer Wolfgang Amadeus Mozart (1756-1791). Mozart was responsible for the construction of the musical dice game. The rules to be observed are quite simple. A minuet, at the time, had an almost standardized form, in 3/4, included a first part (the minuet), followed by a central part, more melodic and relaxed, called a "trio" because originally it was entrusted to three instruments and the conclusion that resumed the starting minuet.

Both the minuet and the trio were 16 beats and in creating their internal structure it was possible to use a precise scheme. Kleine Nachtmusik's minuet, for example, is composed of 4 musical beats (A) followed by a variation of equal length (A'); we then have a second sentence (B), also of 4 beats, which ends with a reprise of A'. In turn, the trio opens with a melody (C) of 8 beats followed by a different motif (D), 4 beats long, which ends with an abbreviated reprise of C. It was structures of this kind that allowed the compositional automation realized by Mozart. The game of musical dice is based on two tables, one for the minuet, for which 2 dice are used, and the other for the trio, for which a single dice is sufficient, then there is a libretto with the list of 176 bars composed by Mozart. Each table has in abscissa the number of bars and in ordinate the result of the roll of the dice. At the intersection of abscissa and ordinate there is the number of beats to be used.

The player simply rolls the dice once for each beat and transcribes the corresponding beats by copying it from the list and placing it after the previous one.

Throwing the dice 16 times gives you the 16 beats that make up the minuet. The table for the minuet, in fact, has 11 rows (the dice go from 2 to 12) for 16 columns and in this case, Mozart has written 11 possible beats for each of the 16 beats of the minuet (11 • 16 = 176).

To "compose" the trio, on the other hand, only one dice is used and the table has only 6 rows. For each beat, therefore, Mozart has written 6 possible beats.

Mozart's skill, in this case, was to write some lines that could in any case logically concatenate without creating great imbalances with each of the previous ones. Being tonal music, the thing is less complex than it may seem at first sight (the tonality is always the same, D for the minuet, G for the trio), but it is still quite difficult if you also consider the fact that a composition of this type does not admit great sequential irregularities.

Finally, it would be interesting to know whether Mozart, when arranging the beats in the tables, took into account the different exit probabilities determined by the roll of the dice. The problem does not arise in the case of the trio where only one dice is used and each number comes out with a probability of 1/6, but for the minuet where 2 dice are used with different probabilities for the 11 possible results. With 2 dice, in fact, the 7 has the greater probabilities because it is obtained with 6 different combinations. Followed by 6 and 8 with 5, 5 and 9 with 4, 4 and 10 with 3, 3 and 11 with 2, while 2 and 12 are obtained with only one.

Ultimately, however, 11^{16} minuets and 6^{16} trios can be generated with this system. Since the minuet as a musical form is composed of a minuet and a trio, we have a huge total of several possible compositions. Obviously, many of them will be almost equal as the difference will be only one or two bars; nevertheless, on such a large total, a large number of interesting combinations remain to be listened to.

The problem is that, even if you leave out the final take of the minuet, each track lasts about a minute, so it would take a disproportionate number of years to listen to them all.

TIC-TAC-TOE

Tic-tac-toe is an extremely popular and quite simple game normally played with paper and pencil. It is played on a square grid of 3×3 squares. In turn, players choose an empty cell and draw their symbol on it (usually one player has an "X" and the other "O" as symbol). The winner is the player who can place three of his symbols in a straight horizontal, vertical, or diagonal line. It can of course be played on a chessboard using different objects such as coins or ... dogs (Figure 2).

An example of a game can be the one shown in Figure 3:

Considering only the state of the board, taking into account symmetries (i.e. rotations and reflections), there are 138 positions. A combinatorial study of the game (Bolon, 2013) shows that when "X" makes the first move, three cases can occur:

a. 91 separate positions are won by (X);
b. 44 separate positions are won by (O);
c. 3 separate positions lead to a tie.

Figure 2. Possible Tic-tac-toe position

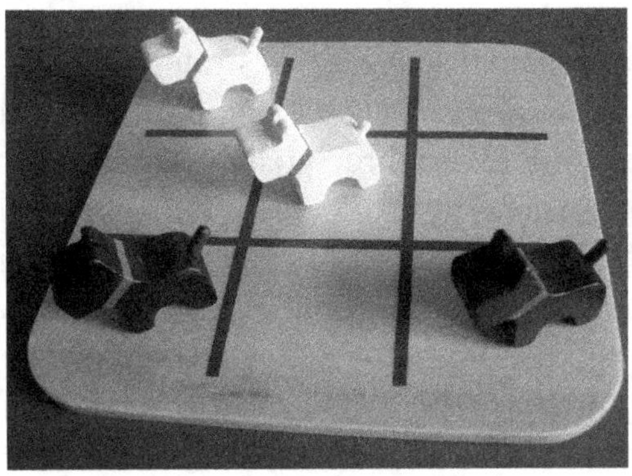

Figure 3. Possible Tic-tac-toe game

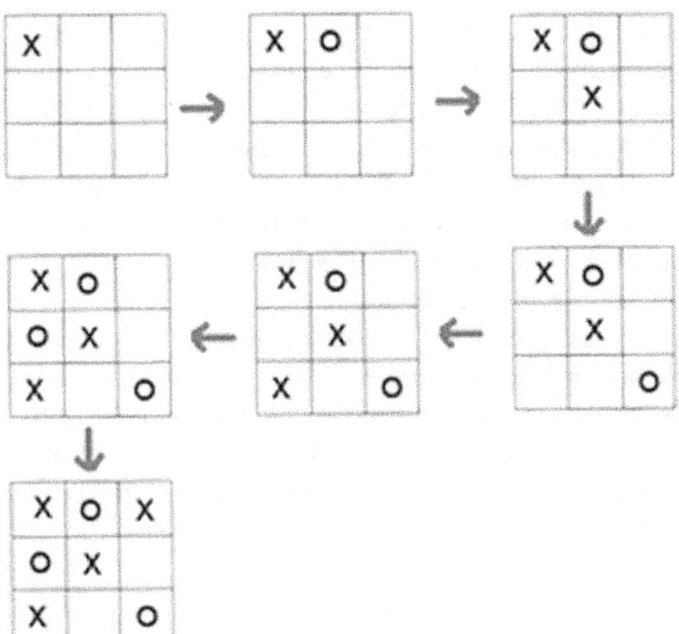

The chances of winning, calculated on the number of possible games, are as follows:

a. p(X) = 0.51;
b. p(O) = 0.31;
c. p(draw) = 0.18.

All this applies in the absence of a strategy. In practice, after a few games, players should find that the best game of both sides leads to a draw. For strategy purposes, there are three possible first occupations for the first player (suppose X): corner, edge, or center. This player can win or force the draw from any of these first occupations; according to some players playing the corner gives the opponent the smallest choice of squares that must be played to avoid losing, according to others the opening move in the middle is the best for X.

SMARTPHONE

A tool that can be used to think about combinatorial possibilities is perhaps the most widely used object in the world: the mobile or, in its most recent version, the smartphone (Figure 4). The most recent models have been enriched with the functionality of multimedia devices able to play music, take photos and videos, connect to the Internet. With the technological development, another innovation brought by smartphones is the possibility to make selfie and then share the image on social networks. For a brief analysis of the relationship between selfie and social see Stefanone (2019).

An amusing example linked to this habit was proposed by an Italian comedian and singer who also graduated in mathematics, Lorenzo Baglioni (2018). Let's follow his reasoning. Shared selfie are characterized by three elements: the photo, a quote, a hashtag. Let's suppose that each element has a finite number of possibilities. The author identifies, for the selfie of a girl on holiday by the sea:

1. five typical poses of a selfie made during the holidays, for example from photos with bent legs and sea in the background to photos of feet with red enamel in transparent water;
2. fifty quotations of phrases and aphorisms, from Jim Morrison to Mother Teresa of Calcutta;

3. ten hashtags, among the most used in this context, such as #beach, #fun, #relax, #sun, etc..

Figure 4. Smartphone

The number of selfies that can be posted is, according to the well-known multiplicative principle, 5 • 50 • 10 = 2500. At this point, Baglioni introduces the temporal aspect and supposes that the girl is at the beach from May 20th to September 10th and that she takes one photo a day (personally I think that these data are not credible, overestimated the first and underestimated the second). The number of summers (S) with different selfie becomes:

$$S = \frac{2500}{113} \approx 22.12$$

After about 22 years the selfie repertoire will have run out.

A second example concerns a question of probability. Suppose I want to phone my friend who lives in London but has lost his address book and can't remember the number: all I know is that the country code is 0044 followed by eight digits, the first of which is 7 and the second is definitely not 0. If I try randomly what is the probability of doing his number?

Since I can choose a number between 1 and 9 for the second digit and all subsequent numbers from 0 to 9, the possible numbers are 9 • 10^6 = 900000000 and therefore the probability is a disarming 1/9000000, which means an even lower chance of winning the second prize at the USA Power Ball lottery (about one in five million)!

The problem of lotteries and other probabilistic problems are deepened in Rosenthal (2006).

One of the relevant issues for smartphone users is protection against malicious intent, so Android users are asked to choose a PIN, a password or to select and remember a graphical template on a grid 3 • 3. The template must consist of at least four points.

Since Android devices block access if you make 20 wrong attempts an attacker must guess the unlock code within 20 attempts at most. A study by Cho and al (2017). has shown which unlocking patterns are most used by users. The theoretical space of the patterns calculated by the researchers is 389112 but the actual space of the patterns chosen by the user is much smaller. The interesting aspect is that the starting points were strongly oriented towards the highest point on the left which has a probability of use of 63.5% and that the most used segment was 2-3 while some diagonal lines were never used. In Figure 5 there are the 20 most used codes according to the study:

Figure 5. Most Used Patterns

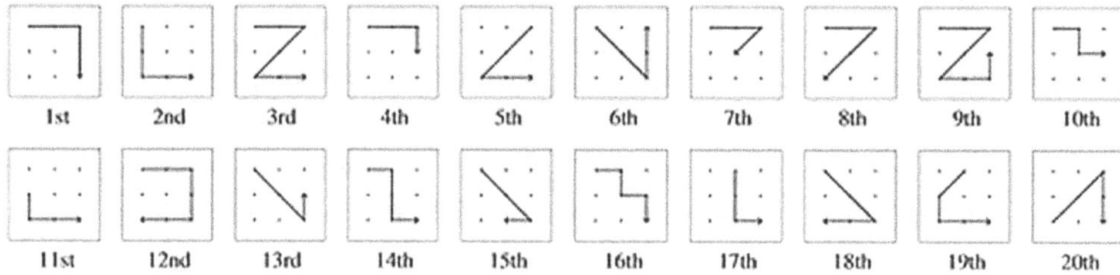

What is interesting is to examine if the points are touched with different probabilities, Table 1 shows with grey scales the probabilities of use for each of the 9 points:

Table 1. Probability of using the 9 points

0.117	0.114	0.113
0.104	0.120	0.105
0.109	0.108	0.110

SNOW WHITE

Snow White (Figure 6) is the title of a European folk tale, also made famous by an animated film from 1937 produced by Walt Disney (1901-1966).

Figure 6. Snow White

In one of the most famous depictions, you can see Snow White standing next to a rectangular table with the dwarves sitting around her. Bashful sits on her right in particular.

Suppose that this event occurred three times in a row, can we think that it was a coincidence or that there was a choice?

Since it is possible to assume that Snow White always sits at the head of the table, the first dwarf has 7 free chairs, the second 6, etc., while the last can only occupy the last remaining chair. In practice the positions are 7 • 6 •...• 1. This is the factorial (already seen previously). The seven diners can then sit around the table in 7! = 5040 different ways.

Let's now examine Bashful's position: he can sit on Snow White's left or right; the other dwarfs can occupy the 6 positions left free. This leads to 2 • 6! = 1440 different ways for the phenomenon to occur.

The probability of the event (X = Bashful sits randomly next to Snow White) is given by the number of cases favorable to proximity over the total ones, so it can be said that p(X) = 1440 / 5040 = 0.286. This is a non-negligible probability, but the probability that the phenomenon repeats itself three times is given by P(X) • P(X) = 0.286^3 = 0.023. The probability that it is a case is therefore just over 2%, which is low enough to think that it was just a chance case. So we can be led to believe that the situation is not reliable and, paraphrasing Shakespeare's Hamlet, we could say "Something is rotten in the state of Denmark".

Does it make any difference if the table is round? If the elements are arranged circularly, there is no point in talking about first and last place, so once one person is seated, the others can be arranged by exchanging the positions of the remaining seats. A drawing (Figure 7) can clarify ideas:

Figure 7. Arrangement of three people on a round table

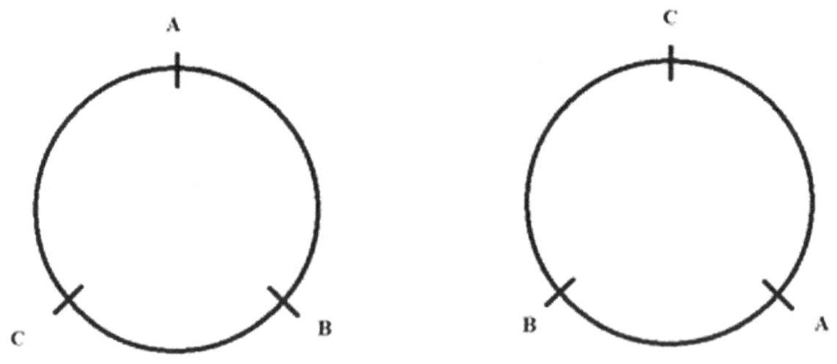

In this case we speak of cyclic or circular permutations; the number of cyclic permutations of *n* must take into account that each diner can take one of the n places, the permutations must be divided by this number and then you will have:

$$\frac{n!}{n} = (n-1)!$$

LANDSCAPES

We have 24 cards of 3 • 6 cm representing an 18th-century landscape (Figure 8). The manufacturer states that it is possible to create 1686553615927922354187720 different landscapes simply by lining up the cards in different ways. An enormity!

Figure 8. Continuous Landscapes

To get closer to understanding the phenomenon, let's try just three tiles. The first one can be chosen between 3, therefore only 2 tiles remain and finally 1. For the principle of multiplication, we will have $3 \cdot 2 \cdot 1 = 6$ possibilities. These possibilities are called *Simple permutations* of a finite set of distinct elements.

A similar example concerns the possible permutations of the four-letter series “ABCD”:

ABCD BACD CABD DABC
ABDC BADC CADB DACB
ACBD BCAD CBAD DBAC
ACDB BCDA CBDA DBCA
ADBC BDAC CDAB DCAB
ADCB BDCA CDBA DCBA

The 24 permutations are calculated using the generalized formula, where n indicates the elements:

$$P_n = n! = n \bullet (n-1) \bullet (n-2) \bullet \ldots \bullet 1$$

The symbol *Pn* is often indicated with *n!*, called factorial, which is the product of the natural numbers *n* to 1. It should be noted that 0!, by definition, is worth 1.

Let’s go back to the problem we started from: if you line up the tiles, all different from each other, you can calculate how many different landscapes you can build. Since there are 24 tiles, we discover that $P_{24} = 24! = 24 \bullet 23 \bullet 22 \bullet \ldots \bullet 1 » 6 \bullet 10^{23}$. Even if all the inhabitants of the earth could form a new grouping every second, it would take almost three million years to build all possible landscapes.

The speech changes if we are content to take only *k* tiles among the possible *n* tiles. For example, choosing each time 3 tiles you get 12144 landscapes, the reasoning is quite simple: initially I can choose one of the 24 tiles, next to this I can put one of the 23 tiles, below I can add one of the 22 remaining tiles; the result is $24 \bullet 23 \bullet 22 = 12144$.

A different meaning of the term permutation refers to those ordered arrangements where no element occurs more than once, but not all elements of a given set are used. These objects are also known as *Partial permutations* or *k-permutations of n* while in Europe the term disposition is preferred.

$$P_{n,k} = n \cdot (n-1) \cdot (n-2) \cdot \ldots \cdot (n-k+1) = \frac{n!}{(n-k)!}$$

Among the permutations many have in common at least one object in the same position; others do not: in this case we will talk about *Derangement*. For example: with the four letters seen before we have $4! = 24$ permutations, but of these only 9 with the desired characteristic.

BADC BCDA BDAC
CADB CDAB CDBA
DABC DCAB DCBA

In the case of n objects, a theorem of Euler comes to our aid. The number of dismutations is:

$$D_n = \sum_{i=0}^{m} (-1)^i \frac{n!}{i!} \approx \frac{n!}{e}$$

Being n! the number of permutations you have that the ratio between permutations and dismutations is:

$$\frac{n!}{D_n}$$

How much is this relationship worth when n tends to infinity?

It is interesting to note that this ratio tends to the number of Euler $e =$ 2.718...

PALETTE

If the elements can be repeated, as in the case of word anagrams containing equal letters, then they are *Permutations with repetition* or *Permutations with Indistinguishable Objects*.

For example, the word ABCA always has four letters like ABCD, but there are only 12 anagrams:

AABC AACB ABAC ABCA ACBA
BAAC BACA BCAA CAAB CABA CBAA

For the calculation of permutations with repetition, where a, b, c, ... represent the number of repetitions of each element, the formula is used:

$$P_n^{a,b,c\ldots} = \frac{n!}{a!b!c!\ldots}$$

In the word probability, composed of 11 letters, there are two *b* and two *i* and therefore the number of distinct arrangements is 9979200.

Let us now analyze the problem with an object, or rather a series of objects. There are 4 palettes used to give votes to a contest, two show the 7, one the 6 and one the 8 as in Figure 9. In how many possible ways can they be arranged?

$$P_n^{2,1,1} = \frac{4!}{2!1!1!\ldots} = \frac{4\cdot3\cdot2\cdot1}{2\cdot1\cdot1\cdot1} = 12$$

Figure 9. Palette

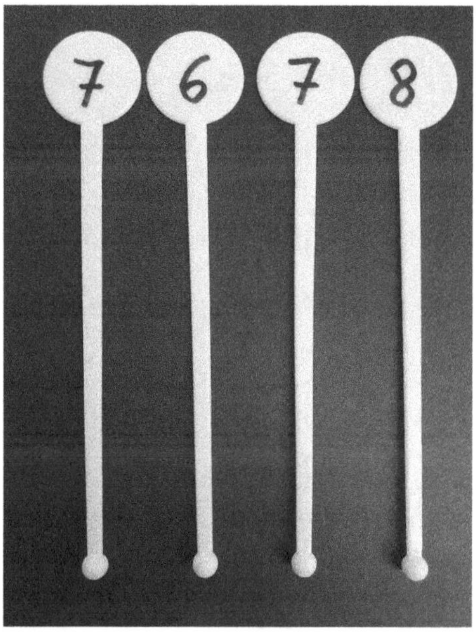

A small final parenthesis. Permutations can be found in areas not immediately imaginable: for example, in mathematics, the development of a binomial of the type $(a + b)^n$ is widely discussed. If we wanted to know what is the coefficient of a^5b^7 in the development of $(a + b)^{12}$ it would be enough to consider the anagrams of aaaaabbbbb. They are permutations with repetition and, applying the formula, we find that their total number is $\frac{12!}{5!\cdot 7!} = 792$.

A similar example is the following: suppose that the city plan is similar to that of Manhattan, i.e. with streets parallel and perpendicular to each other (Figure 10).

Figure 10. Manhattan Grid

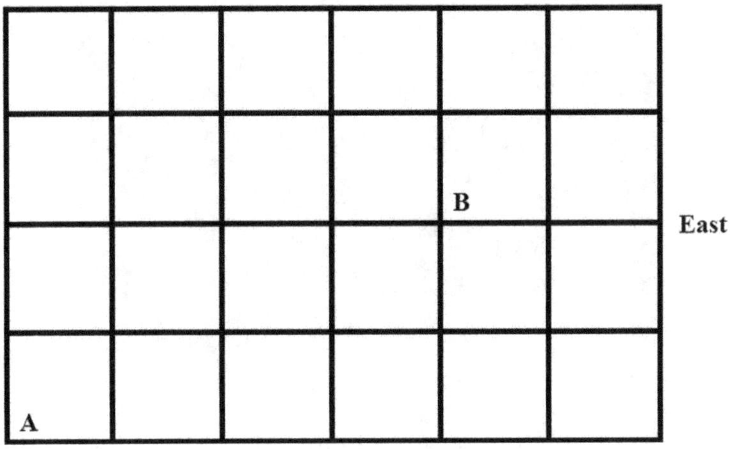

How many ways can you reach point *B* from point *A* by choosing to go up or to the right at each junction?

Evidently to reach *B* you need to do four blocks true east (E) and two blocks north (N), for a total of six. The number of possible paths is given by the permutations of EEEENN therefore $\frac{6!}{4!\cdot 2!} = 15$.

Let us now suppose that *A* points to the school and *B* points to a boy's house. He gets home from school as quickly as possible. If the mother is at the intersection to the left of *B*, what is the probability that Mario, having arrived home, met her on the way? Since the routes from school to his mother are $\frac{5!}{3!\cdot 2!} = 10$ the probability is 10/15 = 2/3.

BEAGLE BOYS MUG

The cup is a round mouth container with a handle. Often the cups are decorated as in the case shown in Figure 11, where a component of the Beagle Boys is reproduced. This artifact gives us the opportunity to reflect on a permutation problem.

Figure 11. Cup with a Beagle Boy

It is a gang of criminals, operating in the city of Duckburg, composed of brothers, whose main objective is to steal money from Uncle Scrooge and not only. We don't know their name, but only their prison freshman number imprinted on the yellow plaque that each of them wears on their red shirt.

In Don Rosa's comic book version, there are seven Beagle Boys. Their serial number is always of the type "176-***" where the three asterisks indicate a permutation of the previous three digits, such as 176-176, 176-671, etc. Is this an adequate way to identify all seven members of the Band? Actually, since the permutations of the three digits are only six ($3! = 3 \cdot 2 \cdot 1 = 6$), it is impossible to provide everyone with a different serial number. The author is aware of the problem and in fact, when all the Beagle Boys appear in a vignette, there is always one of the group whose nameplate is partly covered, so that the last digits are not readable. It is a suitable graphic solution to mask an insoluble mathematical problem.

15 PUZZLES

The 15 puzzle (Figure 12) is a classic game invented by Samuel Loyd (1841-1911) in 1878. The game consists of a square table, usually made of plastic, divided into four rows and four columns, on which are placed 15 square tiles, numbered progressively starting from 1. The tiles can slide horizontally or vertically, but their movement is obviously limited by the existence of a single empty space. The aim of the game is to rearrange the tiles, in an increasing sense, after having "mixed" them randomly.

Figure 12. Game of 15

Beyond the purely playful aspect, the game of 15 represents a mathematical problem that can be solved through permutations. The objective is, as said, starting from an initial (random) configuration of the game, to exchange its elements to place them in the natural order. Let's ask ourselves the question: is it always possible to do this? To answer this question, we need to analyze the game's scheme from a mathematical point of view. Each move, understood as a permutation of two tiles, modifies the parity (or disparity) of the system. If we look at the solved state of the puzzle, the number of inversions is zero, an even number. Any sequence of moves that returns the empty box to its original position in the lower right-hand corner will result in a reordering of the tiles that correspond to an even permutation. Any state of the puzzle with disparity characteristics cannot be solved. Note, incidentally, that even permutations of game tiles are exactly half of all possible permutations: 15! = 1307674368000.

Let us now consider the configuration in Figure 13, partially ordered:

Figure 13. Initial game diagram of the 15 puzzle

1	**2**	**3**	**4**
5	**6**	**7**	**8**
9	**12**	**14**	**15**
13	**10**	**11**	

It can be expressed in the form of a list according to the normal reading pattern from left to right and from top to bottom.

(1, 2, 3, 4, 5, 6, 7, 8, 9, 12, 14, 15, 13, 10, 11)

This arrangement of the cards is a partially ordered set with respect to the operator <, you can assess the degree of disorder in the scheme by checking how many cards are already partially ordered. We extract all pairs that do not meet this criterion:

(1210),(1211),(14,13),(1410),(1411),(1513),(1510),(1511),(1310),(1311)

You have 10 unordered pairs: since it is an even number the game is solvable.

Before starting the game, is it possible, given an initial configuration, to know immediately if it will succeed or not? Yes, for example using a spreadsheet. In this case we will only give a few simple indications of the procedure to follow, assuming that those who use this software have gained sufficient knowledge of this tool.

First of all you type the numbers from 1 to 15 in the first column of the sheet, in the second column these values are flanked by a random number, using the appropriate function =RAND(). After you have selected all the data you make a custom Sort by choosing the second column as sorting criteria. In this way we generate the sequence of numbers of the game whose outcome we want to know. The next step consists in counting how many numbers are lower than the first in the sequence, this implies the construction of a double entry table that allows the comparison between the numbers previously ordered, placed in abscissa and in ordinate. This is done by using the conditional test "if ... then", available as function =IF($column row<=column $row;0;1). The insertion of the $ symbol allows you to copy the formula throughout the table keeping the correct references, but then you have to delete the formulas placed under the diagonal (technically a triangular upper matrix is generated). For each number of the sequence placed in order, count the times when the condition is verified with =SUM (interval). If the sum of this count column gives an even number then the game is bound to succeed. This problem offers the opportunity to reflect on the potential of computer science. With the Excel or Calc macros, it is possible to repeat the simulation in automatic way and to apply it also in other contexts. This procedure can be used in other situations in which it is necessary to carry out modifications in the ordering of objects, for example when we must simulate the shuffling of a deck of cards.

PADLOCK

As is known, the padlock is a portable lock used to protect the contents of suitcases, bags, lockers from theft. The mechanism is controlled by a key or by means of numbered rollers (Figure 14).

An employee takes the subway to the office and the journey time is 35 minutes. His briefcase, in which he holds important documents, is closed with a padlock made up of three rollers with the ten digits. The briefcase opens only if the three digits of the "combination" are aligned. Due to a sudden loss of memory the employee does not remember what it is. Assuming that each attempt takes three seconds: A) Does he have time to try all the combinations

Figure 14. Padlock with rollers

before arriving at the final stop? B) can he still hope to open the suitcase before arriving at his destination?

Before we get to a formalization let's think about the concrete case: the first roll can take one of the ten values, this can be associated with one of the ten values of the second roll and the same goes for the other cases. To open a combination padlock you have to try all possible cases, so, with three rolls, we will have 10 • 10 • 10 = 1000 different positions. The time required is 3 • 1000 = 3000 seconds or 50 minutes. Actually, he can hope to guess the combination before all the possibilities are exhausted, let's say with probability 1/2, so the minutes are only 25.

In the text we talked about "combinations" because it is the term used in common language, actually it would be more correct to use the term *Permutations with repetition*. In this case the order counts (123 cannot open a padlock with combination 231) and the elements are repeatable.

For the calculation, examining the example, we deduce that, indicating with n the number of cases and with k the repetitions, we arrive at the formula:

$$P'_{n,k} = n^k$$

An interesting problem that has similarities with the previous one is due to Stephen Siklos: "In order to get money from a cash dispenser I have to punch in a Personal Identification Number. I have forgotten my PIN, but I do know that it is equally likely to be any one of the integers 1, 2, ..., *n*. I plan to punch in integers in ascending order until I get the right one. I can do this at the rate of *r* integers per minute. As soon as I punch in the first wrong number, the police will be alerted". In reality the problem presents itself in a much more complex way because the arrival of the police depends on a particular probability distribution (a topic discussed in Chapter 6) called exponential distribution that, however, is not dealt with in the text.

T-SHIRT

The T-shirt is the short-sleeved shirt, with a round neckline and a cut that gives it the classic "T" shape. Since the eighties of the last century the T-shirt, initially white, becomes a creative piece of clothing thanks to designs and/or writing.

On the T-shirt of Figure 15 is reproduced a matrix 4 • 4 which is on one of the facades of the Sagrada Familia, a bizarre work by the architect Antoni Gaudì (1852-1926).

The table, created by sculptor Joseph Subirach (1927-2014), constitutes an almost *magical square* with constant 33. A magic square is constructed by arranging all the integers from 1 to n^2 in a grid of n • n boxes so that the sum of the numbers of each row, each column, and each diagonal (called the magic constant) is always the same. As said, it is a fake magic square, in which all the natural numbers from 1 to 16 are not used: 12 and 16 are missing while 10 and 14 are repeated twice. This is due because the constant was supposed to represent the presumed age of Christ at the moment of the crucifixion.

In the past, however, magic squares have been studied in depth by many mathematicians and scholars. The first magic square 3 • 3, called *Lo Shu*, even dates back to Ancient China, in the year two thousand years ago. C.

The most famous example is taken from the work Melancholia I by the painter Albrecht Dürer (1471-1528) exhibited at the Staatliche Kunsthalle in Karlsruhe, Germany (Figure 16).

The work, incredibly famous, is full of esoteric references, but we are interested in the magic square depicted at the top right, as shown by the enlargement of Figure 17.

Figure 15. T-shirt

Figure 16. Melancholia I. (Public domain)

Figure 17. Magic square

This square, which has a constant of 34, is complex. Also, the sum of the numbers of the four main square sectors in which it is divisible and also the four numbers in the center, added together give 34, as well as the four numbers at the corners. Furthermore, if you take the central numbers of the last line, you will find the number 1514, the year in which the work was created. Bernard Frénicle de Bessy (1605-1665), French mathematician, in 1663 calculated the number of perfect magic squares of the fourth-order: 880, with constant sum 34, on rows, columns and diagonals. The use of computers widened the horizon to magic squares of a higher order and thus discovered, in 1973, that the magic squares of order 5 are 275305224. Determining how many are the magic squares of order 6 is still an open question today, however, it is estimated that the number of these magic squares is about 10^{19}. But let's go back to probabilistic problems. What is the probability of guessing one of the 880 perfect order 4 squares at random?

The number of possible permutations is 16!, so the probability is given by the ratio $880/16! = 4 \bullet 10^{-11}$.

POKER

Poker is one of the most famous card games. There are many variations: Draw poker, where each player receives five cards and can decide to change from one to four (Figure 18); Stud poker. where the cards are given one at a time, some face down and some face down; Texas hold'em, where the player receives two face down cards and can use a number of the five community cards face down.

Figure 18. Example of distribution of the five cards

The aim of the game is to get the best combination or get the other players to quit. It is not a simple game of chance because, although there is a probabilistic basis, the skill of the player also counts. They have several options available to them:

1. bet: initially put a certain amount in the pot;
2. check: do not place any bets;
3. call: bet the same amount as the previous player.
4. raise: bet a larger sum;
5. fold: quit the game.

We can examine the main aspects of American-style poker, i.e. played with 52 cards.

The face value of the singular cards is as follows, from the strongest to the weakest: Ace, King, Queen, Jack, Ten, Nine, Eight, Seven, Six, Five, Four, Three, Two. For the seeds the sequence, from the strongest to the weakest, is: Hearts, Diamonds, Clubs, Spades.

The winning combinations in descending order are:

1. Royal flush: a sequence of the 5 highest cards belonging to the same suit
2. Straight Flush: five cards in sequence of the same suit
3. Four of the kind: four cards of the same value
4. Full house: three equal cards and an equal pair.
5. Flush: five cards of the same suit, but not in sequence
6. Straight: five cards in sequence but not of the same suit
7. Three of a kind: three equal cards and two different cards
8. Two pairs: two pairs of cards of the same value, plus a different card
9. One pair: a pair of cards of the same value and the other different ones
10. Higher card: a hand that does not contain any of the above combinations

In order to define the probabilities of each hand, it is necessary to use the study of *Combinations,* i.e. that part of the combinatory calculation where the order of the elements does not matter.

In particular, we will deal with the simple combinations that are indicated with $C_{n,k}$ or $\binom{n}{k}$, where n indicates the total number of elements and k the number of objects that each grouping must contain.

The number of combinations, i.e. the number of subsets of k elements that can be formed from a set containing n elements, often referred to as a binomial coefficient, is given by:

$$\binom{n}{k} = \frac{P_{n.k}}{k!} = \frac{n!}{k! \cdot (n-k)!}$$

Using all 52 cards you can calculate all possible combinations of five cards:

$$C_{52,5} = \frac{52!}{5!\bullet(47)!} = 2598960$$

Once the possible cases have been calculated, the favorable ones remain to be defined. In order not to be protracted, only two cases can be considered:

1. Three of a kind: choosing one card among the 13 possible multiplies by the ways in which you can choose 3 cards in a set of 4 equal cards ($_4C_3$), then you multiply the ways in which you can choose 2 cards in a set of 12 ($_{12}C_2$), finally having for each type four suits, among the two chosen you must multiply 4 for itself twice (4^2): P = 13 • $_4C_3$ • $_{12}C_2$ • 4^2 = 54912;
2. Pair: choosing one card among the 13 possible is multiplied by the ways in which you can choose 2 cards in a set of 4 equal cards ($C_{4,2}$), then you multiply the ways in which you can choose 3 cards in a set of 12 ($_{12}C_3$), finally having for each type four suits, among the three chosen you must multiply 4 for itself three times (4^3): P = 13 • $_4C_2$ • $_{12}C_3$ • 4^3 = 1089240.

To obtain the probabilities of each "point" just divide the favorable combinations by the total ones. Just out of curiosity, we report in Table 2 the combinations and the probabilities of obtaining a certain result at first hand:

Table 2. Combinations and odds in Draw poker

	Combinations	**Probability**
Royal flush	4	0.000002
Straight flush	36	0.000014
Four of the kind	624	0.000240
Full house	3744	0.001441
Flush	5108	0.001965
Straight	10200	0.003925
Three of a kind	54912	0.021128
Two pair	123552	0.047539
One pair	1089240	0.419106
Higher card	1311540	0.504640

So far we have considered traditional poker, but currently Texas Hold'em with all its variations is much more in vogue. In poker we know that mathematics is not the only factor to be taken into account, but being able to calculate the percentage of realization of a project is certainly a good assumption to make correct decisions. In Texas Hold'em to be able to excel there is the need to be able to calculate the probability of achieving a project, i.e. to get a certain point. The first step in this direction is to understand the concept of *outs*. Outs are the cards the player needs to complete a project. Example: my cards are A♥ and 7♥. The first three cards turned on the table (Flop) show: 5♥ K♥ 3♠. At this point if a heart card came out in the following rounds I could flush, a combination that could allow me to win. Outside of mine there are 9 useful hearts cards left, i.e. there are 9 outs available, the higher this number and the higher the chances of completing the game using the two extra common cards available (which will be dealt in the following passes, called Turn and River). The remaining 39 cards do not interest us to close a flush, at most they can allow us to make One pair, Two pair or Three of a kind. The ratio of useless cards to useful cards is 4.33, i.e. about one card out of four is favorable to us. In order for it to be convenient to chase the flush, the value of the pot we play for must be at least 4 times the bet. This is a concept that will be taken up in the next chapter. If you want to know the odds of making a flush at the end of the game, you can consult special tables; in this case the odds in my favor would represent a not despicable 35%.

Another game where you have to take the combinations into account is Powerball, a US lottery that can be played in 44 states. Drawings are held twice a week. You have to guess 5 numbers out of 69 (drawn between as many white balls) plus one number out of 26 (drawn between as many red balls, called Powerballs); hence the probability of winning the maximum prize is 1 out of 292201338. The figure is obtained by taking into account how many possible combinations there are:

$$\binom{69}{5}\cdot\binom{26}{1}=\frac{69!}{64!5!}\cdot\frac{26!}{25!1!}=292201338$$

In Italy there is a version (Lotto) that has a much older history, dating back to the mid-1500s. In this case the game consists of drawing five numbers between 1 and 90. Whoever plays can win by guessing even just one of the five numbers, or by doing an "ambo" (two numbers) or a "terno" (three numbers) and so on. The example is useful to examine a classic misconception widely studied by Tversky & Kahneman (1974): the availability heuristic. It is a cognitive strategy that consists in the search for a satisfactory estimate with a minimum expenditure of mental resources. For example, if one asks to estimate the number of distinct "ambo" that can be formed from five extracted and to compare it with the number of "terno", individuals are led to think that the first number is greater because they believe that there are more possible combinations and do not resort to a calculation using concepts that they may possess. In reality, from the five numbers we can form both 10 couples and 10 distinct triads: ${}_5C_2 = {}_5C_3$.

STAMPS

Before concluding the chapter, we can examine an enumerative problem, in which the calculation of possible combinations depends on geometric reasoning. The stamp is a rectangle of notched paper, printed on one side and gummed on the back, which is used for postage on mail services. The stamps are divided into two main categories: ordinary and commemorative like the one shown in Figure 19 and related to the captain of the legendary Torino football team of the immediate post-war period.

Figure 19. Full sheet of stamps

Suppose we have bought a whole sheet of 12 stamps and want to remove four of them so that they remain united on at least one side.

How many ways can this division be achieved?

According to the English mathematician Henry Dudeney (1917) author of many logical games, the possible cases are the five in Figure 20:

Figure 20. Five separation modes

The enumerative problem, reported by Moscovich (2015), requires knowing in how many ways the figures shown can be realized.

Assuming that the twelve stamps are numbered progressively, the first shape, square, can be achieved in six ways (Table 3):

Table 3. The six square stamps

1	2
5	6

2	3
6	7

3	4
7	8

5	6
9	10

6	7
10	11

7	8
11	12

The other figures can be realized, respectively, in 14, 28, 14 and 3 ways.

It is interesting to note that the pieces are those of Tetris (Figure 21), a video game of logic, skill and reasoning invented in 1985 by Russian researcher Alexey Pazhitnov. The chosen polygons, in fact, have a long history, prior to the invention of the game. The work of the famous English author and mathematician Henry Ernest Dudeney (1857-1930), who specialized in logical puzzles and mathematical games, appeared at the beginning of the century. Later they were used by Solomon Golomb (1932-2016), then a Harvard student, to propose a series of geometric problems. The term Tetris comes from tetramino, a figure composed of four blocks. The pieces fall one at a time and, by means of translations and rotations, an attempt is made to compose a full line. When the line is created, the "bricks" disappear.

Figure 21. Five pieces of Tetris

Actually there are seven pieces, but two more can be obtained by symmetry from the second and the third shown in Figure 20. The seven possible combinations of Tetris are called as the corresponding letters of the alphabet that come closest to the shape of the piece: O, Z (S), L (J), T and I. Tetris shapes are the result of all the possible combinations that can be obtained by arranging 4 squares, each of which has at least one side in common with at least one of the other three squares.

Over the years, many mathematicians have dedicated themselves to the analysis of the game. The best-known document of this kind is due to Demaine (2003) who have shown that all Tetris games certainly end with a defeat of the player, as an alternating sequence of S and Z pieces prevents the formation of a complete line.

KEYCHAIN

A common object is the keyring, a rigid metal ring to which an ornamental pendant is often added (Figure 22).

Figure 22. Keychain

Taking inspiration from this simple object you can propose a combinatorial problem proposed in a math competition in 2016.

Four friends got tired of their key chains and decided to redistribute them, so that each of them had a different one from the one they had before. How many different ways can they exchange keychains?

The solution of the problem is much simplified if you assign each keychain a number. Then it is necessary to exchange the numbers 1, 2, 3, 4 so that the i number does not appear at position i. Being a Derangement we can easily calculate the nine possibilities to do so. In this case the verification is simple, the cases are: 2143, 2341, 2413, 3142, 3421, 3412, 4123, 4312, 4321.

CONCLUSION

One of the leading experts in the field, Gian Carlo Rota (1973) wrote: "Almost all classical mathematics, from elementary algebra to the theory of differential equations, is applicable to the real world only if it consists of objects and events of a continuous character. However, in many common situations in physics and chemistry and other sciences, one can realistically speak only of a collection of discrete objects, which act in combination, one step at a time; mathematics applied to such situations is called combinatory analysis. Many of the most interesting problems of combinatory analysis have been presented in the form of ingenious riddles, challenging mathematicians and non-mathematicians together: at first sight, some of them may even seem frivolous, yet almost all of them have immediate and important applications to concrete scientific problems".

I think there is little to add to this consideration except that the presence of concrete objects on which to reflect can be a good starting point. Formulas are fundamental to allow what has been defined as "the art of counting ... without counting", but they must be acquired logically and not instrumentally.

FUTURE RESEARCH DIRECTIONS

One of the questions we could ask is whether there are models that can define how combinatorial thinking is characterized in students. When students are faced with a counting problem, which mechanisms are activated? Do we see the use of patterns that we could define as algorithms, in the sense that there is a sequence of questions about the characteristics of the problem that come to the use of a known relationship? In the absence of standard notions already acquired, what is the approach? Are there recurrent errors?

These (and other) questions should be the basis of the reflection of those involved in training and researchers. Some studies already exist, but it would be useful if they were expanded.

REFERENCES

Baglioni, L. (2018). È tutto calcolato. *Mondadori*, 71-81.

Bolon, T. (2013). *How to never lose at Tic-Tac-Toe*. BookCountry.

Cho, G., Huh, J. H., Cho, J., Oh, S., Song, Y., & Kim, H. (2017, May). Syspal: System-guided pattern locks for android. In *2017 IEEE Symposium on Security and Privacy (SP)* (pp. 338-356). IEEE. 10.1109/SP.2017.61

Demaine, E. D., Hohenberger, S., & Liben-Nowell, D. (2003, July). Tetris is hard, even to approximate. In *International Computing and Combinatorics Conference* (pp. 351-363). Springer.

Dudeney, H. (1917). *Amusements in Mathematics*. Nelson & Sons.

Halbeisen, L. (2011). Combinatorial Set Theory: With a Gentle Introduction to Forcing. Springer.

Loehr, N. (2017). *Combinatorics*. Chapman and Hall/CRC.

Moscovich, I. (2015). *The puzzle universe: A history of mathematics in 315 puzzles*. Firefly Books.

Rosenthal, J. S. (2006). *Struck by Lightning: The Curious World of Probabilities*. Joseph Henry Press.

Rota, G. C. (1973). *Analisi combinatoria*. Le Scienze Matematiche -UMI-Zanichelli.

Siklos, S. (2016). PIN guessing. In *Advanced Problems in Mathematics: Preparing for University* (pp. 152–153). Open Book. doi:10.11647/OBP.0075

Stefanone, M. A., Yue, Z., & Toh, Z. (2019). A social cognitive approach to traditional media content and social media use: Selfie-related behavior as competitive strategy. *New Media & Society, 21*(2), 317-335.

Tversky, A., & Kahneman, D. (1974). Judgment under uncertainty: Heuristics and biases. *Science, 185*(4157), 1124-1131.

ADDITIONAL READING

Albini, G. (2018). Combinatorics, Probability and Choice in Music Composition: Towards an Aesthetics of Composing Systems for Non-Musicians. In Bridges 2018 Conference Proceedings (pp. 395-398). Tessellations Publishing.

Anzola, J., & Falk, M. (2017). Avances en la caracterización del pensamiento combinatorio.

Chen, C. M. (2020). *Introduction to Probability: Bison Hold'em Poker Game*. INFORMS Transactions on Education.

Mazur, D. R. (2020). *Combinatorics: a guided tour* (Vol. 55). American Mathematical Soc. doi:10.1090/text/055

Nugraha, G. D., Mulyana, E., & Suryadi, D. (2018, December). Lets reveal it: what are learning obstacles in combinatorial problems. In *International Conference on Mathematics and Science Education of Universitas Pendidikan Indonesia* (Vol. 3, pp. 639-644).

KEY TERMS AND DEFINITIONS

Combinations: Each subset of elements extracted from a set.

Derangement: A permutation whereby no element appears in its original position.

Partial Permutation: Ordered subset of elements such that subsets differ at least in one element.

Permutation With Repetition: Ordered arrangements of the elements where repetition is allowed.

Pigeonhole Principle: The principle states that if there are more pigeons than boxes, then at least two pigeons will enter some boxes.

Simple Permutations: A way of ordering distinct objects in succession.

Tree Model: Classification scheme with ramifications.

Chapter 5
Probability and Game

ABSTRACT

Probability is generally concerned with dealing with problems of a random or uncertain nature. The fact that it arises and develops from the analysis of gambling is something that cannot be overlooked. From the point of view of teaching, in addition to historical aspects, it is important to point out the importance of putting students in front of situations that, if not known, can lead to incorrect behavior and pathological attitudes. For this reason, the authors tried to emphasize not only the theoretical aspects, but above all the certainty that you always play "against the dealer" with an expected loss assessable for the various games.

INTRODUCTION

The probable is what usually happens. ~ Aristotle

Probability has historically been created precisely to study the game and in particular gambling, in general to deal with problems of a random or uncertain nature, the so-called aleatory problems. It is not surprising that the etymology of the word gambling and the word aleatory are similar: both derive from the word dice (*az-zahr* in Arabic and *alea* in Latin). Dice and betting were also well known in ancient Greece. An example is provided by mythology: playing dice, the gods divide the universe: Zeus gets the sky, Poseidon gets the sea, Hades loses and finds himself in the underworld.

DOI: 10.4018/978-1-7998-3871-5.ch005

Even the Romans can be regarded as remarkable gamblers. It is known that they bet almost everything, from gladiator fights to chariot races, from the launch of astragals to lottery tablets during the holidays dedicated to the god Saturn.

One of the most famous treatises on the game dates back to 1283. Alfonso X (2007), King of Castilla and Leon (Spain) wrote the *Libro de los Juegos*. In the text it is said that an Indian king had proposed to three wise men to reflect on what is worth more between luck and intelligence. One wise man takes the side of intelligence, another of luck, and the third argues the importance of both. The king asks to demonstrate these positions concretely, and the three wise men take chess, dice and the last one games based on a board (such as backgammon) as proof.

In the Middle Ages, gambling became compulsive in Europe. The chronicles of the time speak of numerous edicts aimed at limiting the game, which, in places as bad as taverns, but also special gambling dens had taken on an incredible dimension. The Italian term *baratteria* (barratry) indicated precisely who kept a gambling dealer; the term was born in the 13th-14th century and the chronicles tell us that many municipalities taxed the game. *Barattiere* was also the one who, having a public role, was bribed for money. The accusation of barratry could be profoundly serious and also used to strike political enemies. At the State Archives of Florence is preserved the book where all the court sentences issued in that period were recorded. On March 10, 1302 it reads: "Alighieri Dante is condemned for barratry, fraud, falsehood, malice, unfair extortion practices, illicit proceeds, pederasty, and is sentenced to 5000 florins of fine, perpetual disqualification from public office, perpetual exile, and if you take him, at the stake, so that he dies. In the fourteenth century, with the introduction of modern cards, an adaptation of the previous ones coming from the East, new forms of game appear.

As time goes by, the games find an accommodation in small pavilions, called *casine*, a term that then evolves into "casino". The first real modern casino was built in 1861 in the Principality of Monaco and only in 1931 did the state of Nevada legalize gambling, making Las Vegas the gaming city par excellence. Other states followed suit and today gambling is widespread almost everywhere.

These few considerations would be enough to highlight how rich the history of gaming is and the space that would deserve analysis in the school environment. As Siew Pei Hwa (2018) says: "Play-way method involves act of seeing, hearing, saying and doing, which makes it easier to remember". Some difficulties, however, arise because many teachers do not have a clear picture of the spread of the phenomenon while students are frequently the protagonists, especially since the possibilities of online gaming became more popular. The study of gambling must consider both mathematical and psychological aspects. The study of gambling must consider both mathematical and psychological aspects. It is now well established among scholars that the accuracy of probability judgments is poor, as is familiarity with numbers. Some scholars such as Cosmides & Tooby came to explain the fact in evolutionary terms: according to them evolution has not consolidated *ad hoc* processing tools. Another interesting aspect is that, according to Drazen Prelec (1988), the human mind tends to overestimate small probabilities and underestimate high ones, which determines that the point of coincidence between probability and its estimate is for a value that is not, as one might suppose, 0.5 but coincides with a mathematically more interesting value: p = 1/e. The author has constructed a non-linear transformation of probabilities into "decision weights" w(p). If we try to represent this characteristic we get the following graph (Figure 1):

Figure 1. Prelec diagram

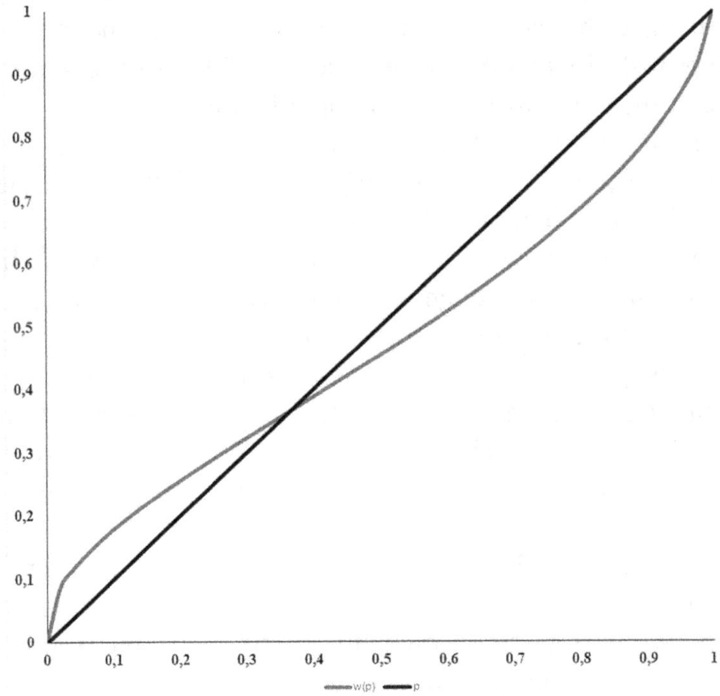

This mechanism makes the individual think he can win with greater probability than the real one, plus the *sunk cost bias* that consists of continuing in the game because you have already invested a lot of money, energy, time.

An interesting field of research concerns a behavior that involves our cognitive system. Let's just think about the difference in attitude between finding a four-leaf clover in the meadow (unlikely event) and finding a poppy (much more likely event). In the first case the event, linked to luck, creates a short circuit between improbability and mental representation of desirability, probable events are instead easily structured in terms of rationality. A similar mechanism occurs in the game, if one wins, he thinks he is "lucky" and therefore can continue to play without taking into account that the probability of continuing to win may be extremely low. It is this behavior that contributes to what has been called *gambler's ruin*, widely described in literature. To give an idea we use a simple example.

Two players compete for heads and tails, if heads *A* wins \$1, if tails *B* wins \$1, the game ends when one of them has exhausted his capital. Apparently it is a fair game. But let's suppose that the available capital is not the same: *A* has \$10 and *B* \$50.

We can represent it all as the random journey of a placeholder on a flat route with at the extremities two bars, the first place 10 steps to the left of the initial position and the second to the right 50 steps from the initial position. What is the probability that the placeholder will fall into one of the two bars, i.e. that one of the two players will run out of capital?

When the game develops a hypothetical gain and loss line oscillates around the zero-gain line (being the game fair) but the oscillations can be more or less wide: if the line exceeds a player's availability, he has lost everything. Those who have less money available are more likely to lose first. In general, the probability of winning depends on the ratio between the initial capital of each player and the sum of the two players' capital. So the probability of winning for *A* is given by $p_a = 10/60$ and for *B* by $p_b = 50/60$.

The problem becomes even more dramatic if you participate in a game where the dealer has a better chance of winning. In this case it is interesting to compare the "loss coefficients" of some extremely popular games according to Canova and Rizzuto (2016) (Figure 2):

Underlying this is the difficulty for the mind to manage uncertainty. It is no coincidence that many, not finding a rational causal explanation for an event, resort to almost magical concepts. Everything could be enclosed in a question: can superstition influence events?

Figure 2. How much you can lose in various games

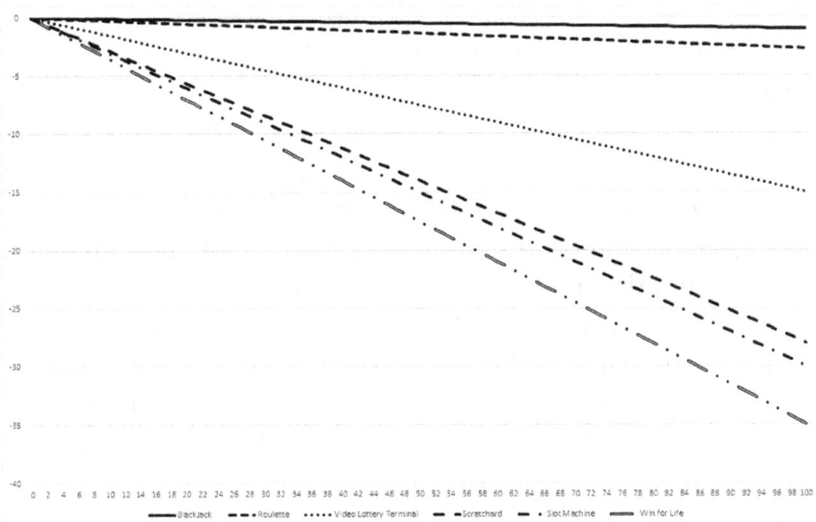

The term superstition refers to practices that are intended to reject supposed negative influences. Many actions are avoided by players: never knock before rolling the dice, never cross your fingers, never touch the cards before they are all served at the table, etc..

These brief considerations are just the tip of the iceberg of an issue that affects the daily life of the population, not just the players. For this reason it would be essential to provide all the tools for understanding random phenomena, starting from school.

Returning to didactics, according to Siller & Maaß (2012), teach probability through gambling: "is a very authentic one, so that the students can discuss a real-life problem which could be found in their near environment".

All educational experiences in this sense have shown that betting is an emotionally charged activity, involving the individual as a whole. Leveraging this can stimulate the student to reflect on these issues, also exploiting the mechanism of playful adhesion.

In the chapter will be presented, mixed, both board games and some particularly significant games of chance.

ORCHARD

The first game presented is certainly the simplest, but not trivial in terms of probabilistic reasoning.

It is a cooperative game aimed at pre-schoolers. The board has four trees loaded with fruit that must be picked by the players: apples, pears, cherries and plums. A raven wants to take advantage of the opportunity to seize the fruits. The game is regulated by throwing a dice that shows the colour of the fruits on four sides, on the fifth side a basket (if it comes out you can pick two fruits of your choice) and on the sixth side the image of the raven. Depending on the result of the throw the children collect one or two fruits, but if the raven comes out one of the nine pieces of the puzzle representing the animal is placed on the board to compose the drawing of the raven (Figure 3).

Figure 3. Orchard

The game ends when the players have collected all the fruits or the puzzle is completed.

The game is simple, but it is less simple to answer the question: what is the probability that the children are winners?

The modeling of the game itself is not difficult and a tree graph can help us, each launch corresponds to 6 different outcomes, but after n throws the possible results are 6n. Even eliminating the basket so as not to complicate the analysis too much (if its symbol comes out, the die is thrown again and therefore it is as if we had a five-sided die), there are too many cases. Let's try a simplified version of the game (Bibbona et al,2014) tree with two fruits and a puzzle consisting of three tiles and using a coin; if heads come out the children collect the fruit, if tails come out they put a crow tile on the board..

Now we can build a simplified tree diagram (Figure 4).

Figure 4. Simplified Orchard Graph

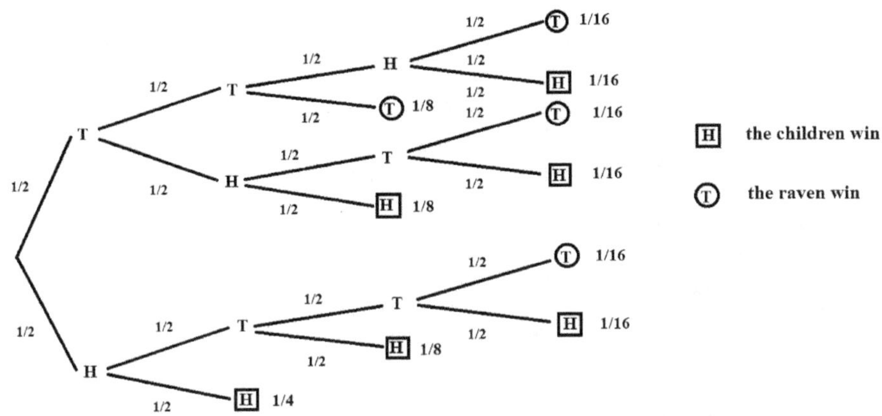

The maximum number of rolls is 4, i.e. $n_{fruits} + n_{raven} -1$. As you can see the probability of the children winning is 11/16 while the crow is left with a more miserable 5/16.

Even using an old programming language such as BASIC you can achieve the theoretically expected result. Here is a possible program:

```
FRUITS = 2
RAVEN = 0
RANDOMIZE
INPUT "# GAMES    ", n
FOR I = 1 TO n
    DO
        COIN = FIX(RND * 2) + 1
        IF COIN = 1 AND FRUITS > 0 THEN FRUITS = FRUITS - 1
ELSE FRUITS = 2
        IF COIN = 2 AND RAVEN < 3 THEN RAVEN = RAVEN + 1 ELSE
RAVEN = 0
    LOOP UNTIL FRUITS = 0 OR RAVEN = 3
        IF FRUITS = 0 THEN win = win + 1
        IF RAVEN = 3 THEN lose = lose + 1
NEXT I
```

```
PRINT "% WIN", win / n * 100, "% LOSE", lose / n * 100
The VBA version for Excel does not differ much:
Sub OrchardGame()
fruits = 2
raven = 0
n = InputBox("# games")
For i = 1 To n
      Do
            dado = Int(Rnd * 2) + 1
            If (fruits > 0 And dado = 1) Then fruits = fruits -
1 Else fruits = 2
            If (raven < 3 And dado = 2) Then raven = raven + 1
Else raven = 0
      Loop Until fruits = 0 Or raven = 3
            If fruits = 0 Then win = win + 1
            If raven = 3 Then lose = lose + 1
Next i
Range("A2") = "% win"
Range("A3") = "% lose"
Range("B2") = win / n * 100
Range("B3") = lose / n * 100
End Sub
```

Generalizing, an interesting case assuming an unbalanced coin where the probability of heads is p and the probability of tails $q = 1 - p$, you will have $P(T) = p^2 + 2\,p^2q + 3\,p^2q^2$.

The problem becomes more complicated when you switch to regular play. In this case the number of trees becomes 4 with 40 fruits, the pieces of the puzzle rise to 9 and in addition there is the variant of the basket. For the analysis of this type of problem with multiple objectives, see an article by Paul J. Campbell (2007). A simplified version that is easier to program could include the elimination of the basket and the presence of a single tree with 40 fruits.

RISK

It is a board strategy game, published in 1957 under the title *La conquête du monde* and later in the United States under the name *Risk*. In Italy it was published in the variant called *RisiKo!* (Figure 5).

The goal of the game is to achieve a predefined but secret and different goal for each player. It may consist in the conquest of a number of states, the conquest of two or more continents, the entire planisphere or the annihilation

Figure 5. Italian variant of Risk

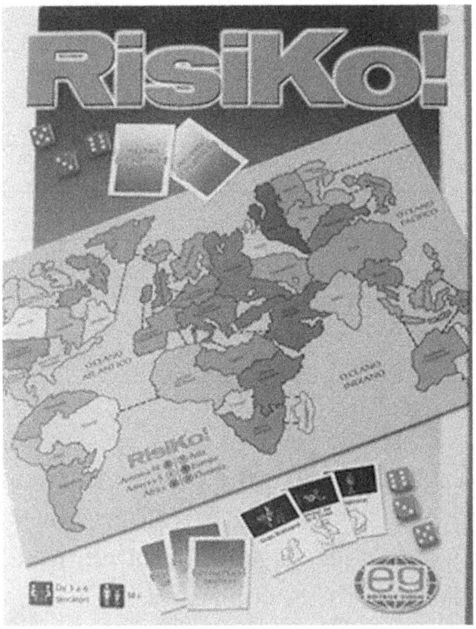

of an opponent. The scoreboard on which a game develops represents the world divided into 42 territories grouped in 6 continents. At each moment of the game each territory is occupied by one player's armies. Each participant commands a group of armies, identified by the color of the tanks, with which they occupy their board territories, attack the territories occupied by the other players and defend themselves against attacks. When a goal is achieved the game ends and the player who has achieved it has won. The rule foresees that the attacker (A) and the defender (D) roll from 1 to 3 dice, to be compared in pairs in order of value, in case of equal dice, the defender wins.

The simplest situation is when *A* and *D* compete with one die each: *D* wins if the number obtained by throwing his die is greater than or equal to that of *A*.

Built the Table 1 of the 36 possibilities, which represents the *sample space*, you can see that *A* wins with probability 15/36 = 5/12 = 41.7% while the chances are obviously greater for the defender.

Let's suppose instead that *A* attacks with two dice, while *D* defends with 1 dice: in this case *A* wins if at least one of his results is greater than D's. The calculation of A's probability of winning is now a bit more complicated as the possible cases are already $6^3 = 216$. What happens can be displayed with a graph in which the attacker collides head-to-head with the defender and, in case of defeat, uses the second die.

Table 1. Results in the toss of a dice

		Striker					
		1	2	3	4	5	6
Defender	1	D	A	A	A	A	A
	2	D	D	A	A	A	A
	3	D	D	D	A	A	A
	4	D	D	D	D	A	A
	5	D	D	D	D	D	A
	6	D	D	D	D	D	D

It should be noted that if *A* loses the first fight (with probability 21/36), this affects what happens afterwards; in other words, the events are not independent: it is therefore necessary to use conditional probability: that is, we must calculate p(A wins the second shot | A loses the first shot). Once the calculations have been made, we can see that the probability that *A* wins the second fight is only 10/36. At this point, the tree (Figure 6) becomes:

Figure 6. Risk Graph 2 vs 1

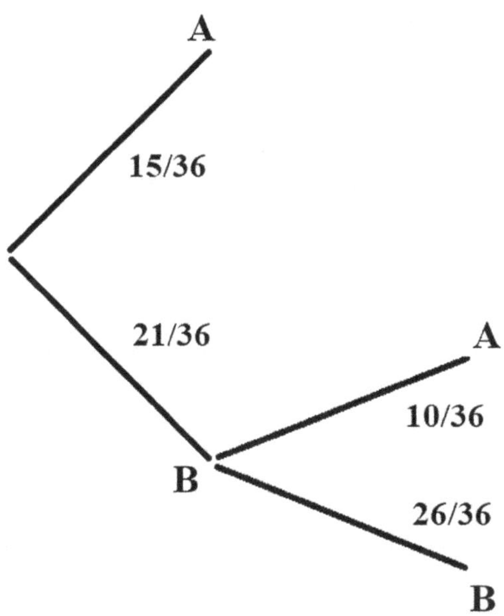

What happens in general if *A* and *D* play with a different number of dice or are repeated attacks to be considered?

From a strategic point of view it is always advantageous for *A* to roll the maximum number of dice. So when the attacker rolls three dice against two, three against one or two against one, he has a slight advantage, otherwise the advantage is all of the defender. The analysis is not trivial, it is a typical case where it is convenient to use a program to simulate a large number of lots and empirically verify the probability. The method (already seen) used to draw estimates through simulations is called *Monte Carlo Method*; it is used to find solutions to mathematical problems that are difficult to deal with. To give an idea of the odds involved, the results of 20000 virtual fights can be examined (Table 2).

Table 2. Summary table of simulations

<table>
<tr><td colspan="2" rowspan="2"></td><td colspan="3">Striker</td></tr>
<tr><td>1</td><td>2</td><td>3</td></tr>
<tr><td rowspan="3">Defender</td><td>1</td><td>1-0 = 0.417
0-1 = 0.583</td><td>2-0 = 0.579
0-1 = 0.221
1-0 = 0.200</td><td>3-0 = 0.658
2-0 = 0.203
0-1 = 0.072
1-0 = 0.067</td></tr>
<tr><td>2</td><td>0-2 = 0.741
0-1 = 0.162
1-0 = 0.097</td><td>0-2 = 0.444
2-0 = 0.225
0-1 = 0.177
1-0 = 0.154</td><td>3-0 = 0.375
0-2 = 0.208
2-0 = 0.180
0-1 = 0.130
1-0 = 0.107</td></tr>
<tr><td>3</td><td>0-3 = 0.825
0-2 = 0.130
0-1 = 0.028
1-0 = 0.017</td><td>0-3 = 0.615
0-2 = 0.176
0-1 = 0.079
2-0 = 0.069
1-0 = 0.061</td><td>0-3 = 0.384
0-2 = 0.173
3-0 = 0.138
2-0 = 0.115
0-1 = 0.108
1-0 = 0.082</td></tr>
</table>

CLOCK SOLITAIRE

In this case the artifact consists of a deck of cards and therefore the topic does not require a particular introduction to the object. In this game you take a deck of 52 cards, divide it into 13 decks, of four covered cards, arranged according to the numbers of the clock with the last deck in the middle. You take one of these middle cards and place it under the deck corresponding to the value of the card itself; then the first card of this latter pile is uncovered and the previous operations are repeated. If you find a King, you discard it and uncover the next card in the middle deck. The game is successful if all the cards are revealed before the fourth king is found.

This is an interesting example where you can compare *Monte Carlo Method* with algorithmic reasoning. The analysis is taken from a text on games edited by Lucia Salomone (1979).

Since there are 52 ways in which cards can be dealt, it is not possible to systematically examine possible distributions. Then you can play a certain number of games on the computer, assuming the fraction of successful solitaires in the sample as an estimate of the probability. The order of the cards is due to chance and is influenced only by the initial distribution. The probability of winning, on average, is 1 in 13. If you do not have a computer to check the result, you can use a different solution based on reasoning. To predict from the beginning if the game will succeed, simply know the value of the last card of each bundle! Figure 7 shows a possible configuration.

Figure 7. Final configuration of the solitaire

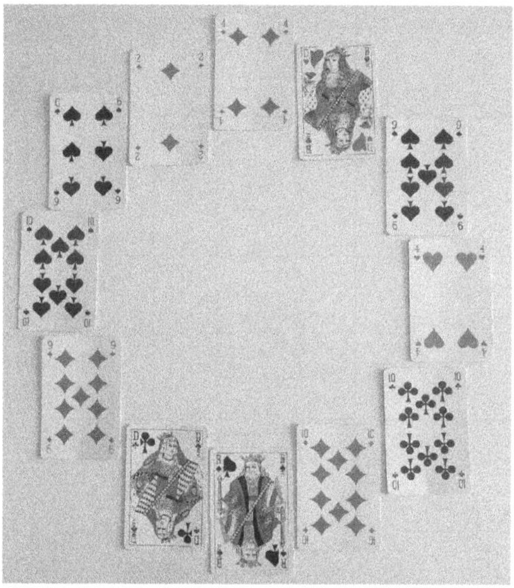

Each position of the clock is matched to a corresponding value of the cards, so one coincides with the Ace, two with 2, ..., eleven with Jack, twelve with Queen.

Then you draw a line from each of the 12 card values to the card that occupies the corresponding place, for example: the card that is placed on the one is a Queen and therefore must be connected with Q. The card placed on the two is a nine and then connects with 9, the card placed on the three is a four and then the connection must be made with the next card, which is a four, and so on. In practice, for each card, the connections are indicated as shown in Figure 8.

Figure 8. Solitaire connections

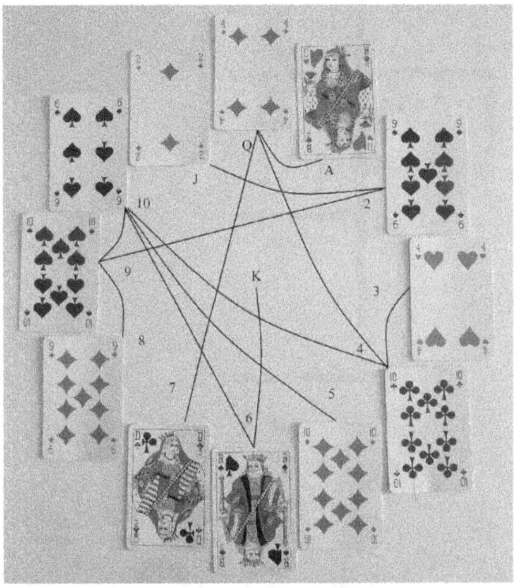

The representation of the connections is then transformed into a graph (Figure 9).

The game is won if and only if the graph of the connections between the last cards is a tree that includes all thirteen decks.

For example in this case the solitaire will succeed because the graph is connected.

SLOT MACHINE

A slot machine is a gambling machine, now widespread not only in casinos but also in many Internet sites. Even a slot/money box (Figure 10) can serve to introduce different concepts.

Figure 9. Solitaire graph

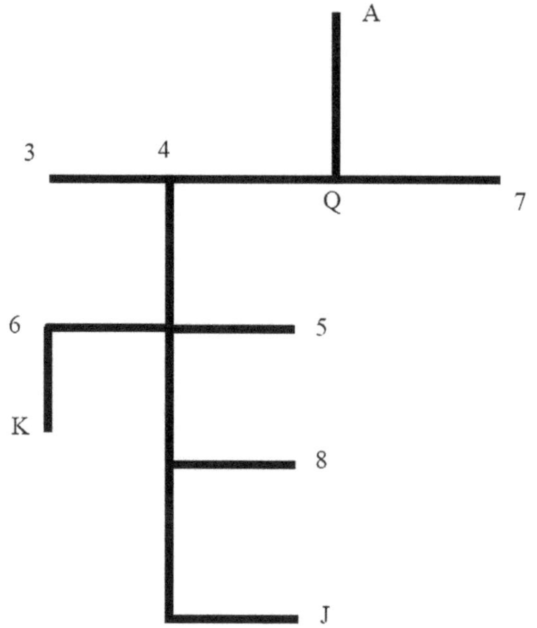

Figure 10. Slot machine – money box

The aim of the game is to achieve a winning combination of the figures that appear on the reels. To play the game you have to insert a coin and then pull the lever that operates the rollers, wait for the rollers to stop and check if the combination is winning or not. The odds of winning vary depending on the type of slot. Taking a cue from the slot machine in the figure we can see that there are three wheels and five symbols: BAR, cherry, apple, 7, \$. Lowering the arm the three wheels slide independently to form an arrangement between the $5 \bullet 5 \bullet 5 = 5^3$ possible. Suppose you win if you get a triple BAR or a triple 7 or two BAR and a 7 or two 7 and a BAR. In total you have eight winning combinations because the triple 7 and the triple BAR have only one occurrence while the other two can each be presented in three different ways. For example: 7 7 BAR, 7 BAR 7, BAR 7 7.

The probability of winning is therefore $8/125 = 0.064$. How much should you get from a bet if the game was fair, i.e. both the player and the machine operator did not have an advantage?

From this consideration we can take a fundamental concept, that of *Mathematical Expectation* or, more simply, *Expectation*.

The Expected value is the product between the sum S to be won and the probability p to win:

$$Expectation = S \bullet p$$

Returning to the case of the slot machine, if the bet is \$1, the $S \bullet 0.064 = 1$ should apply in case of victory.

Obtaining S, we have:

$$S = \frac{1}{0.064} \approx 15.6$$

In reality the slot manager does not give this amount to the player but retains part of it as remuneration, so the game is unfair.

Following an example set by Jean-Louis Boursin (1986), we now shift our interest to a situation where there are two slot machines. Let's first take as a reference the previously calculated probability of winning (0.0064). Next to it (without our knowledge) there is a slot machine which, due to a malfunction, sees this probability increased by a factor of ten (0.064). What is the probability that, choosing at random one of the two, the win (V) or the loss ($\overline{V}$) depends on the use of the faulty one (F) or the not faulty one ($\overline{F}$)? The answer depends on the type of question being asked, for example we can ask ourselves what the probability is:

1. If I play a game and I win it;
2. If I play a game and lose it;
3. If I play two games and win them both;
4. If I play two games and lose both;
5. If I play two games and lose one, I win the other.

The answers can be found using a tree graph (Figure 11):

Figure 11. Tree of the two slot machines

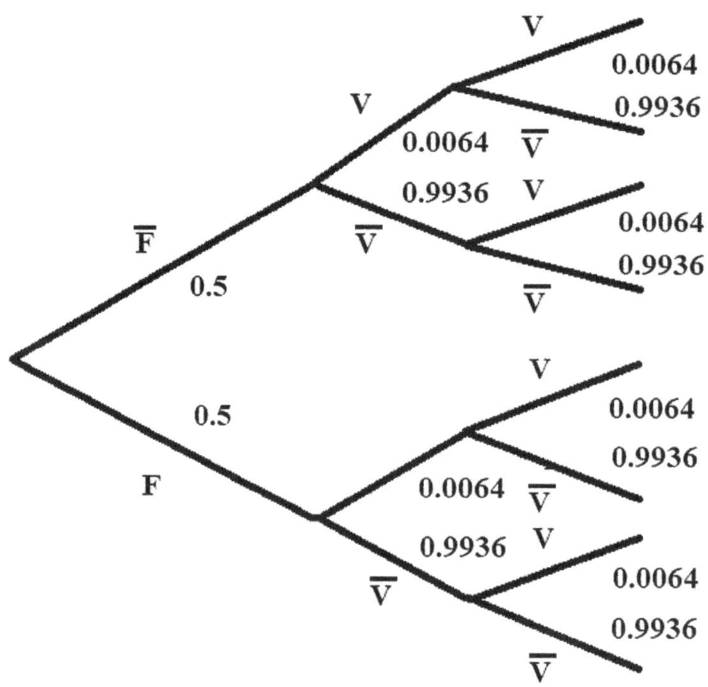

The solution is to be found by applying Bayes' Theorem. If I play a game and I win, the probability that this victory is due to the use of the broken machine is remarkably high:

$$P(V|F)=\frac{P(\bar{F}\cap V)}{P(\bar{F}\cap V)+P(F\cap V)}=\frac{0.0032}{0.0032+0.032}=0.909$$

If I play a game and lose it, then I get:

$$P(\bar{V}|F)=\frac{P(\bar{F}\cap \bar{V})}{P(\bar{F}\cap \bar{V})+P(F\cap \bar{V})}=\frac{0.4968}{0.4968+0.468}=0.485$$

If I play two games it is necessary to continue the calculations by going further through the branches but the reasoning does not change. Summarizing the results you will have (Table 3):

Table 3. Probability of the two slot machines

Question	Probability
1. If I play a game and I win it;	0.909
2. If I play a game and lose it;	0.485
3. If I play two games and win them both;	0.990
4. If I play two games and lose both;	0.470
5. If I play two games and lose one, I win the other.	0.904

ROULETTE

Roulette is a gambling game of Italian or, according to other sources, French origin. It consists of a disc, divided into 37 sectors (38, in American roulette with 00) numbered from 0 to 36 and colored alternately in red and black, while the zero (0), like the double zero (00) when present, is normally colored green; the disc is rotated in its seat by an attendant (the croupier) who then throws a ball at it. The ball, rotated in the opposite direction to that of the roulette, stops falling into one of the numbered sectors, determining the number and the winning combinations. The presence of some rhombuses (diamonds) placed on the roulette wheel that make the ball bounce randomly makes the result unpredictable. Then there is the tableau, a green cloth on which all the possible bets are shown. Roulette and tableau are shown together, as in the actual playing table, in Figure 12.

Figure 12. Roulette and tableaux

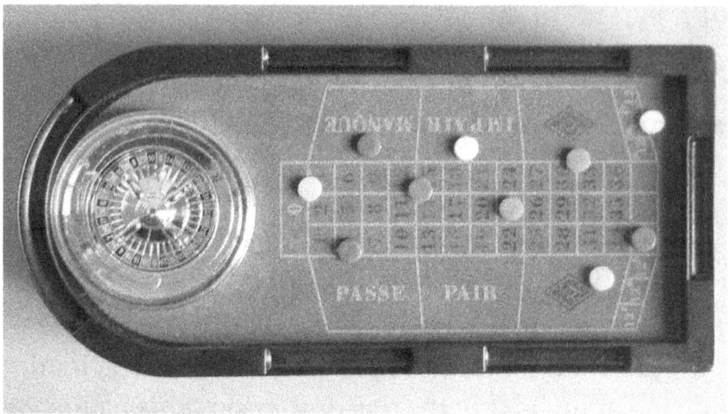

While the slot machine, seen above, is an object known to most students, typical casino games do not often have the same notoriety and therefore it would be useful to show how they work. There are excellent virtual reproductions of the roulette game on the net. The problem is that, almost always, online simulations offer, in addition to the possibility of playing for free, also that of playing for real money with the risk of sponsoring certain sites, albeit in good faith. To avoid this, it is preferable to turn to the few versions not explicitly linked to betting sites or, better, to use the equipment readily available in any toy store in the classroom.

The combinations and payments per unit of bet are those in Table 4:

Table 4. Table of bets

Name	Payments
Straight/Single (single number)	35
Split (contiguous pair)	17
Street (transverse)	11
Corner/Square (four contiguous numbers)	8
Six Line/Double Street (two contiguous transversal)	5
Dozen bet (first, second or third)	2
Column bet (first, second or third)	2
Even or Odd	1
Low or High (1-18 or 19-36)	1
Red or Black	1

Roulette, like all games of chance, guarantees the dealer a mathematical percentage of advantage over the player, which in the case of French roulette is about 2.70%. If you bet on a number the probability of winning is 1/37 and that of losing 36/37, betting a $ the expected "gain" is 35 • 1/37 - 1 • 36/37 = -0.027. If roulette is American the losses double because the numbers are 38.

Repeating the calculation for the other combinations can verify that the result does not change. For example, if you bet on red or black, the relationship 1 • 18/37 - 1 • 19/37 = -0.027 applies. As a result, there are no systems or techniques that guarantee a secure win.

Many players thought they could beat roulette: the best-known method is doubling the bet (martingale). The system is, at least in appearance, simple. Suppose we use the minimum bet, which we suppose is $10, on the red. If it comes out red I win $20 and then I have a balance of +$10. If it comes out black or the zero that is green (it's worse for roulette with double zero) I bet red again, doubling the bet to $20, so I bet $30. If it comes out red I win double, i.e. $40, and then my balance is positive again. If unfortunately this color doesn't come out I keep betting on red, doubling the bet again, and so on, until finally it comes out red and I end up with a balance of +$10. So I can think that, sooner or later, I will be able to gain from this strategy and in any case not to be at a loss. Too bad the method only works if I have unlimited money available. In fact at some point I may not have enough money to make continuous doubles. Also, casinos impose a cap on bets (suppose it is $2000). After seven bets I have to bet $1280 and if I don't win I've lost $2550 but I can't keep playing doubles. The probability of such a threatening event is not as low as you might think:

$$P = \left(\frac{19}{37}\right)^8 = 0.0048$$

So the question arises: am I willing to risk losing $ 2,550, about once every 200, to earn $ 80, if it comes out red at most eight times?

Despite the broad and motivated mathematical justifications that advise against believing in the possibility of beating the dealer in the long run, there is still a proliferation of sites on the Internet where they sell out amazing strategies to win. Many players have tried to perfect a game system thinking they are skilled enough to beat the wheel. In this regard I would like to recall an anecdote that happened years ago in a casino, I observed a player who did not place bets but noted the results in a booklet building a fairly complicated scheme. After a certain time a smile appeared on his face, he put all his chips on the green table and relaxed in his chair. Needless to say, he lost everything and, more than disconsolate, he was stunned by the failure.

A more scientific approach is based on an analysis that takes into account some fundamental parameters such as knowledge of the initial position, speed and acceleration of the ball (Small & Tse), the problem lies both in the definition of the initial position (as we will see in chapter dedicated to Chaos) and the need to mount a digital camera above the wheel, a procedure obviously prohibited by casinos.

An interesting study on the link between skill and chance is due to Dreef et al (20010. According to the authors the outcome of any game depends on learning and chance. The ability to improve, with learning, one's performance allows an expert to increase the winnings compared to a beginner.

The random aspect affects the increase in winnings of a player who not only knows the rules of the game, but also knows what may depend on the case. Synthetically: the skill (S) can then be expressed as the ratio between learning (L) and the sum of learning and randomness (R):

$$S = \frac{L}{L+R}$$

Poker is a typical example: any player who only knows the rules of the game but has no knowledge of the odds, will not improve his winnings by playing. So learning is unlikely to significantly increase your winnings; poker is a game that mixes skill and chance. If we then consider the game of roulette then the skill, defined as above, gives us as result 0, i.e. the game of roulette is a game of pure chance.

Speaking more generally of the *gambler's ruin* we must mention Abraham de Moivre (1667- 1754) who managed to solve this classic problem by demonstrating that in a fair game the chances of ruin are inversely proportional to the capital available to play. Even in a fair game against a dealer with unlimited monetary availability, ruin is certain; this is even more true in a game like roulette, which is clearly not fair.

For a lot of players, this is a discouraging news (though hardly accepted). All gambling games involve a loss that represents the dealer's remuneration. So there's no chance? Two losing games, according to the physicist Juan Parrondo, can be set so that when played one after the other, they become winners. This is a surprising result in game theory called Parrondo's paradox, named after its discoverer. For the explanation we follow (with some modifications) the indications contained in an article by Harmer and Abbott (1999). Game *A* consists of flipping a coin that has a chance of winning a little less than half, let's say with a small bias like 0.005; the probability of winning (p = 0.4995) is such that it is a long losing game. The *B* game consists of playing with two particular dice. The rule is that, if our capital is a multiple of a whole number (Harmer and Abbott (1999) use n =3), we play with an octahedron and we win if face 1 comes out so with low probability of victory (p = 1/8); we play with a tetrahedron and we lose if face 1 comes out so with favorable probability of victory (p = 3/4). It should be easy to convince yourself that the combination of these two strategies leads overall to a loss, the advantage of playing with the tetrahedron does not compensate the losses of when I use the octahedron. Here the paradox comes into play: both games are losers, but if we alternate them, for example playing AABB, this periodic change leads to a rapid increase in capital. Starting with a certain capital, after a large number of games we see a positive *Average final capital.*

Presented as a game the paradox may seem a futile exercise, in fact it is widely used in game theory, engineering, the study of population dynamics, the diversification of financial risk, etc..

BLACKJACK

Although there is no total agreement, many scholars in the history of blackjack believe that this game dates back to Vingt Et Un ("21" in French) played in 17th century casinos in France. The only The only certain thing is that there is official news of Blackjack in 1623, in a local press there is an article dedicated to gambling that mentions it.

Players are dealt cards whose value is equal to the number that appears on the card except for figures that are worth 10. The Ace is the only card that can have two values: it can be worth 11 or 1, in this case it is said that the hand is soft. The goal is to get a score as close to 21 as possible. This value must not be exceeded, because in that case the player, or the dealer, "bust". Often it is thought that the goal of the game is to get as close as possible to 21, without exceeding this number, actually the goal of Blackjack is first of all to beat the dealer without exceeding 21. This aspect affects the game strategy.

To play you must place the bet in the appropriate box. After the players have made their bets, the dealer distributes two cards to all those sitting at the table and also gives two cards to himself (Figure 13). One of the dealer's cards is always uncovered. The other card, covered, is called the "hole card". Most casinos around the world prefer an "uncovered game" where the cards are placed face up next to the bet, there are other casinos that prefer the "covered game" where the cards are placed face down and the player has the right to look at them.

Figure 13. Blackjack table

Once the distribution has taken place, players can exercise different options. If with the first two cards you do not have a "Blackjack" (21) in your hand you can:

1. Stay: you reject other cards and keep your score;
2. Ask for card: you can ask for "card" as many times as you want;
3. Split: if the first two cards are of the same value, you can split them, to play two separate hands, automatically betting the same amount;
4. Doubling: you can double the bet but in this case the dealer only needs one card.

At this point the dealer must pull paper with a score less than or equal to 16 and stay with 17 or more. The possibilities for the dealer are closely tied to its first card value, as shown in Table 5:

Some statistical calculations have made it possible to elaborate a "basic strategy" of play which, considering the uncovered card of the dealer and the two initial cards of the player, suggests what is more convenient to do. A different technique is based on counting cards that have already been issued.

When the deck is "full" of aces and cards with a ten value, the player places high stakes. When the deck has more low-value cards such as two to six, it is advisable to make small bets or leave the table.

To understand how this strategy works, we can see a film, 21: based on the story of a group of MIT students who, between 1980 and 1990, bankrupted numerous casinos by counting cards.

Table 5. Advantages and disadvantages based on the dealer's first card

Dealer Show Card	% Of Dealer Bust	Player Advantage %
2	35.30	9.8
3	37.56	13.4
4	40.28	18.0
5	42.89	23.2
6	42.28	23.9
7	25.99	14.3
8	23.86	5.4

REVOLVER

A game not to be experienced as potentially lethal, but very well known, is called Russian roulette. The rules are simple: a six-shot revolver has only one bullet in the drum, in turn the participants aim the weapon at their temple and pull the trigger risking, of course, to die. A famous Russian roulette scene is present in the film *The Deer Hunter* with Robert De Niro. The film features three American soldiers captured during the Vietnam War and forced to play this type of roulette. The game goes on until all but one of the competitors are killed.

In our case the object/stimulus (Figure 14) is represented by a card of the game known in Italy as the *Merchant at the fair* which is played with two decks of special cards with typical figures.

Leaving aside the bloody and senseless aspect of the game, you can formulate some questions whose answers depend on the conditions in which the game takes place.

Assuming there are six players taking turns at shooting each time the drum:

1. What are the chances of dying on the first try?

2. Is it better to try your luck first, second... or last?

The answers are a typical example of the application of conditional probabilities.

Figure 14. Paper with a revolver depicted

By indicating with $P(E_i)$ the probability of the first player to die at gunshot *i*, you obviously have $P(E_1) = 1/6$. Furthermore $P(E_2|\bar{E}_1) = 1/5$; $P(E_3|\bar{E}_1\bar{E}_2) = 1/4$; ...; $P(E_6|\bar{E}_1 \ldots \bar{E}_5) = 1$. Note that $E_2 = \bar{E}_1E_2$; $E_3 = \bar{E}_1\bar{E}_2E_3$; ...; $E_6 = \bar{E}_1 \ldots \bar{E}_5E_6$. Applying the composite probability theorem you have $P(E_1) = 1/6$, $P(E_2) = P(\bar{E}_1)P(E_2|\bar{E}_1) = 5/6 \bullet 1/5 = 1/6$; $P(E_3) = P(\bar{E}_1) \bullet P(\bar{E}_2|\bar{E}_1) \bullet P(E_3|\bar{E}_1\bar{E}_2) = 5/6 \bullet 4/5 \bullet 1/4 = 1/6$; $P(E_6) = P(\bar{E}_1) \bullet P(\bar{E}_2|\bar{E}_1) \bullet P(\bar{E}_3|\bar{E}_1\bar{E}_2) \bullet \ldots \bullet P(\bar{E}_5|\bar{E}_1 \ldots \bar{E}_4)\, \bar{E}_4$ $P(E_6|\bar{E}_1 \ldots \bar{E}_5) = 5/6 \bullet 4/5 \bullet 3/4 \bullet 2/3 \bullet 1/2 \bullet 1 = 1/6$.

As you can see, it is indifferent to try your luck in any position.

If you consider only two players the problem changes. By indicating with P(A) the probability that *A* dies, and with P(B) the probability that *B* dies occurs:

1. 1st round $P(A) = 1/6 = 0.17$ (ends the game); $P(B) = 5/6 \bullet 1/6 = 0.14$; probability that nobody dies $5/6 \bullet 5/6 = 0.69$
2. 2nd round $P(A) = 5/6 \bullet 5/6 \bullet 1/6 = 0.11$; $P(B) = (5/6)^3 \bullet 1/6 = 0.09$
3. 3rd round $P(A) = (5/6)^4 \bullet 1/6 = 0.08$; $P(B) = (5/6)^5 \bullet 1/6 = 0.06$.

At this point we can generalize the reasoning; we can say that P(B) = 5/6 • P(B). This happens at every turn, because, to win, *A* must pass his turn, whose probability is 5/6. If no one dies, the conditions return identical to the starting situation: it is as if he hadn't played. Moreover, since one of the two must die P(A) + P(B) = 1.

Solving the simple system in the two equations:

$$\begin{cases} P(B) = 5/6 \cdot P(A) \\ P(A) + P(B) = 1 \end{cases}$$

you get

P(a) = 6/11 and P(b)= 5/11.

An additional question could be the following: if there are two bullets is it better to shoot straight away or make a spin?

The problem of Russian roulette does not end with the consideration that it is a crazy game, in fact mathematicians and psychologists have tried to deepen the subject in terms of *Expected utility*. The Expected utility theory is based on the hypothesis that, under conditions of uncertainty, the utility can be calculated as a weighted average of utilities in every possible state, using as weights the estimated probabilities of the occurrence of possible events. We follow the reasoning of Kahneman & Tversky (1979): "Suppose you are compelled to play Russian roulette, but are given the opportunity to purchase the removal of one bullet from the loaded gun. Would you pay as much to reduce the number of bullets from four to three as you would to reduce the number of bullets from one to zero? Most people feel that they would be willing to pay much more for a reduction of the probability of death from 1/6 to zero than for a reduction from 4/6 to 3/6. Economic considerations would lead one to pay more in the latter case, where the value of money is presumably reduced by the considerable probability that one will not live to enjoy it". Experimental research (Wu & Gonzale, 1996.) based on risky decision-making processes shows that people systematically violate the principle of expected utility.

Remaining always in the field of firearms one can remember what is considered the paradox of the three gunslingers.

Three gunfighters *A*, *B*, *C* challenge each other to a duel by placing themselves at the vertexes of an equilateral triangle. They will shoot one at a time until there is only one survivor; each can choose who to shoot. What is interesting, from a probabilistic point of view, is that the skills of the three gunfighters are not the same: *A* when he shoots hits 100% of the time, *B* 80% of the time and *C* only 50% of the time.

To help the weakest are supposed to shoot first C and then B to shoot. What does *C* have to do to have the best chance of winning? Paradoxically, he has to shoot high!

It's not good for him to shoot the strongest of the gunfighters because, assuming he can eliminate him, he only has a 20% chance of surviving when it's B's turn. In the second phase *B* won't shoot him, he'll shoot *A*. There are two cases: if he hits him it will be *C* who will then shoot B with a 50% chance of hitting him; if he doesn't hit *A* he will obviously shoot B and kill him with certainty. At this point the turn is up to *C* who can hope to hit *A*.

Summing up, you can see that *C*, firing in the air the first time, has a higher probability of survival.

BACKGAMMON

One of the oldest games is Backgammon, whose origin dates back more than 5000 years. A similar game was also present among the Greeks. The current rules, dating back to 1931, can be traced back to the president of the Racquet and Tennis Clubs of New York. The players have 15 checkers that move (one clockwise and the other counterclockwise), based on the throw of two dice, along 24 arrows (Figure 15). Players must, if possible, move the checkers according to the number indicated on each dice, the same checker can be moved twice, if the throw generates a double result the player must play each dice twice. The game's aim is to be the first to move all their checkers off the board. If you play a money stake, there is also a doubling dice that allows you to win (or lose) a double amount every time a player uses it.

This is a probabilistic optimization game that requires careful evaluation of positions and is influenced by the possibility of doubling the stakes before each shot. Also in this game enters the concept of Mathematical Expectation.

This can be clarified with an example, suppose that the final position, where the move touches the grey, is the one shown in Figure 16. The grey has two checkers on arrows 2 and 3 while the white has two checkers on points

23 and 24, remember that the white checkers move clockwise and the grey ones counterclockwise.

Let's consider what the probability is for grey to let his checkers out with a throw. He will win unless one of the dice comes up with face 1. Since the possible results in the throw of two dice are 36 and 25 are the combinations that do not present the one, the probability of victory is 25/36 = 0.694. Now, before launching, he can choose to double or double the $ 1 stake. If he doesn't double, although the odds are almost 70%, his expected earnings per game is 25/36 • 1 - 11/36 • 1 = 24/36 = $ 0.39. If it doubles, two things can happen: white gives up and grey's equity becomes $ 1, given that it wins; if white accepts the doubling, the hope of grey doubles becoming 0.39 • 2 = 0.78 $. This is certainly not a good situation for the white that must decide how to minimize losses; especially assuming multiple games, it is not convenient for him to be renouncing.

Figure 15. Backgammon

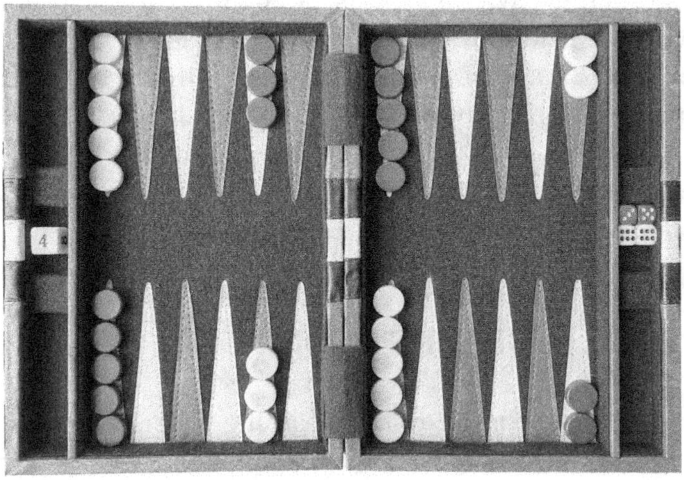

CONCLUSION

As we have already had occasion to clarify, we must distinguish between games and gambling. In the first case we can talk about situations in which a real (or at least realistic) situation is simulated and therefore can play an important role both for training and learning. War-gaming, business gaming, backgammon, etc. are effective tools for teaching. Already in the 1970s, Jerome Bruner (1965) highlighted the usefulness of simulation in the school field.

Figure 16. Final position of a Backgammon game

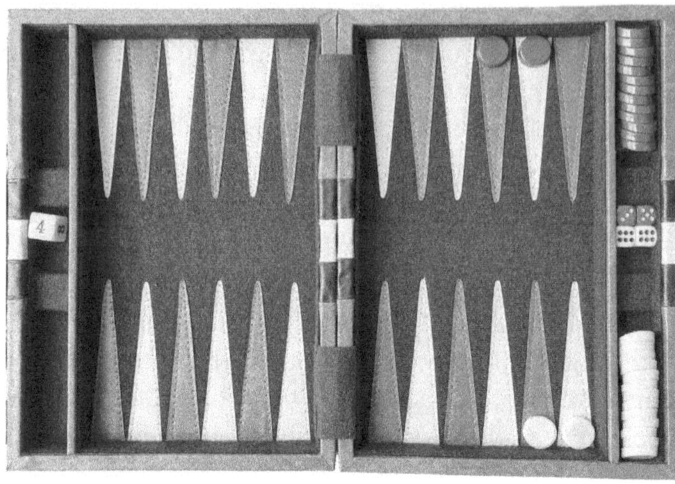

Gambling requires a different kind of consideration. This is because there are important behavioral aspects. A 2018 research carried out in Italy by the ISS (Higher Institute of Health) showed that 18 million adult Italians have gambled (out of a total population of 60 million): of these, more than 13 million play in a 'social' way, 2 million have a low-risk profile, 1400000 have a moderate risk, 1500000 are problem gamblers. Leaving aside the pathological game which represents a real disease, one is struck by people's total ignorance of the main probabilistic rudiments.

FUTURE RESEARCH DIRECTIONS

Games can play an important role in learning probability as students can test themselves by concretely discovering numerous concepts. Immersed in an engaging role (the kids want to win) you can measure an incomparably higher involvement than that obtainable in a normal lesson. "There are some studies revealing that game-based learning enriches the learning environment, improves the students 'performance, increases the students' motivation, provides the opportunity to work with the group and provides a fun learning environment". (Koparan, 2019).

In this sense, a fruitful research path could guide us towards the operational verification of the introduction of these supports in teaching or, at least, in the proposition of elements capable of arousing greater awareness of the implications of this issue.

Seeing this approach simply in a playful way is wrong, just think of the tumultuous development, in the field of artificial intelligence, of genetic algorithms (stochastic optimization techniques based on evolution). Well, all the texts that describe applications in industry, medicine, genetics, etc. cite the Roulette-wheel selection as a randomized generative basis.

REFERENCES

Alfonso X (King of Castile and Leon), & Calderón, R. O. (2007). *Libro de los juegos: acedrex, dados e tablas; Ordenamiento de las tafurerías*. Fundación José Antonio de Castro.

Bibbona, E., Boggiatto, P., Carypis, E., De Simone, M., & Panero, M. (2014). *Gare e giochi Matematici: Studenti all'opera*. Ledizioni.

Boursin, J. L. (1986). *Les structures du hasard: les probabilités et leurs usages*. Le Seuil.

Bruner, J. S. (1965). *Man: A course of study*. Educational Services, Incorporated.

Campbell, P. J. (2007). Games of chance with multiple objectives. *Metrika, 66*(3), 305–313. doi:10.100700184-006-0112-5

Canova, P., & Rizzuto, D. (2016). *Fate il nostro gioco*. Add Editore.

Cosmides, L., & Tooby, J. (1996). Are humans good intuitive statisticians after all? Rethinking some conclusions from the literature on judgment under uncertainty. *Cognition, 58*(1), 1-73.

Dreef, M., Borm, P., & van der Genugten, B. B. (2001). *A New Relative Skill Measure for Games with Chance Elements*. Tilburg University.

Eastaway, R. (2008). *How Many Socks Make a Pair?* Surprisingly Interesting Everyday Maths. JR Books Ltd.

Harmer, G. P., & Abbott, D. (1999). Losing strategies can win by Parrondo's paradox. *Nature, 402*(6764), 864–864. doi:10.1038/47220

Hwa, S. (2018). Pedagogical Change in Mathematics Learning: Harnessing the Power of Digital Game-Based Learning. *Journal of Educational Technology & Society*, *21*(4), 259–276.

Koparan, T. (2019). Teaching Game and Simulation Based Probability. *International Journal of Assessment Tools in Education*, *6*(2), 235–258. doi:10.21449/ijate.566563

Li, L. B., He, S. H., Li, S., Xu, J. H., & Rao, L. L. (2009). A closer look at the Russian roulette problem: A re-examination of the nonlinearity of the prospect theory's decision weight π. *International Journal of Approximate Reasoning*, *50*(3), 515–520. doi:10.1016/j.ijar.2008.10.004

Prelec, D. (1998). The probability weighting function. *Econometrica*, *66*(3), 497–527. doi:10.2307/2998573

Roberts, F., & Tesman, B. (2009). *Applied combinatorics*. CRC Press. doi:10.1201/b12335

Salomone, L. (1979). *Raccolta di giochi a due*. A.P.E. Mursia.

Siller, H. S., & Maaß, J. (2012). Learning mathematics or losing money - betting as a topic for mathematics education. *Teaching Mathematics Applications*, *31*(2), 65–83. doi:10.1093/teamat/hrs005

Small, M., & Tse, C. K. (2012). Predicting the outcome of roulette. *Chaos (Woodbury, N.Y.)*, *22*(3), 033150. doi:10.1063/1.4753920

Tversky, A., & Kahneman, D. (1979). Prospect theory: An analysis of decision under risk. *Econometrica*, *47*(2), 263–291. doi:10.2307/1914185

Wu, G., & Gonzalez, R. (1996). Curvature of the probability weighting function. *Management Science*, *42*(12), 1676–1690. doi:10.1287/mnsc.42.12.1676

ADDITIONAL READING

Hays, R. T. (2005). *The effectiveness of instructional games: A literature review and discussion (No. NAWCTSD-TR-2005-004)*. Naval Air Warfare Center Training Systems Div Orlando Fl. doi:10.21236/ADA441935

Kordaki, M., & Gousiou, A. (2016). Computer Card Games in Computer Science Education: A 10-Year Review. *Journal of Educational Technology & Society*, *19*(4), 11–21.

Lipowski, A., & Lipowska, D. (2012). Roulette-wheel selection via stochastic acceptance. *Physica A*, *391*(6), 2193–2196. doi:10.1016/j.physa.2011.12.004

Monaghan, J. (2016). Games: Artefacts in gameplay. In *Tools and Mathematics* (pp. 417–431). Springer. doi:10.1007/978-3-319-02396-0_18

KEY TERMS AND DEFINITIONS

Expectation: The arithmetic mean of a large number of independent realizations.

Expected Utility: The attractiveness of an economic opportunity.

Gambler's Ruin: A gambler with finite wealth, playing a fair game will eventually and inevitably go broke against an opponent with infinite wealth.

Monte Carlo Method: Algorithm that rely on repeated random sampling to obtain numerical results.

Sunk Cost Bias: Expenses paid for previously that are not affected by current or future decisions.

Chapter 6
Probability and Distribution

ABSTRACT

The concept of probability distribution is one of the most relevant but, in some ways, among the most complex concepts to be treated at school. If we exclude some fairly simple cases, there is a variety of distributions whose formulation and use is not trivial. In this chapter, many distributions have been omitted (e.g., exponential, t student, Weibull, etc.) because they are too specialized and reserved for a higher level of study. The choice has privileged distributions that can be analyzed starting from objects with which students are, or maybe, familiar. However, the distributions addressed are sufficient to give a fairly broad idea of the topic.

INTRODUCTION

Isn't it extraordinary how difficult it is to get a sample really random? ~ George Yule

A premise: "mathematical knowledge in the classroom is no professional and disciplinary knowledge. [...]. School mathematics is not abstract and universal, but local and exemplary". (Steinbring, 1984).

This is one of the reasons that guided the choice of topics.

A probability distribution, a term already encountered in the text, is a mathematical model that links the values of a variable to the probabilities that these values can be observed. The first distribution that is normally dealt with in didactics is the *Binomial Distribution*. The coin toss is often brought

DOI: 10.4018/978-1-7998-3871-5.ch006

as an example of the probability of obtaining k success in n independent tests according to Bernoulli's well-known formula:

$$P(X=k)=\binom{n}{k}p^k(1-p)^{n-k}$$

An example of a different and quite particular application is the following: the same type of engine equips two aircraft: a twin-engine and a four-engine. The twin-engine flies even with only one engine, the four-engine flies even with only two engines. Calling *p* the probability of an engine failure, which is the safest aircraft? Let's assume that the failure of one engine does not affect the operation of the others and that it is therefore an independent event. In this case we have two probability functions that depend on an independent variable that is in turn a probability (of engine failure).

The probability of remaining in flight for the two types of aircraft can be determined by considering the possible combinations. The twin-engine aircraft will arrive at its destination if at least one of its engines is functioning:

$$\text{P(twin-engine)} = C_{2,1}\ p\ (1\text{-}p) + C_{2,2}\ (1\text{-}p)^2 = 1 - p^2$$

The aircraft equipped with four engines arrives at its destination if the failures do not involve more than half of the engines:

$$\text{P(four-engine)} = C_{4,2}\ p^2\ (1\text{-}p)^2 + C_{4,3}\ p\ (1\text{-}p)^3 + C_{4,4}\ (1\text{-}p)^4 = 1 - 4p^3 + 3p^4$$

The graph (Figure 1) corresponding to the different values that the breaking-point probability could assume shows clearly that there is a breaking point where the probabilities are equal. All this is obviously unrealistic because the probability of failure is extremely low, especially when compared with the number of car and rail accidents due to breakdowns.

Based on the scale of the variable of interest, two types of probability distributions can be distinguished:

1. discrete distributions: the variable is measured with integer numerical values, e.g. the number of people served by a supermarket checkout over a certain period or the longevity of a person expressed in years;

Figure 1. Comparison between twin engine and four engine

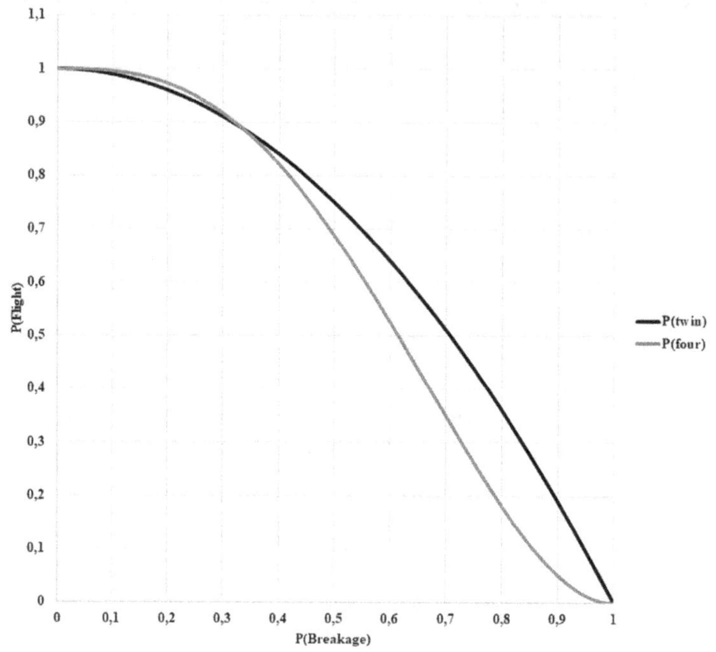

2. continuous distributions: the variable can assume a theoretically infinite number of values within a finite or infinite range, for example the heights of a group of people or the duration of operation of an electronic component.

LOTTERY

A lottery is a form of gambling based on the extraction of numbers, with cash prizes. It is one of the games with an ancient historical tradition. The common aspect of the various lotteries is that in the past they have often been used to finance public works. Traces of this game date back to the Han Dynasty in China, around 200 BC. Already in use in Imperial Rome, it partially declined in the Middle Ages. In 1500 the king of France, Francis I, officially legalized the lotteries with an edict, shortly after these were authorized by Queen Elizabeth I. A lot of news can be obtained from a text, not recent, by Alan Wykes. Currently the countries that manage this activity are distinguished by the different ways of execution and therefore it is not

Figure 2. Number extraction

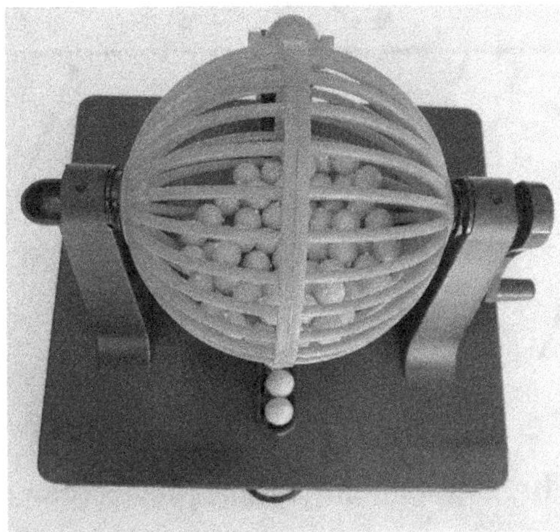

easy a unitary treatment. The common aspect is the extraction of numbers using some form of device (Figure 2).

Before going on, a apparently trivial question: suppose that someone has eliminated from the device all the balls that have a number greater than N, should we bet on the output of an even or odd number?

It shouldn't be difficult to understand that it's better to bet on odd because the chances are equal, or there's an extra odd ball. For example if N=3 the numbers would be 1, 2, and 3 and the probability of an odd number coming out would be 2/3; for N=4 the probability is 1/2. Obviously, the bigger N is, the closer the probability is to equity values.

We now return to our lottery.

In a way it would have been quite logical to include the topic in the Game chapter, but it is useful to look at this topic now, because it offers the opportunity to study a particular type of distribution: *Geometric Distribution*.

A discrete random variable X follows the geometric distribution under certain conditions: independent tests, two possible results, constant probability. The variable we are interested in is the number of trials required for an event to occur for the first time at n-th trial.

Let's examine the simple model of the Italian Lotto: 5 numbered balls are drawn within a group of 90 and the player makes a bet assuming the exit of a certain number. The probability to get the number right is p = 5/90 = 1/18 and the opposite probability is q = 17/18. The assumed number can

Figure 3. Possible routes

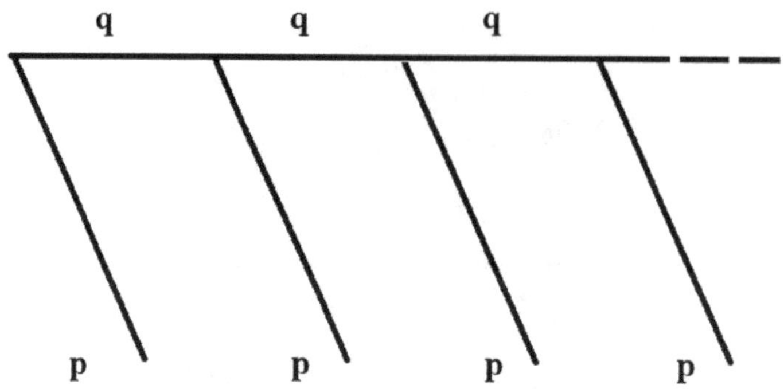

exit at the first draw or after *n* attempts. The process can be well represented graphically (Figure 3):

The probability of having no success in n-1 trials is q^{n-1}, while the probability of success at n-th extraction is *p*. Consequently the probability of success is p:

$$P(X=n)=p\bullet q^{n-1}$$

Figure 4. Geometric distribution function graph

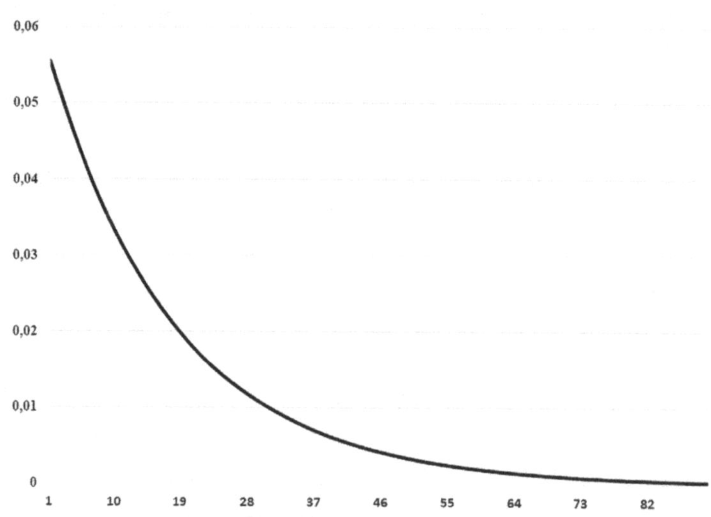

How does this distribution behave? What is the probability that a certain number will come out for the first time during 1, 2, 10, 100 draws? The trend of the function (Figure 4) is of a decreasing exponential type.

However, you must be careful not to make an error of assessment: the probability remains the same, it does not decrease or increase, as many bettors believe. The distribution tells us that, starting from any time t, it is more likely that the event will occur at time t+1 than at time t+2. In other words, the event is more likely to occur sooner than later.

PIN

Figure 5. Pin with Pi symbol

Pins are common and often funny because of the writing that distinguishes them. The one under examination (Figure 5) shows the symbol π.

This symbol has already been mentioned earlier, talking about Buffon's needle and its relationship with probability. It is perhaps the most famous symbol in the history of mathematics, so famous that in honor of its approximation 3.14 there is the Pi Day which is celebrated on March 14. In reality, being Pi an irrational number, it has an infinite number of digits and the race to lengthen the sequence is always alive but hopeless. For this reason we are content with an approximation, even if this effort has led in the past to excessively inaccurate results. In the Old Testament we read about the altar built in Solomon's temple: "ten cubits from one side to the other, that is completely round; its height was five cubits and a rope of thirty cubits surrounded it all around". This passage indicates that the ratio between circumference and diameter is 3. As a more recent but certainly striking case, we remember that the bill n. 246 of 1897 of the law of the State of Indiana fixed the value of π to a clearly wrong rational approximation (3.2).

Normally students associate Pi with geometric calculations but here we want to consider a problem: can we recognize a distribution of decimals?

The figures of π have no apparent pattern and have passed the statistical randomness tests. For a very approximate idea but of considerable visual impact we can associate each cipher from 0..9 to a color and observe that in

Figure 6. Pi chromatic

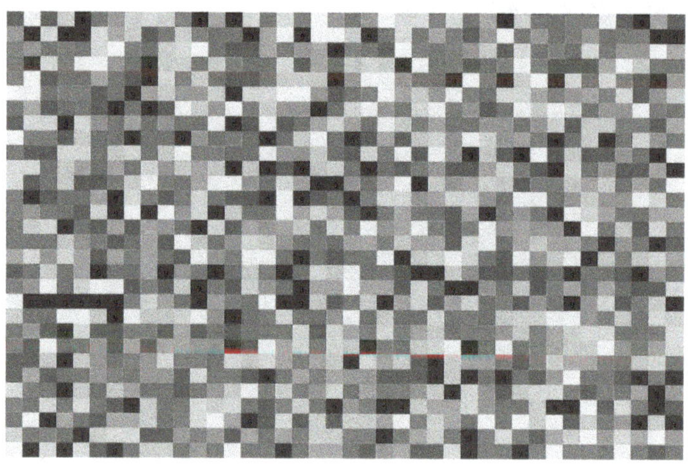

the resulting patchwork (Figure 6) it seems impossible to find a recurring pattern:

Normality tests were also carried out on π figures. Let us remember that a number of infinite length is called normal, according to the notion introduced by the French mathematician and politician Émile Borel (1871-1956), when all possible sequences of digits appear with the same frequency. Unfortunately, the conjecture that π is normal has not been proved (but neither denied). On the other hand, an analysis of the frequencies with which all possible combinations within the known digits of the development of π are repeated fits very well with the statistical hypothesis that it is. In a non-specialist text, but intended for teaching purposes, it may be sufficient to give an idea of what is meant by *Uniform Distribution* and what the concept of normality is.

Figure 7. Distribution of the first million digits of Pi

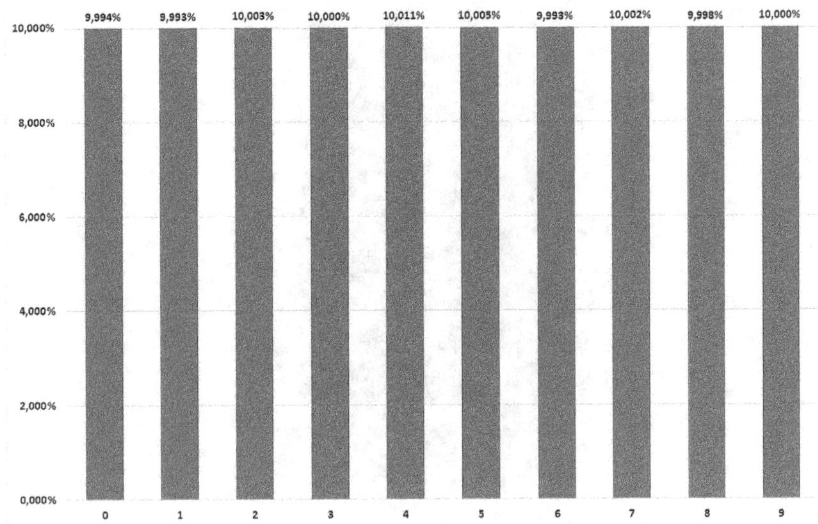

Taking into account even the decimal digits of up to one million, we obtain a distribution that suggests a fair distribution (Figure 7).

The second noteworthy aspect is that you can trace any number within the sequence. For example, a sequence of six 9 already appears in 762 position, the date of Independence Day can be found from position 21134 while the telephone of the Prado Museum in Madrid (without prefix) appears after 56887995 digits.

SPAGHETTI

Spaghetti (Figure 8) is a particular type of pasta with a long and thin shape and represents a fundamental element of the so-called Mediterranean diet.

In this case, however, we are not interested in the culinary aspects, but in a probabilistic problem: a random variable distributed evenly over a set consisting of a segment.

The problem can be divided into two parts. The first refers to the study of the French Audoly & Neukirch (2005). They built a mathematical model of the dynamics of a spaghetti. This model can be extended to rigid and elongated objects. What is interesting is that this break rarely generates two pieces but normally three. Thanks to their work we know that the spaghetti

Figure 8. Spaghetti

breaks in this way because by bending it, it is subjected to an effort that is concentrated in the middle point. This causes the first break, the next one happens because part of the elastic energy is discharged on the remaining pieces of spaghetti.

We are not able to predict how many fragments of dough will be generated, but we assume that with remarkably high probability there will be three; this allows us to introduce the second part of the problem.

If a spaghetti breaks into three parts in a completely unpredictable way what is the probability that a triangle can be formed with the three parts? Let's assume a truly random rupture, even if there are strong doubts that the probability of rupture is the same at the center or the extremes, but let's pretend that this is possible.

Considering the length L of the string, the two breaking points x and y are two random variables evenly distributed over [0, L]. Geometrically, the three pieces form a triangle only if the triangular inequalities apply; in other words, the sum of the lengths of any two pieces must be greater than the length of the third:

$$x < \frac{L}{2}; y < \frac{L}{2}; x + y > \frac{L}{2}$$

obviously with $x \geq 0$; $y \geq 0$; $x + y \leq L$

There are several graphical ways to show that the probability sought is surprisingly equal to 1/4. Perhaps the most elegant one involves drawing an equilateral triangle and connecting the central points of the sides to form an internal equilateral triangle (Figure 9). Starting from any point of the large

Figure 9. Broken spaghetti in the triangle

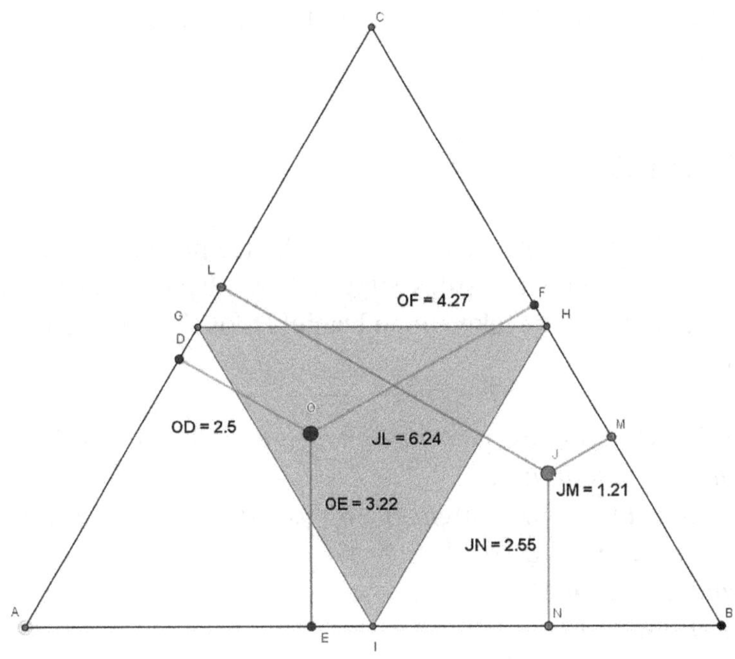

triangle, if we draw the perpendiculars to the sides, the sum of these three segments will be constant and equal to the height *L* of the large triangle, as demonstrated by Vincenzo Viviani (1622-1703).

Only for the points inside the small triangle the previous inequalities apply, in fact OD, OE and OF respect the conditions while this does not happen for JL, JM and JN; moreover it is evident that the area is a quarter of that of the big triangle. The geometric analogy with the problem of broken spaghetti is clear. The sum of the three perpendiculars on the sides corresponds to the length of the spaghetti. Each point of the large triangle represents a unique way of dividing the spaghetti with the three perpendicular segments corresponding to the three pieces. The probability of breaking it favorably is the same as the probability of choosing a random point in the central triangle.

Definitely an interesting discovery, but the probability always reserves surprises. If you change the breaking conditions, the problem can take on a different aspect. Let's suppose that the object (not necessarily a spaghetti) is first broken randomly into two parts and then one of the two is broken again randomly into two parts. Is the result the same? As Martin Gardner (1914-2010) stated if we choose the shortest piece, no triangle is possible. If we take the longer piece the probability is measured between the three triangles below and then it is 1/3. Since the probability of taking this piece is 1/2, we have 1/3 • 1/2 = 1/6.

CHESSBOARD

" Life can best be understood as a game of chance played on the chessboard on space and time with the rules supplied by physics and mathematics" (Hoffmann, 2012). Theme taken up in Daniel Hooks' poem "Life Is A Game Of Chance":

Every time I shuffle a pack of cards I change the outcome of the game.
What if I do the same in life?
A girl I meet for the first time might end up as a friend, an enemy, or my wife!
The random things you say or do
Open or shut doors for you.
That is the nature of life.
Like you throw a six on a dice when all you need is a two
Life deals out blows not always to you.
No one knows where they will fall.

A chessboard of size 6 • 6 and black and white checkers offer the opportunity to examine how the numerical consistency of a population evolves. The game starts with the two players randomly placing their checkers on the board, occupying all the boxes. The game also requires a coin or two dice, which are intended to generate randomness, and rules that can influence the behavior of populations (Figure 10).

Figure 10. Chessboard 6x6

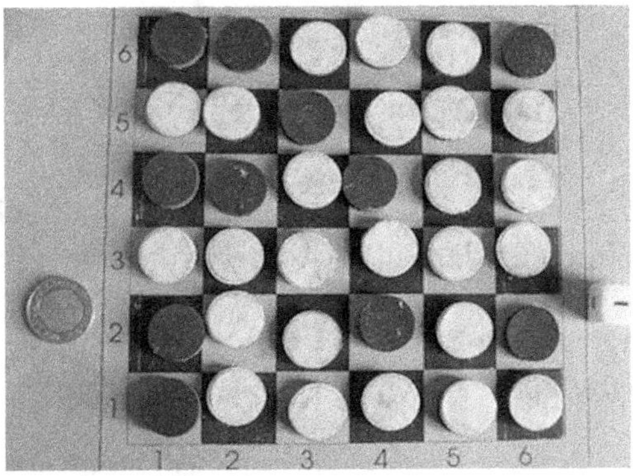

According to Eigen and Winkler (1993), creators of the game, there can be three different models that lead to different distributions. The fundamental behaviors that a population can follow are:

1. Stable: birth and mortality rates are balanced. A change in the population in a positive or negative direction will influence the rates to compensate for this change. We are faced with self-regulation.
2. Indifferent: the population can take on all possible dimensions. The average values of birth and mortality rates are the same. Each value assumed by the population at a certain point in time is a starting point for further fluctuations.
3. Unstable: a small variation influences one or both rates so that this variation is further strengthened. The final effect will be a catastrophe of either species.

The development of a population can therefore lead to three different models.

1. Drift: you flip the coin. If the white comes head on, he removes any black token and replaces it with a white one taken from his reserve. If a white checker is crossed, it is replaced by a black one.
2. Balance: the dice are rolled: the checker in the square corresponding to the coordinates resulting from the roll is removed and replaced with a spare checker of the opponent.
3. All or nothing: roll the dice: the token in the square corresponding to the coordinates resulting from the roll is doubled at the expense of the opponent's color. So if you draw a square with a white token, you remove

any black token and replace it with another white one taken from the reserve.

In the three cases, what will be the distributions after a sufficiently high number of launches?

In the first case the chances of the number of checkers of each color increasing or decreasing are identical. No particular configuration is privileged

Figure 11. Drift distribution

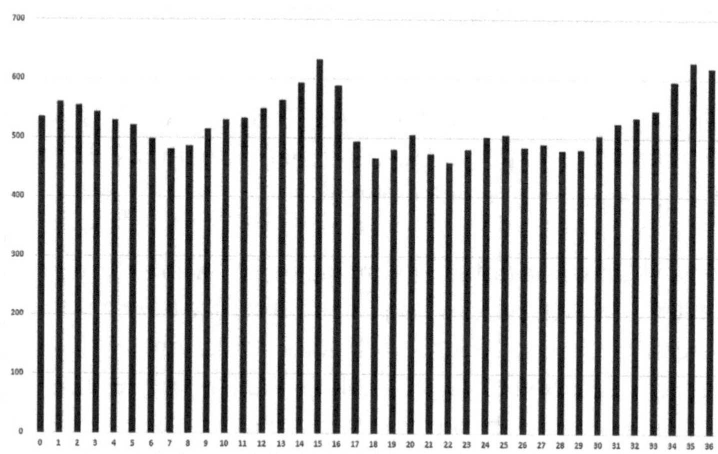

even if particular configurations are reached only after some time. For example, the graphical representation of a simulation of 20000 matches is shown as in Figure 11:

This type of representation can be associated, from a theoretical point of view, to the *Uniform Distribution* in which, said *n* the number of results, we have that for the single event *x* (for example the initial distribution 18-18) is P(x) = 1/n. In other words: a distribution is uniform when all of the outcomes have the same probability. As Sebastian Kwiatkowski says: "Uniform distributions have maximum uncertainty. If your goal is to minimize uncertainty, stay away from uniform probability distributions".

In the second case an equilibrium is established even in the presence of fluctuations, the graph takes the form of a curve called a bell curve that recalls a continuous distribution (which we will see later). It has an average deviation proportional to the square root of the number of checkers. For example, with even just 100 throws you can get the distribution shown in Figure 12:

In the third case the doubling leads to an imbalance between the two contenders, in this game who has more checkers receives more and, in the end, reaches the victory. In this case we can well say that "When it rains it pours". If you examine a simulation (Figure 13) you can see, in this case, that the white man loses, at a certain point, all the checkers.

Figure 12. Balance distribution

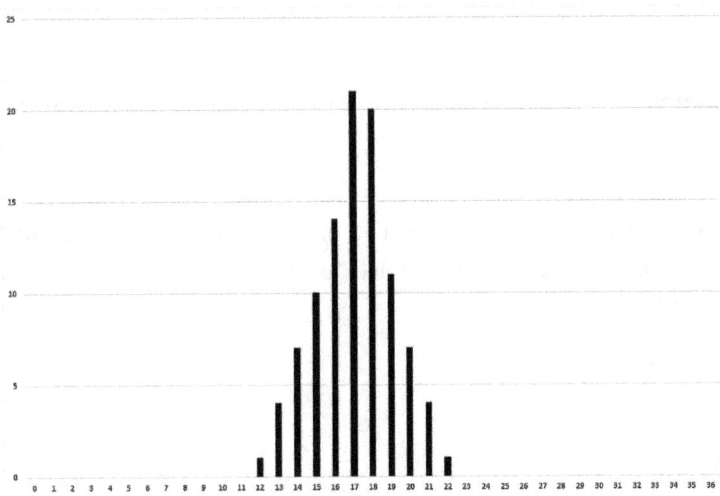

Figure 13 Number of white checkers in play

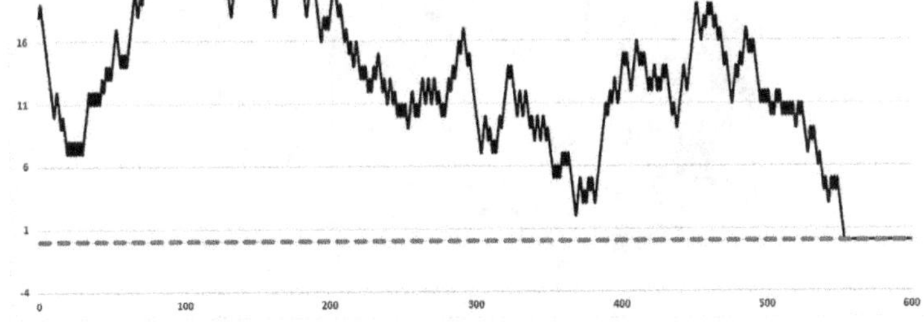

Figure 14. Distribution "All or Nothing".

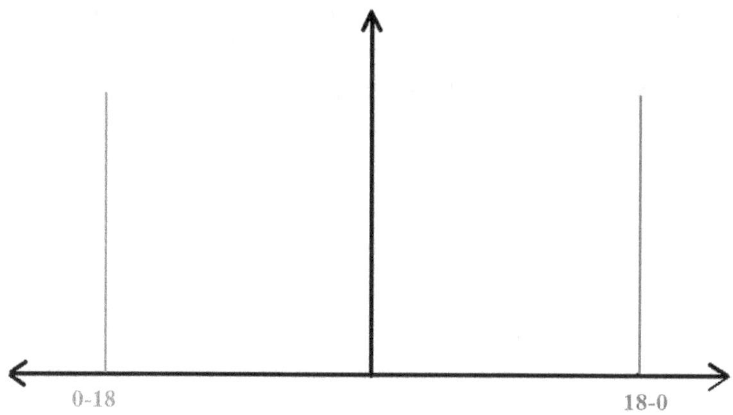

If you look at the possibilities of the game you can understand that it is a Singular Distribution. It is a very particular distribution (Figure 14) and is not frequently used.

Figure 15. Quincunx

QUINCUNX

Figure 15, a sort of foosball table, refers to a device developed by Francis Galton (1822-1911) for the study of a particular type of distribution.

The apparatus consists of a vertical plane with nails perpendicular to the plane, and a series of collection cells. Galton's machine is also called Quincunx because the arrangement of the nails resembles the five imprinted on the faces of the dice. The operation of the device is widely covered by many authors, among them we mention only Alex Bellos (2010).

The experiment consists in dropping balls from the vertex of the "triangle" formed by the nails to see how they are arranged in the cells below. The premises of the experiment are that the spaces between the nails are slightly larger than the falling ball and that the ball, on its way, make only one rebound per line, with the only possibility of falling to the right or left of the nail.

We ask ourselves: is there a connection between the number of rows of nails/grooves and the number of balls falling into the individual grooves?

If we do an experiment we notice that the balls tend to "prefer" the central grooves, why? Is there a mathematical model to describe how the balls fall?

Let's suppose that the deviation suffered by the ball when it hits a nail happens randomly: the probability calculation allows to determine the distribution of the balls according to the number of deviations to the right (p is the probability that the ball "turns" to the right).

There are many situations in which solving a problem admits only two possible results: true-false, good-defective, favorable-opposite, etc. In general we will talk about success (1) or failure (0). Examples of cases are the results of a series of flips of the same coin or of a series of extractions with reintroduction in the urn, each of which can provide only two results: success with probability p and failure with probability q = 1- p. In this case we are talking about *Binomial Distribution* and the probability of achieving x success in n trials will be:

$$P(x) = \frac{n!}{x! \bullet (n-x)!} p^x q^{n-x}$$

Let's go back to our Quincunx: n means the rows of nails and x *means* the final position. The probability that the ball, at the end of its run between the 6 rows of nails, ends up in the central groove (the third) will be:

$$P(3) = \frac{6!}{3! \bullet (6-3)!} 0.5^3 0.5^3 = 0.3125$$

The problem posed by this game has links to situations that are easily found in everyday life.

You could list many cases, but let's settle for one for all: the problem of overbooking.

Overbooking stems from an airline's need to get the most out of a flight in order to get the most out of it. Most of the time no one notices anything because there is almost always someone who gives up or misses the flight they booked. The average number of passengers affected by overbooking in the United States is estimated to be 9 out of 10000. In these cases the airline has to reschedule the passenger on another flight and pay a penalty by offering him/her a voucher that can be spent later.

We try to quantify the probability that there are more passengers than the capacity of the plane. Let's take a 170-seat Boeing 737-800 as a model and assume that the airline may have sold 171, 172, ... tickets. What is the probability that at least one person cannot be boarded?

We can schematize the situation using the binomial model. We must, of course, accept two assumptions: the behavior of one passenger is independent of that of the others and the probability of showing up or not showing up for boarding is always the same. We can outline the situation in this way:

1. number of trials = number of tickets sold, $n > 170$;
2. probability of success = probability of showing up for boarding, $p = 0.99$.

We want to calculate the probability that at least one passenger will not find a seat on the plane (overbooking).

$$P(X > 170) = \sum_{k=171}^{n>170} \binom{n}{k} 0.99^k \bullet 0.01^{n-k}$$

Everything is summarized in Table 1:

Conclusion: overbooking is an advantageous practice for airlines, for two more tickets sold the probability of leaving someone on the ground, with the consequences seen above, does not reach 50%.

Table 1. Calculation of overbooking probabilities

# tickets sold	# Present at boarding	Binomial Distribution	p (at least one on the ground)
171	171	0.179	0.179
172	171	0.308	0.486
172	172	0.178	
173	171	0.267	0.750
173	172	0.307	
173	173	0.176	
174	171	0.155	0.902
174	172	0.267	
174	173	0.306	
174	174	0.174	
Aircraft capacity = 170 seats			

WIN FOR LIFE

One of the relatively newer lotteries is the Win for Life. Launched for the first time in the U.S., it provided that players could choose six numbers between 1 and 42, then seven numbers were drawn including the one called "Free Ball". The game was also introduced in Italy in 2009 and we will discuss this version (Figure 16).

Playing is simple, just choose at least 10 numbers out of 20 plus the "Numerone" (Free ball).

The minimum cost of the minimum bet, equal to a combination, is €1 and allows you to participate in the winning categories: 9+1, 8+1, 7+1 alongside the classic: 10+1, 10, 9, 8 and 7.

Playing 2€ you can also participate in the categories: 9+1 and 1+1, 8+1 and 2+1, 7+1 and 3+1 alongside the classic: 10+1 and 0+1, 10 and 0, 9 and 1, 8 and 2, 7 and 3.

By guessing 10 or 0 numbers plus the "Numerone" you win the special prize, up to €3000 per month for 20 years. As this is a totalizer game, the prizes are divided by the total number of winners.

Like all betting games, there is an obvious link to the calculation of odds. The law of probability governing the outcome of a bet ("Numerone" excluded) is *Hypergeometric Distribution.*

Figure 16. Italian Win for Life

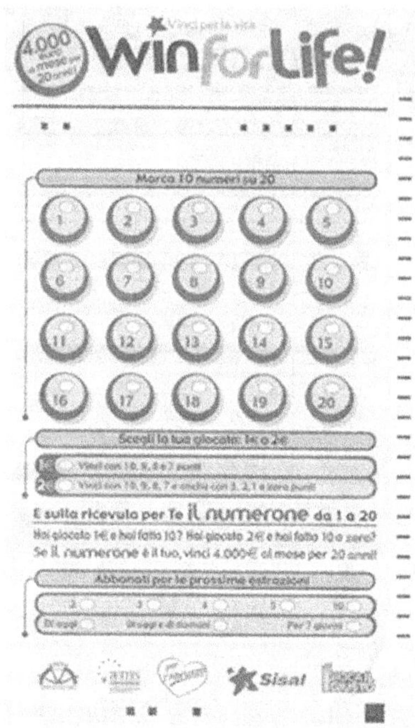

The hypergeometric distribution is a discrete probability distribution that describes the extraction without reinsertion of a number, a card, a ball, etc.. In the formula, indicating with *n* the elements, *h* those chosen, *r those* extracted and *k* those guessed:

$$P(k) = \frac{\binom{h}{k}\binom{n-h}{r-k}}{\binom{n}{r}}$$

The denominator represents the number of possible cases, while the favorable cases are given by the product between the number of possible extractions of *k* elements between *h* and the number of possible extractions of the remaining *r-k* elements between *n-h*. In the game, where out of a total of n = 20 available numbers, h = 10 are chosen by the player and r = 10 are drawn; the probability of guessing for example 4 is:

$$P(4)=\frac{\binom{20}{4}\binom{20-10}{10-4}}{\binom{20}{10}}=0.239$$

Table 2 shows the probability values for the first cases, those from 6 to 10 have symmetrical values.

Table 2. Probability distribution in Win for Life

Result	0	1	2	3	4	5
Probability	1	100	2025	14400	44100	63504
	184756	184756	184756	184756	184756	184756

Guessing "ten plus the Numerone" or "zero plus the Numerone" is 20 times more difficult. Playing € 2 the probability of winning the biggest prize is 1 / 1847560 which is equivalent to placing 1847560 tickets (each 22 cm long) on the road that connects Kansas City to Saint Louis in Missouri and choose the winning ticket by taking one at random.

We can find this distribution in many other contexts. Take a company that produces surveillance cameras, for example. Out of a sample of 150 pieces tested, 8 showed defects. What is the probability that there are no defective cameras on a batch of 50 pieces? The parameters will be n = 150, h = 8, r = 50 and k = 0. So:

$$P(0)=\frac{\binom{8}{0}\binom{150-8}{50-0}}{\binom{150}{50}}=0.035$$

We now wish to add a stimulus related to the problem of probability using the concept of distribution. In this case we apply it to a situation that in 2020 has presented itself in all its drama. In the first chapter we talked about Covid-19 in relation to the Bayes theorem, now the problem will be examined from a different angle. The focus is on the concept of risk. Apparently it is a term with obvious meaning, but things are not so simple. For example, the philosopher Sven Ove Hansson (2007) distinguishes five types of definition of risk:

1. an unwanted event which may or may not occur;
2. the cause of an unwanted event which may or may not occur;
3. the probability of an unwanted event which may or may not occur;
4. the statistical expectation value of an unwanted event which may or may not occur;
5. the fact that a decision is made under conditions of known [rather than unknown] probabilities ("decision under risk") as opposed to decision under uncertainty.

The reference to the fourth definition is the one that best approaches the problem we want to deal with: what is the probability that a person participating in an event is Covid-19 positive?

If, in the United States, 8 million out of 330 were positive, the probability of an individual not being affected by the virus is 8 / 330 = 0.976. Now, what is the probability that in an environment with 25 people (the size of a class of students) there are no positives? From a probability calculation point of view, this means winning 25 bets with that probability; calculating 0.976^{25} gives 0.545, so the probability of finding at least one positive is 1 - 0.545 = 0.455. We underline the fact that the probability that a certain number of virus positive people participate in a given event is described by a *hypergeometric distribution*; however, the use of a *binomial distribution* provides a particularly good approximation, since the number of participants in the event is much smaller than the total population. The data found obviously represents an average value. More precise analysis must take into account that there is a geographical heterogeneity that varies from county to county as shown in Figure 17. It refers to the Covid-19 Event Risk Assessment Planning Tool developed by Joshua Weitz (Professor of Biological Sciences at Georgia Tech). The source is https://covid19risk.biosci.gatech.edu/.

After choosing the group size (25) and placing the Ascertainment Bias (evaluation error that occurs when results are distorted due to factors leading to unrepresentative samples) at 5, when the mouse is hovered over the map, a popup shows the name of the county and the probability (in terms of percentage) that an individual is infected. As you can see from the different colors we have Garfield County, MT with a value <1% and, not far away, Glacier County, MT with an estimate >99%. New York City has a value of 11%.

Figure 17. Covid-19 Risk Map

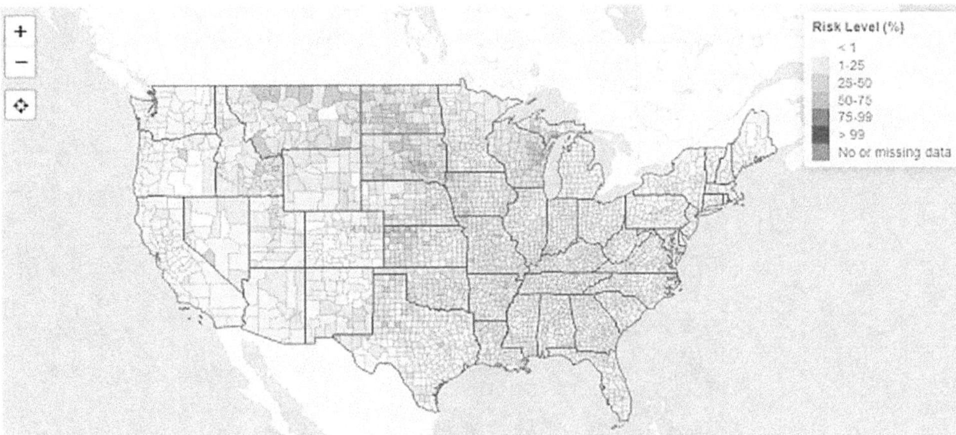

It is important to emphasize that the probability reported does not express the risk of contracting the disease, it then depends on factors that consider the environment of the event, the duration of contact and the precautionary measures taken such as the use of masks and respect for physical distancing.

BOTTLE BULBS

It is well known that the bulb is an electrical device designed to produce light. In this case it is a small bottle in the shape of a bulb with a straw attached (Figure 18). Bulbs are a classic example of a product used in probability issues. In particular, these are examples related to a particular type of distribution and the Bayes theorem.

There are phenomena that rarely occur, in a certain period of time, such as the simultaneous arrival of several people at a counter, the granting of more penalties in the football league, or the duration of a light bulb. A similar argument can be made by taking space into account, for example the number of large hailstones falling on one m^2 of land or the presence of several predatory fish in a river divided into 100-meter sections.

Situations of this type require appropriate modeling. Said $\lambda = n \bullet p$ a positive parameter equal to the average of the successes while *e* represents the already known number of Euler which is 2.718.... we can define the *Poisson Distribution*:

Figure 18. Bottle in the shape of a light bulb

$$P(x)=\lambda^x \frac{e^{-\lambda}}{x!}$$

The distribution is named after the mathematician Siméon-Denis Poisson (1781-1840). It refers to independent repeated tests when the results of each test are only two, the number of tests n is extremely high and the probability p is small. This distribution represents a particularly useful approximation of the binomial distribution. This distribution is also known, as Hand (2015) says, as a *large number of rare events*. The most interesting aspect is that the historically most famous and cited examples refer to very particular situations: the number of soldiers of 10 Prussian regiments who died annually for the kick of a mule in 20 years, reported by Ladislaus von Bortkiewicz (1898) and the distribution of V1 fallen over London during the Second World War. The examples are so well known that it is not worth repeating them. Here we deal with two typical school examples because they have as their object the light bulbs, one referring specifically to Poisson and the other that recalls the Bayes Theorem.

Example 1: A large company wants to replace all the bulbs in its offices. It turns to a factory that has a percentage of defective bulbs of 3%. What is the probability that out of 500 bulbs purchased, 10 are not defective?

First of all you can calculate the average of defective bulbs: $\lambda = n \bullet p = 500 \bullet 0.03 = 15$. Applying the formula will result in the following:

$$P(10) = 15^{10} \frac{e^{-15}}{10!} = 0.05$$

Example 2: the factory in the previous example has two production lines: the first one produces 60% of the bulbs produced and has a percentage of defective bulbs of 2%; the second one supplies the remaining 40% with a percentage of defective bulbs of 4.5%. What is the probability that one of the defective bulbs purchased by the company comes from the first line (1)?

Applying Bayes you have:

$$p(1|d) = \frac{p(1) \bullet p(d|1)}{p(1) \bullet p(d|1) + p(2) \bullet p(d|2)} = 0.4$$

To go deeper into the Bayesian approach we can examine a classic problem that has a light bulb as its protagonist. Suppose that a detector can identify particles with an efficiency of 95% and a probability of confusing a particle of 2%. If a particle is identified as a muon (μ) a light bulb (L) lights up. Knowing that the particle beam contains 10% of μ, we ask ourselves: What is the probability that a muon has passed through if the light bulb is lit ?

$P(L|\mu)=0.98$ is the probability that the bulb will turn on if it passes μ;

$P(L|\bar{\mu}) = 0,01$ is the probability that the light bulb will turn on by mistake;

$P_0(\mu)=0.10$ is the a priori probability;

$P(\mu|L)$ is the probability that the particle is μ if the bulb is lit, applying the Bayes Theorem:

$$P(\mu|L)=\frac{0.98 \bullet 0.10}{0.98 \bullet 0.10+0.01 \bullet 0.10}=0.99$$

Poisson's distribution is applicable in many situations, for example in football. There is a betting method that refers to the famous mathematician.

On the net you can find a simple calculation tool that, based on the *goal expectancy of* two teams in a particular match and using the Poisson function, calculates the percentage of probability of victory for the two teams. If you take for example a match of the Italian Serie A football championship the procedure is quite long but not particularly complex, having the statistics available.

Poisson's betting method is built from 4 coefficients: home attack, away attack, home defense and away defense, i.e. the general estimate of the offensive propensity (attack coefficient) and defensive strength (defense coefficient) of the two teams.

The figures for the year under consideration are, for all teams in the league, taking into account that 380 matches were played:

1. Coefficient of attack at home = total goals scored at home / no. of matches = 564 / 380 = 1.48
2. Coefficient of attack away from home = total goals scored away from home / no. of matches = 455 / 380 = 1.20
3. Home defense coefficient = total goals conceded at home / no. of matches = 455 / 380 = 1.20
4. Defense coefficient away from home = total goals conceded away from home / no. of matches = 564 / 380 = 1.48

Now it is a question of estimating the goals that the home team (A) against the team (B) will presumably score, taking into account that 19 games were played at home and 19 away games:

1. Average goals scored by A at home = home goals / no. of home games played = 32 / 19 = 1.68
2. Average goals conceded by B away from home = goals conceded away from home / no. of matches played away from home = 17 / 19 = 0.89
3. Attack value of A = average home goals / home attack coefficient = 1.68 / 1.48 = 1.14

4. Defense value of B = average goals conceded away from home / defense coefficient away from home = 0.89 / 1.48 = 0.60

These values allow to determine the *Home Goal Expectancy*: attack value of A • defense value of B • overall home attack coefficient = 1.14 • 0.60 • 1.48 = 1.01

Now it is a question of estimating the goals that the visiting team (B) opposite the home team (A) will presumably score.

1. Average goals scored by B away from home = goals scored away from home / no. of matches played away from home = 31 / 19 = 1.63
2. Average goals conceded by A at home = goals conceded at home / No. of home games played = 19 / 19 = 1.00
3. Attack value of B = average goals scored outside the home / coefficient of attack outside the home = 1.63 / 1.20 = 1.36
4. Defense value of A = average home goals conceded / home defense coefficient = 1 / 1.20 = 0.83

These values allow to determine the *Away Goal Expectancy*: attack value of B • defense value of A • average away goal = 1.36 • 0.83 • 1.20 = 1.35.

As you can see, the difference between the two goal expectancy values suggests a superiority of the visiting team.. After determining the values you can calculate the probabilities of the three possible results using, for example, a Poisson calculator tool (Retrieved from https://sinceawin.com/data/tools/poisson): Victory of A 28.0; victory of B 45.6; tie 27.4.

Let us now examine a second problem.

If you consider the Premier League 2018-19 there was an average of 2.82 goals per game. Taking into account the duration of a match, the probability of seeing a goal in a given minute is 0.03. Using this information we can try to superimpose Poisson's distribution on the histogram of the goals scored (Figure 19). We can notice the interesting correspondence between the model and reality. The football match is a random event but it is possible to determine a precise pattern. That it is not a case is given by the fact that also in previous years and in other leagues you can notice the same pattern.

Figure 19. Comparison between model and reality

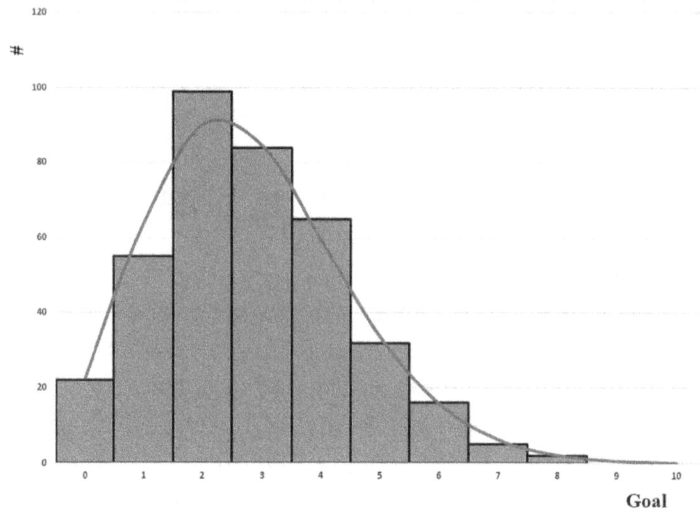

TOOTHBRUSH

The toothbrush is an oral hygiene tool consisting of a small synthetic fiber brush fixed on a handle. Toothbrushes are available in different sizes and shapes. In Figure 20 the toothbrush consists of two faces.

Figure 20. Two-sided toothbrush

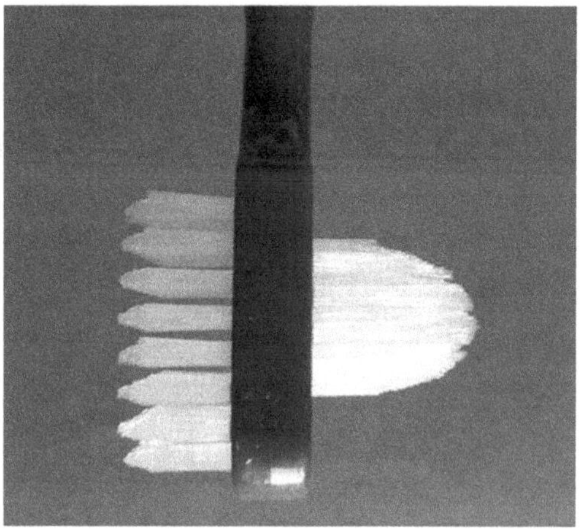

What is interesting is that the shape of the bristles reminds us of two probability distributions: on the left the uniform distribution that, with a forcing and reasoning in terms of continuity, can be approached to the *Rectangular distribution*, and on the right the *Normal distribution*.

A small consideration about Rectangular Distribution. It is certainly an important type of distribution that, in some practical cases, can be usefully simplified in terms of calculation, but it should be analyzed carefully. For example, if you consider calls to a call center throughout the day, the assumption that the probability distribution is uniform appears to be poorly credible unless the call center operates on a global scale.

The continuous probability distributions are expressed by three factors: a mathematical law called probability density function, *mean* μ and *variance* σ^2.

The mean is a central trend measure too well known to require definition; variance is a measure of the variability of the values assumed by the variable itself; it indicates how far these values deviate from the arithmetic mean. Since the variance is a second-degree quantity and this poses problems of homogeneity in the measurements, it is often preferred to use the standard deviation which is the square root of the variance and is indicated with σ.

This information is summarized in Table 3:

Table 3. Parameters of the two distributions

Distribution	Density function	Media	Variance
Rectangular	$\frac{1}{b-a}$	$\frac{a+b}{2}$	$\frac{(b-a)^2}{12}$
Normal	$\frac{1}{\sqrt{2\pi\sigma^2}}e^{-\frac{1}{2}(\frac{(x-\mu)^2}{\sigma^2}}$	μ	σ^2

Using the generation of random numbers from a spreadsheet, the two distributions can be verified. For example, for n = 10000 you could have the following rectangular distribution (Figure 21):

The next picture shows the typical bell shape of the normal distribution.

This is the most widely used continuous distribution, identified for the first time by the aforementioned De Moivre as a means of roughly assessing the binomial. Since it was rediscovered by Carl Friedrich Gauss (1777-

1855) within the error theory, it is often referred to as Gaussian distribution. Many physical, biological, social, and economic situations have a frequency distribution that fits well with the theoretical distribution (Figure 22).

Figure 21. Rectangular distribution

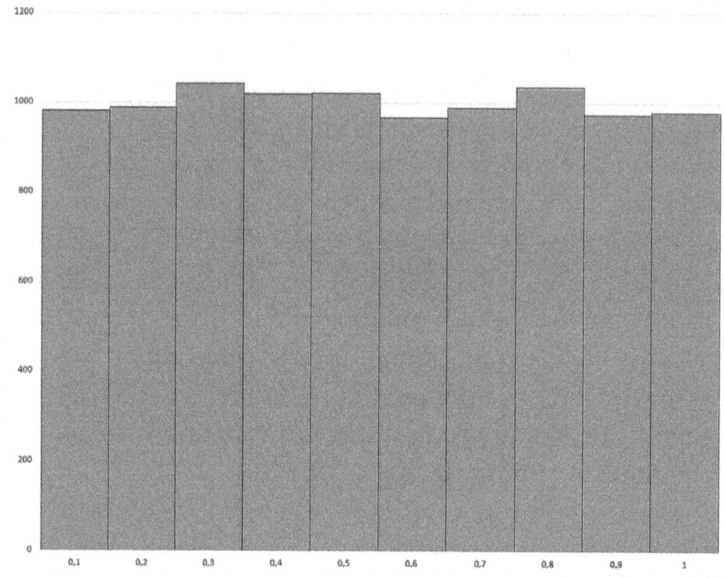

Figure 22. Normal distribution

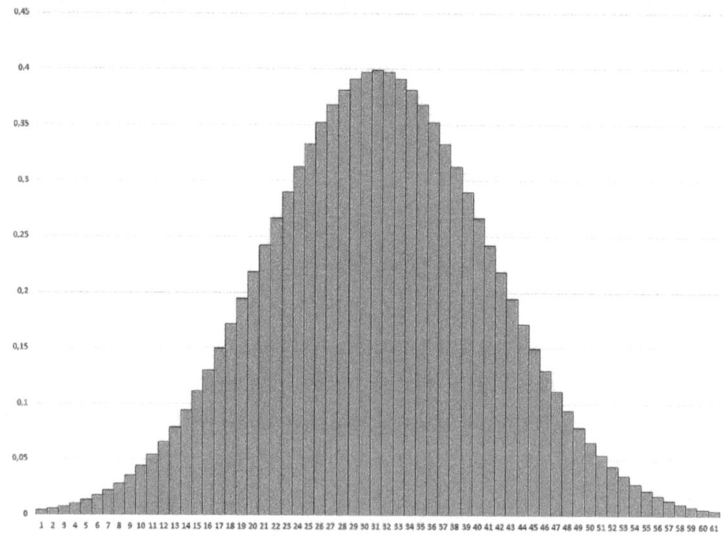

The curve, also known as a bell curve, represents a pattern of random error distribution. The life of Poincaré gives us a curious application example (Bellos, 2010). Poincaré went to the same bakery every day to buy a loaf of one kilo, but one day he began to suspect that the baker was deceiving him and so he carefully weighed the bread for a year. The form of bread he bought every day weighed only about 950 grams, with an approximate Gaussian distribution and a standard deviation of about 50 grams. After being reported to the authorities, the average weight of the daily loaves purchased became one kilo. Not content, Poincaré continued to weigh the loaves and observed that the shape of the distribution was quite different from the expected curve; the lighter loaves were missing although the average value was actually one kilo. The baker had quietly continued to defraud, but taking care to provide Poincaré with the heaviest form of bread!

One area in which the distributions are studied is financial: the hypothesis is that the phenomena can be expressed with a Gaussian curve. This assumes that the price of a share is well defined around the average value and with a low probability that large deviations will occur. Unfortunately, the painful experience of crises has shown that these deviations are not so unlikely and that surprises are always just around the corner. Nassim Nicholas Taleb (2007) has condensed the phenomenon into the *black swan* term. The black swan theory is a metaphor that describes an unforeseen event, with an important effect, and is only rationalized in retrospect. The expression is based on the fact that until the 17th century all known swans were white and therefore the statement "all swans are white" was normal. Only in 1697 were black swans discovered in Australia. According to Taleb (2007), a black swan cannot be predicted or imagined and when it arrives, it is often not even recognized for what it is.

After this digression, we return to the Gaussian. To determine the probability in a given range it is necessary to calculate the area under the curve. However, the mathematical expression is very laborious to calculate, so it would be useful to have tables with the required probabilities. Unfortunately, there is an infinite number of combinations of the parameters μ and σ, but fortunately this can be remedied with an operation called standardization.

Standard normal distribution is a normal distribution obtained by expressing the variable x in terms of standard units z:

$$z = \frac{x - \mu}{\sigma}$$

These linear transformations retain the area below the curve, but allow each distribution to be referred to a single numerical table.

The importance of standardization can be clarified by an example. Let's resume the examination by going back to talking about light bulbs. Let's imagine that the life of certain bulbs is normally distributed with $\mu = 100$ hours and $\sigma = 8$ hours. What is the probability that a lamp taken at random will last between 110 and 120 hours?

The apparently difficult problem is solved as follows: the values of x 110 and 120 are transformed into z 1.25 and 2.5. From special tables we obtain that $P(z_a) = 0.8944$ and $P(z_b) = 0.9938$. Then we can calculate the probability value of the desired area by operating by difference between the two areas

$$P(110 \leq x \leq 120) = 0.9938 - 0.8944 = 0.0994$$

The calculations appear more understandable when looking at Figure 23.

Figure 23. Normalized graph

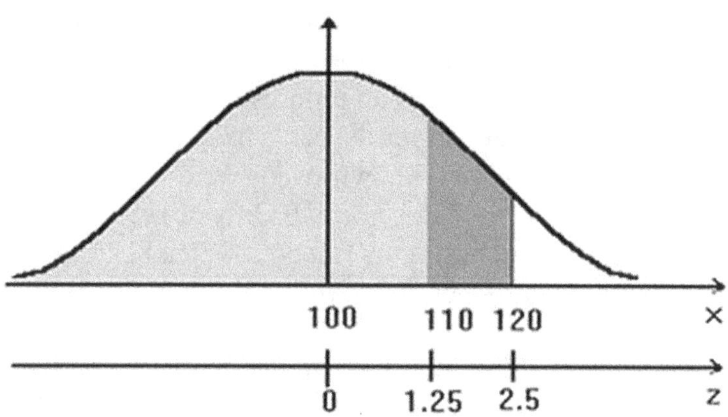

CONCLUSION

In the chapter, the term random variable has returned several times. This word may not be well understood for several reasons: "the random variable is assimilated with a random experiment, the random variable is confused with its image universe (the most common confusion), the random variable refers to an arbitrary variable, the random variable is confused with its law of probabilities". (Amrani & Zaki, 2015). The two researchers examined some wrong answers (in whole or in part) and two seem particularly interesting. The first one refers to a student's sentence stating a partial truth as "The random variable is a continuous function" and cites instead a correct example "The diameters of the tubes in a factory". The second sentence, decidedly false, is "A random variable is a variable that helps us calculate a particular probability" and the example reported in support " Let *X* be a random variable that determines the number of natural childbirths among women" is also incorrect.

These, and other conceptual errors are probably caused by overly formal teaching that is not adequately supported by an intuitive approach. In this sense the production of concrete examples can be of great help in correcting some misunderstandings.

FUTURE RESEARCH DIRECTIONS

The objects present in the chapter have favored the introduction of the most common distributions; in the future we could expand the discourse from the normal distribution. The already mentioned Taleb (2007) defines the bell curve as a great intellectual fraud. This is because, in his opinion, social phenomena are impossible to model with the normal Gaussian distribution because of feedback cycles. The extraordinarily rich are not only a little bit richer, but they are so abnormally rich that they cause a distortion of the distribution. A widespread idea is that one could, in certain cases, replace the normal distribution of Cauchy (1789-1857) which has more pronounced tails. Of course we are hypothesizing a simple qualitative reflection since this distribution, used in many fields including financial analysis, goes beyond the limits of the book.

REFERENCES

Amrani, H., & Zaki, M. (2015). Student's Conceptual Difficulties With respect To the Notion of Random Variable. *International Journal of Education, Learning and Development*, *3*(9), 65–81.

Audoly, B., & Neukirch, S. (2005). Fragmentation of rods by cascading cracks: Why spaghetti does not break in half. *Physical Review Letters*, *95*(9), 095505. doi:10.1103/PhysRevLett.95.095505 PMID:16197227

Bellos, A. (2020). *Alex's adventures in Numberland*. Bloomsbury Publishing PLC.

Bortkiewicz, L. (1898). *The law of small numbers*. BG Teubner.

Chande, A., Lee, S., Harris, M., Hilley, T., Andris, C., & Weitz, J. S. (2020). *Real-time, interactive website for US-county level Covid-19 event risk assessment*. medRxiv.

Eigen, M., & Winkler, R. (1993). *Laws of the Game: How the Principles of Nature Govern Chance*. Princeton University Press.

Hand, D. J. (2015). *The Improbability Principle: Why coincidences, miracles and rare events happen all the time*. Transworld.

Hansson, S. O. (2007). Risk. In E. N. Zalta (Ed.), Stanford Encyclopedia of Science. Academic Press.

Hoffmann, P. M. (2012). *Life's ratchet: how molecular machines extract order from chaos*. Basic Books.

Steinbring, H. (1984). *Mathematical concepts in didactical situations as complex systems: The case of probability*. Theory of mathematics education (TME: ICME 5): Occasional paper, 54.

Taleb, N. N. (2007). *The black swan: The impact of the highly improbable* (Vol. 2). Random house.

Wykes, A. (1964). *The complete illustrated guide to gambling*. Doubleday Books.

ADDITIONAL READING

Papoulis, A., & Pillai, S. U. (2002). *Probability, random variables, and stochastic processes*. Tata McGraw-Hill Education.

Sánchez, E., & Landín, P. R. (2014). Levels of probabilistic reasoning of high school students about binomial problems. In *Probabilistic Thinking* (pp. 581–597). Springer. doi:10.1007/978-94-007-7155-0_31

KEY TERMS AND DEFINITIONS

Binomial Distribution: Discrete distribution of the number of successes in a sequence of independent experiments yes-no.

Geometric Distribution: Distribution of the number of failures before the first success.

Hypergeometric Distribution: Discrete distribution describing the extraction without reinsertion of certain objects.

Large Number of Rare Events: Frequency distribution estimation when rare events are common.

Normal Distribution: The most famous and used of the continuous distributions.

Poisson Distribution: Discrete distribution expressing the probabilities for the number of events occurring successively and independently in a given time interval.

Rectangular Distribution: Continuous distribution attributing the same probability to all points belonging to a given range.

Singular Distribution: Assumes values only in particular measurement sets.

Standard Normal Distribution: Is obtained by a transformation of variable X of a normal curve.

Uniform Distribution: Discrete distribution attributing the same probability to all points belonging to a given range.

Chapter 7
Probability and Strategies

ABSTRACT

When you participate in a game, but in general when you compete with one or more opponents, it is essential to define strategic behavior. Because probability often plays a central role, it is important to learn how to profitably manage the information we have. The chapter presents a multiplicity of games, more or less known. The whole allows us to take up some already known concepts and to suggest new ways of deepening them. It should be noted that the center of the analysis is on probability and we do not want to focus only on games of chance, even if they constitute a field historically very studied.

INTRODUCTION

Always be a little improbable. ~ Oscar Wilde

The concept of strategy, understood as the ability to achieve a given objective by providing the means to do so, is fundamental in all aspects of daily life; all the more so when not all the elements in our possession can be determined with certainty. Numerous experiments have shown that mice also adopt a refined behavioral model based on the calculation of probabilities when it comes to waiting for a reward. The literature on the subject of strategies related to random games is very extensive, for a particular study of Stochastic and Bayesian Games you can refer to the text by Leyton-Brown, & Shoham (2008).

Strategy plays an important role during the games. In a game there are three key concepts:

DOI: 10.4018/978-1-7998-3871-5.ch007

1. the scenario, i.e., the environment in which the game takes place;
2. the randomness, which intervenes to a greater or lesser extent in most games;
3. the bet, which may be absent or present as in gambling.

Two cases can be distinguished here: *Pure strategy* completely defines the way a player plays a game, so what choice he will make in each situation he can find; *Mixed strategy* is a distribution of probability over all the available ones. In practice, a player can choose any pure strategy with $0 \leq p \leq 1$.

A particular strategic game in its simplicity is *skunk*, proposed in a didactic key by Dan Brutlag (1994) but originally launched in 1953 by William Herbert Schaper as a board game in a box. The term indicates the animal but, in slang, means to defeat heavily, annihilate the opponent. Before starting the game each competitor has a sheet with five columns on which to write the results (Table 1).

Table 1. Initial skunk table

S	K	U	N	K

The object of the game is to accumulate points by rolling two dice. The points are accumulated by making several throws, but by choosing to stop before the face with the value one comes out on one of the dice (a skunk in the original version). In this situation, the points of the turn are deleted. If a double one is rolled, then all the points accumulated in the previous columns are deleted.

The game itself is not complex, but it can be a starting point to answer some questions that imply a reflection on some not secondary aspects. What are the odds of throwing two dice, subject to certain constraints? Was the winner just lucky? Which strategy should be used in the game?

If we try to analyze the space of events, it is easy to see that the probability of double one is 1/36, that of simple one is 10/36 and that a favorable throw has a probability of 25/36. If we multiply each positive result by the probability of its achievement we see that the average score is 8 (Table 2).

Table 2. Calculating the average score for the skunk game

	1	2	3	4	5	6						
1	**0**	**0**	**0**	**0**	**0**	**0**						
2	**0**	**4**	5	**6**	**7**	**8**	0.16	0.20	0.24	0.28	0.32	
3	**0**	**5**	**6**	**7**	**8**	**9**	0.20	0.24	0.28	0.32	0.36	
4	**0**	**6**	**7**	**8**	**9**	**10**	0.24	0.28	0.32	0.36	0.40	
5	**0**	**7**	**8**	**9**	**10**	**11**	0.28	0.32	0.36	0.40	0.44	
6	**0**	**8**	**9**	**10**	**11**	**12**	0.32	0.36	0.40	0.44	0.48	
											Σ	**8**

The cumulative gain is $G = (25/36)^n \cdot 8n$. This gain begins to decrease when you pass the three throws as shown in Figure 1.

Figure 1. Skunk earnings graph

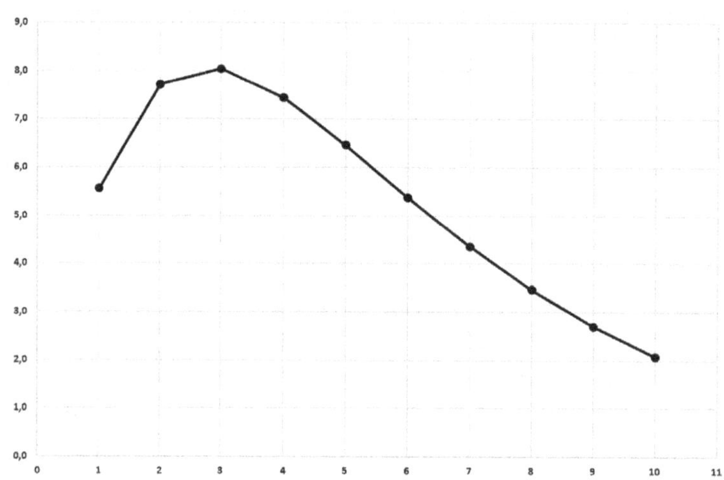

The best strategy, unless you believe in your shameless luck, is to stop after the third launch.

Strategy problems often arise in real life. An example that occurs quite frequently is the following: given the situation shown in Figure 2, where is it best to park the car to reduce the probability of having the door dented by the car door next to it?

Figure 2. Parking problem

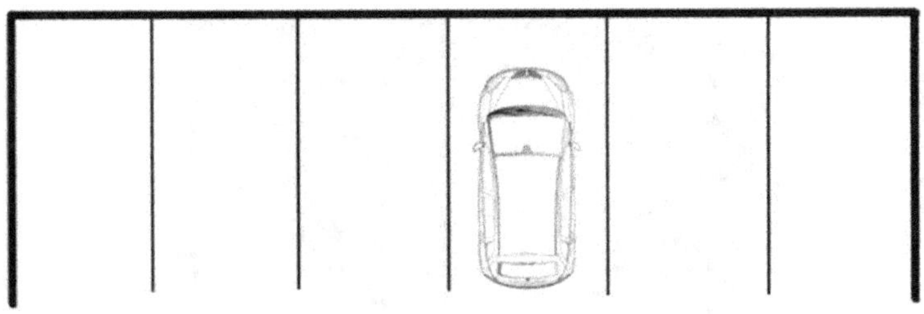

This question can be seen as a problem of strategy to be adopted when faced with a multitude of possibilities. In order to be able to answer it, it is necessary to bear in mind an essential element: the probability that the car next to ours has a passenger in addition to the driver. According to some Italian statistics, car occupancy is 1.33; this obviously does not mean that in every car there is a third passenger, but that there are more cars with just the driver. At this point the answer to the question is clear, if you occupy the most right-hand stall the probability that there is a passenger drops to 1/3 and therefore the risks are lower.

PERUDO

Perudo is an ancient Peruvian dice game, dating back to the Incas. The game became famous because it was featured in the movie Pirates of the Caribbean: *Dead Man's Chest*. It looks like a game for children, but it's a complex strategy game with a probabilistic basis.

Each player has 5 dice and one glass (Figure 3); all players shake the dice in their glass and place the latter upside down on the table: they can look at their own dice without showing them to the others. The goal is to keep their own dice: whoever rolls them first loses.

The first player chooses any value of the dice and declares the number of dice in play with that value, for example "on the table there are five dice of value 3". The next player has two choices: if he thinks it is true, he must raise it. In this case, he can bid either the same number of dice with the highest value, or a number of dice higher than any value, then he could bid "six dice with value 3" or "five dice with value 5". If a player does not believe the previous competitor's bid says *Dudo*, that is, I doubt it. In this case everyone discovers their dice and a check is made: if the bid is correct

Figure 3. Perudo materials

(number of dice equal or higher), the doubted player loses a dice; if it does not match (lower number of dice), the player who made the last bid loses a dice. There is one aspect that makes the game more complicated and makes it more difficult to formulate winning hypotheses: the value 1 of the dice is replaced by an image (the *Paco* toucan), it acts as a Joker in the counting of the dice declared and therefore can significantly change the sum of the faces. Normally the number of dice can only be increased, but Paco allows a variant. It is possible to raise by declaring Paco and halving (with rounding up) the number of dice declared by its predecessor. Example: one player bids "six 5", the next player can raise with "three Paco", the next player can choose to increase the number of Paco or change the value of the dice. There are other rules but they do not substantially change the concept behind the game: adopt a strategy that takes into account both the theoretical probability distribution of the dice and the influence of the *Paco*. The most likely declaration from the probabilistic point of view is the expected value of the faces, taking into account the conditioning of the declarations of the other players. Of course you also have to suspect bluffing and this brings the game closer to poker.

The researchers Bartolomei, Giambalvo & Macaluso (2018) presented this game highlighting the criteria that should guide the players in the declaration and evaluation of the declarations of others in the framework of an educational activity, held in Palermo (Italy), included in the National Scientific Graduation Plan (PNLS). Particularly interesting is the quantitative evaluation, based on the number of dice present, of the declarations to be doubted in order to obtain a maximum probability of error of 5% or to be accepted to obtain the maximum probability of victory. In the first case the average value is almost 97% while in the second case the value drops to 76%, so it is more convenient to doubt. An interesting example of mixing probability and strategy.

ELEPHANT

The elephant, of which we can admire a reproduction in Figure 4, is certainly a much-loved animal, the images of a herd of elephants that cares for their young softens us, as well as the story of the flying elephant Dumbo (the reference is to the Disney cartoon) is full of pathos. The elephant is also synonymous with an animal with a proverbial memory, which implies that it can also be extremely vindictive.

Figure 4. Elephant

This animal can lead us to a game that can study the mechanisms of cooperation or conflict. The theoretical reference is the famous *Prisoner's dilemma*. The dilemma can be described as a kind of paradox involving two criminals arrested on charges of committing a crime. The criminals, who cannot communicate, are given two choices: to cooperate or not to cooperate. They are also explained that:

1. if only one of them cooperates by accusing the other, the other one is sentenced to many years in prison;
2. if they both charge the other, they're both sentenced to a lower sentence;
3. if neither collaborates, both are sentenced to 1 year, because the charge is reduced to an illegal port of arms.

This dilemma has been analyzed by numerous scholars, many of whom have posed the problem of how behavior could evolve in case of not a single event, but a possible repetition. Among these scholars, the contribution of Robert Axelrod (1981) is particularly relevant, he highlighted how a strategy called *Tit for Tat* is amazingly effective. This strategy involves cooperating initially and then copying the opponent's behavior in the immediately preceding period. In this way one adjusts one's behavior to the actions of the other without voluntarily taking an initiative of betrayal.

Let us now adapt the problem to the animal world by introducing the possibility of random choices following a suggestion of Delahaye (1988). Four animals are taken into consideration, each one has a characteristic that induces it to pursue a strategy (Table 3) that can be of collaboration (C) or betrayal (B):

Table 3. Animal behavior

ANIMAL	CHARACTER	STRATEGY
rhino	moody	Cooperates, then chooses C (1) or B (0) randomly
donkey	methodical	The first choice is random, then makes the opposite move to the previous one for 10 times, then starts again
monkey	quite cooperative	Cooperate, Cooperate, Betrays alternately
elephant	vindictive	Cooperates in the first three steps; later, if the opponent cooperated in steps 2 and 3, Cooperate, otherwise Betray

By indicating with 1 the collaboration and 0 the betrayal, as if it were heads or tails, you can observe the results in the long run. This competition can be seen as a game: if the behavior is identical (C-C, T-T) both participants do not score points, otherwise whoever betrays wins 1 point while the other loses it.

A spreadsheet can serve the purpose (Figure 5).

Figure 5. Spreadsheet for C-B simulation

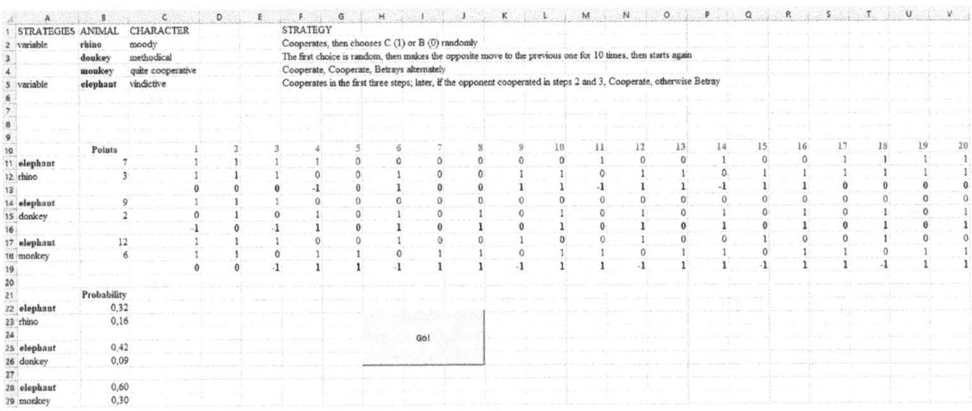

A small macro is sufficient to obtain the desired results:

```
Sub Macro1()
n = InputBox("# of games")
    For i = 1 To n
        Calculate
        el1 = el1 + Range("B11")
        rhinoceros = rhino + Range("B12")
        el2 = el2 + Range("B14")
        donkey = donkey + Range("B15")
        el3 = el3 + Range("B17")
        monkey = monkey + Range("B18")
    Next i
        Range("B22") = (el1 / 100) / n
        Range("B23") = (rhino / 100) / n
        Range("B25") = (el2 / 100) / n
        Range("B26") = (donkey / 100) / n
        Range("B28") = (el3 / 100) / n
        Range("B29") = (monkey / 100) / n
End Sub
```

Assuming that each clash between the elephant and one of the other animals lasts 20 times and that the event is repeated 1000 times, you can get results like those shown in the Figure above (Table 4):

Table 4. Probable animal behavior

	Probability of Collaboration
elephant	0.32
rhino	0.16
elephant	0.42
donkey	0.09
elephant	0.60
monkey	0.30

This introduced the concept of simulation, which will be discussed in the next chapter.

ROCK, PAPER, SCISSORS

This is a classic game of Chinese origin. There are three symbols available: Rock symbolized by the hand closed with a fist; Paper symbolized by an open hand; Scissors, a closed hand with index and middle finger to form a "V" (Figure 6). The aim is to choose a sign capable of beating that of the opponent, according to four simple rules:

Figure 6. Rock, paper, and scissor on the fingers

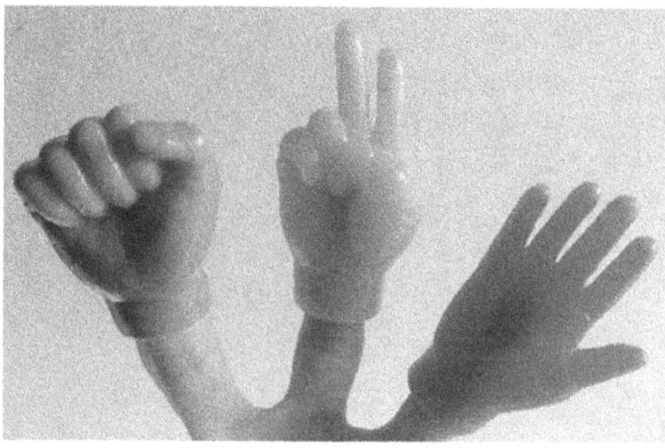

1. The rock breaks the scissors;
2. Scissors cut paper;
3. Paper wraps rock;
4. If the two players choose the same object, the game is even.

The game is, at least in theory, aleatory: every turn the odds of success against the opponent are reset to zero. If the two players adopt a purely random behavior, it is not possible to establish a strategy. However, various studies show that, usually, players are not robots guided by chance, but apply their game strategy. It is quite clear that if an opponent adopts a truly random strategy, it is impossible to look for an advantage, but a sly player can exploit the psychological attitudes of the opponents in order to achieve better results. Someone has observed that scissors are played less often than paper and rock: precisely 29.6% of the time, while it should happen a third of the time, i.e. 33.3%. According to research by the University of Zhejiang there is a good chance that if a player wins one round, in the next round he will play the same thing; if he loses, it is likely that he will play the sign that he would have won against the one he has just played. In practice, if he wins with scissors, he will play scissors again; if he loses with scissors, he will play rock. Based on these considerations, several algorithms have been proposed that are basically based on the analysis of past frequencies. A different approach has been proposed by other researchers. They defined as "increase" the passage in the sequence paper → scissors → rock, that is the movement from left to right; as "reduction" if you move in the opposite direction, that is from right to left.

The best strategy against a human would seem to be the following:

1. If I lose, I increase his move (I'll use scissors if I've been beaten by paper against rock);
2. If I win, I use his losing move (I'll use rock if I won paper against rock);
3. If it ends up tied, I'll decrease the move in the next round compared to the one played before (I'll use paper if it ended up tied with scissors against scissors).

Dance & Jackson proposed in 2010 in the New York Times a site where you can challenge a computer that adopts two different modes: novice, i.e. the computer learns to play from scratch and uses a strategy on a statistical basis; veteran, i.e. the computer acts considering the sequences played based on the experience of 200,000 games. Even if the computer doesn't always win, it can use a person's patterns to gain an advantage.

MATCHBOX

The match is a simple object, now little used, used to get fire safely: it is a stick with one end covered with a flammable substance that ignites by rubbing on a rough surface. The matches are contained in a box.

Matchboxes remind us of a problem known as the *Banach problem*. A smoker holds two matchboxes in two different pockets. To light a cigarette, he chooses one pocket at random. The question is: in each box there are initially *n* matches, what is the probability that, finding one empty box, the other contains *k* matches? (Frosini, 1990).

The number of extractions that an empty box produces while the other still has matches is the sum of n+1 extractions for the empty box and n-k extractions for the other. In total we can say that we are faced with 2n+1-k extractions. By indicating with p(k) the probability to stay with *k* matches in the other box, when you take for the first time an empty box you have:

$$p(k) = \frac{(2n-k)!}{n!(n-k)! \bullet 2^{2n-k}}$$

If we take into consideration the random variable that assumes the values given by the remaining matches, we can write Table 5, for n = 20 and considering only the first 10 values of k:

Table 5. Distribution of the random variable k in the match problem

k	0	1	2	3	4	5	6	7	8	9	10
p(k)	0.125	0.125	0.122	0.116	0.106	0.095	0.081	0.067	0.053	0.039	0.028

To which the graph in Figure 7 corresponds:

Let's examine a strange game between a competitor and a host. Suppose we have a matchbox divided into three compartments (Figure 8). In one there is a matchbox and the other two are empty. The goal is to find the match. The player chooses, without opening it, a drawer: the host opens another one, showing that it is empty. At this point the competitor can, if he wishes, change his choice. What should he do?

Figure 7. Graph of the random variable k in the match problem

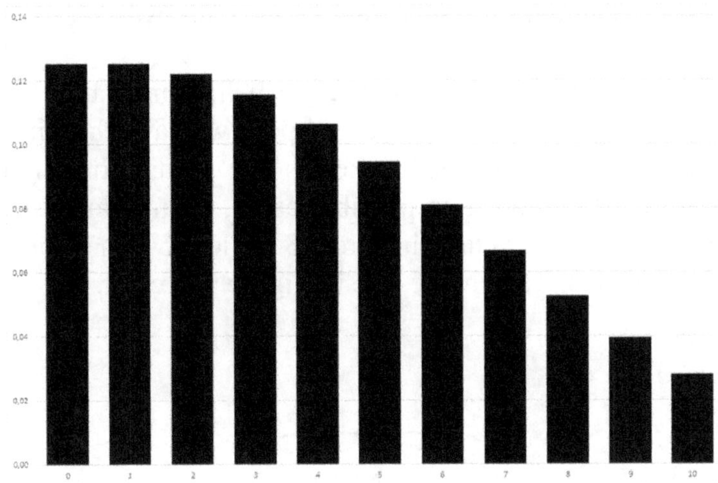

Figure 8. Box with a match as a prize

The game is based on a famous problem of probability theory, related to an American game of chance. It is named after the show's host, Maurice Halprin, known under the pseudonym Monty Hall. In the game three closed doors were shown to the competitor; behind one was a car, while each of the other two was hiding a goat (Isaac, 2013). The contestant selected one door, without opening it; at this point the show host (knowing what is behind each door) opened one of the other two, revealing a goat, and offered the player the chance to change his initial choice.

What is more convenient to do: keep your choice or change it?

Marilyn vos Savant, a writer with an extremely high Intelligence Quotient (IQ) solved the problem correctly but her explanation was not recognized as correct by several experts.

Let us examine the problem with the help of a diagram that well represents the solution proposed by Savant. Let us suppose we have chosen the drawer with the match. In this case change makes us lose. If instead we have chosen an empty box, and there are two possible cases, change makes us win. In practice the odds in favor of changing are 2/3 while those in favor of keeping the first choice are 1/3. The choice tree in Figure 9 is clear.

Figure 9. Choice tree

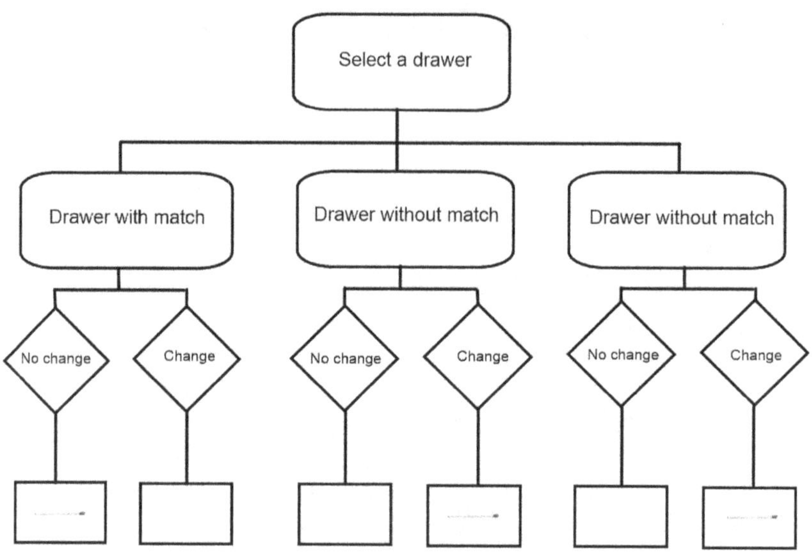

For all those with a normal IQ it can be a relief to discover that even the smartest and most educated people can take big blunders. In fact, in the December 2011 issue of Parade Magazine, this letter and its response appeared in your December 2011 column.

Question: "I manage a drug-testing program for an organization with 400 employees. Every three months, a random-number generator selects 100 names for testing. Afterward, these names go back into the selection pool. Obviously, the probability of an employee being chosen in one quarter is 25 percent. But what's the likelihood of being chosen over the course of a year? "

Answer: "The probability remains 25 percent, despite the repeated testing. One might think that as the number of tests grows, the likelihood of being chosen increases, but as long as the size of the pool remains the same, so does the probability. Goes against your intuition, doesn't it?".

Actually the correct answer is 68% as he explains in one of his books, with sharp irony, Paul Nahin (2013).

BATTLESHIP

The battleship (Figure 10) is a two-player game. To play each participant has two tables of equal size (typically 10 • 10). The table squares are identified by pairs of coordinates, corresponding to row and column; at the beginning, players must, by marking them on one of their two grids, position their ships: an aircraft carrier (size 5), a battleship (4), a submarine (3), a cruiser (3), a destroyer (2). The player on duty "fires a shot", declaring a coordinate and marking it in the second grid with the result. If he guessed the position of an opposing vessel the result is "hit", if he failed it is "miss". Whoever destroys the opposing fleet first wins.

Figure 10. Naval battle

The game is simple but lends itself to several considerations. First of all it proves that choosing a point at random is a losing strategy: the probability of playing a perfect game by randomly pulling on a grid 10 • 10 is extremely small, in the order of 10^{-7}.

Let's look at some tips for a strategically correct game:

1. Shoot in the middle of the board because it's more likely to hit a ship. The reason is simple: there are only two ways to put the ship in a corner but in the center are more;
2. A second consideration takes into account that the smallest ship covers two squares: we can see the board as a chessboard and then "shoot" only on the white (or black) squares, which halves the possible shots;
3. A less secure strategy but one that proves quite efficient is to fire one shot every three squares (instead of two); this makes us lose a few small ships but covers the grid in 34 moves instead of 50;
4. Once you get the first shot, you only have to fire in the squares immediately adjacent to it.

DARTS

The game consists of throwing darts stabilized in the flight from flaps to a target (Figure 11) so that the tip of the darts hits one of the partitions. The target is divided into 20 sectors of equal size, numbered from 1 to 20 not in order, with a center divided into 2 concentric sectors. The goal is to score the highest score.

The value of each throw is determined by the endpoint of the arrow:

1. A dart that ends outside the target is not worth any points;
2. The dart that ends in one of the colored black or white portions makes the player score the number of points indicated on the edge of the sector;
3. If you hit the outer ring, colored green or red, the score on the outer edge is doubled;
4. By hitting instead the inner ring, also colored green or red, the points of the segment are tripled;
5. The area in the center of the target, colored green, is worth 25 points, the "heart" colored red guarantees 50 points.

Figure 11. Target

This game raises several questions:

1. Why are the scores so distributed?
2. The probability of hitting a certain sector depends on what?
3. Where do you aim in the darts game to maximize your score?

The first two questions have simpler answers. In fact, it's easy to realize that:

1. The numbers are arranged in that order to reward accuracy and to "punish" poor aim. Placing low numbers next to high numbers (e.g.: 5, 20, 1) prevents inaccurate shots from producing high scores.
2. The probability is related to the width of each sector and/or ring.

More complex is to deal with the third issue, as throwing darts does not correspond to a totally random process but the player tries to maximize the result by "directing" the dart. If a good player can hope to reach the maximum score of 60 points on each throw, a less experienced player is content to aim at the center so as not to miss the target, an average player can concentrate on certain areas that could prove to be productive.

If we think we are able to target a dart within three sectors, we need to determine which portion of the target is most convenient. In fact, it is a matter of calculating the average value of the bet, for example by targeting the three sectors that are worth 10, 15 and 2 points you can total (if the dart scores) an average score of $\frac{10+15+2}{3}=9$. Brief analysis indicates that the three most favorable sectors are those with score 16, 7, 19.

MONOPOLY

Monopoly is a classic board game. Players compete to earn money through an economic activity involving the purchase, rent and trade of land properties using fake money. Players, in turn, move on the game board according to the result of the roll of two dice. There are three possibilities. The placeholder arrives on a square that:

1. does not belong to anyone; in this case it can be purchased by paying the indicated amount or auctioned;
2. belongs to the player himself, in this case he can, if he has the necessary money, build some buildings;
3. belongs to another player, in this case you have to pay to the owner as provided.

The aim of the game is to purchase as many lots as possible so that players who stop over the property will pay and go bankrupt (Figure 12).

You play with a board on which are printed 40 boxes divided by colors in various streets, courses, and corporate enterprises. Money can also be gained or lost through Chance and Community Chest cards. There is also a jail. One may wonder whether there is a strategy based on the probability of ending up on certain boxes. Some principles are simple: remembering that it is more likely to do 7 than 2 or 12 and, since there are 28 properties, normally four will be touched in a turn; it takes on average six turns to do a full turn (40/7); on average it will happen once every turn to get a double shot (17% probability).

However, all of this must be supplemented by the calculation of the probability of happening on one of the 40 boxes. The best method is to carry out a simulation. In Figure 13 each box is associated with the respective transit probability.

Figure 12. Monopoly board

Figure 13. Monopoly simulation

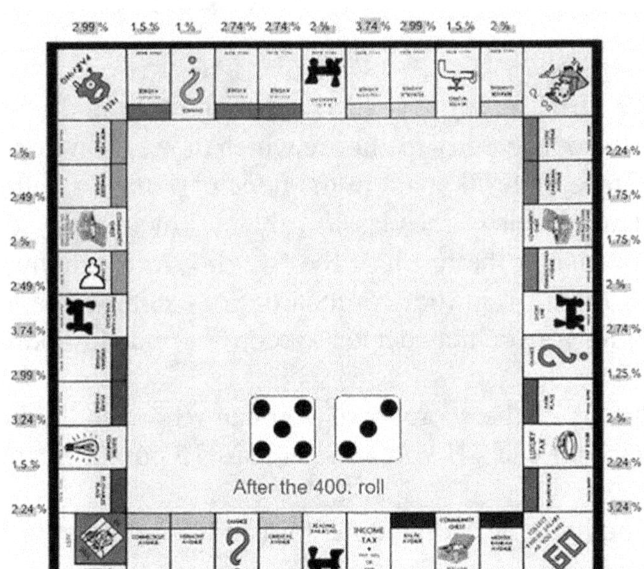

POKER DICE

Dice poker is played with five dice bearing on the six faces Ace, King, Queen, Jack, Ten, and Nine (Figure 14). The purpose of the game is to make the best combination of poker.

Figure 14. Poker dice

Players take turns rolling five dice. After the first and after the second roll, you have the opportunity to choose which dice to leave on the table and which ones to take and roll again in the hope of getting a better result. The game is very reminiscent of the classic Yahtzee which is played with normal dice. You can consider the result of the five dice as single numbers or as a poker hand, and mark it in the scoreboard. For example if after rolling the dice you get K-K-K-9 you may decide to score 4 points for four K or 4 points for K Poker.

In the first half of the scoreboard you can record individual results by getting one point for each. If you score at least 15 points you get a bonus of three points.

In the second half you score points for hands equivalent to poker hands:

1. Pair: 1 point;
2. Double pair: 2 points;

3. Three points: 3 points;
4. Scale: 4 points for 9-10-J-Q-K or 10-J-Q-K-A;
5. Full house: 4 points;
6. Poker: 4 points;
7. Five equals (Yahtzee): 5 points.

Each of these categories can only be registered once during the match. If you cannot record the score you must score 0 points in a category that is still free. Actually, there are other variants of the game with other counting modes. The game is influenced by luck, but also by a correct strategy that takes into account the odds using combinatorics.

We keep in mind that from the first throw all possible combinations can result, they amount to $6^5 = 7776$. The Yahtzee can only be done in 6 ways, the poker comes in $6 \bullet C_{5,4} \bullet 5 = 6 \bullet 5 = 150$ ways, the full house in $6 \bullet C_{5,3} \bullet C_{5,2} / 2 = 6 \bullet 10 \bullet 10/2 = 300$ modes, the scale comes in $6! / 2 = 720 / 2 = 360$ ways, the tris $6 \bullet C_{5,3} \bullet 5 \bullet 4 = 6 \bullet 10 \bullet 5 \bullet 4 = 1200$ occurrences, the double pair is formed in $6 \bullet C_{5,2} \bullet C_{5,2} \bullet 3 = 6 \bullet 10 \bullet 10 \bullet 3 = 1800$, the pair has $6 \bullet C_{5,2} \bullet 5 \bullet 4 = 6 \bullet 10 \bullet 5 \bullet 4 \bullet 3 = 3600$ combinations, finally remain the instances where no point is obtained. Odds, as we know, are obtained by dividing favorable combinations by total ones. To calculate the respective probabilities, simply divide the combinations of each result by the number of total combinations so, for example, the probability of making Yahtzee at the first throw is a modest $6/7776 \approx 0.0008$. It is interesting to compare theoretical and empirical probabilities calculated using an applet available on the net.

Applet (Retrieved from http://www.randomservices.org/random/apps/PokerDiceExperiment.html) in which the following values are shown on the abscissae:

0: all distinct
1: one pair
2: two pairs
3: three of a kind
4: full house
5: four of a kind
6: five of a kind

A simulation with 10,000 launches leads to the following graph (Figure 15):

Figure 15. Comparison of theoretical and empirical odds in Poker Says

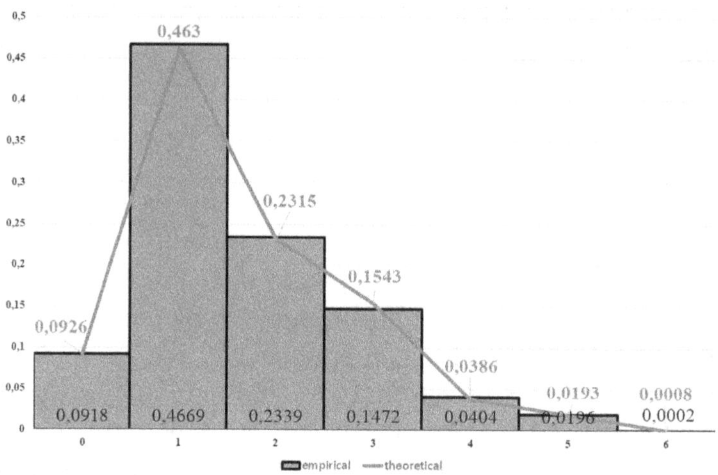

HEXAPAWN

There are problems where no quick procedure is known to identify a winning strategy. One example is the game of chess: it has been estimated that the number of strategies exceeds the number of atoms in the universe, so we will have to limit ourselves to a partial analysis. Calculators can estimate the best moves thanks to the technique known as "tree research".

A representation of this technique can be found with an automaton that simulates a reduced version of the game of chess: the Hexapawn. On a chessboard 3 x 3 there are three white and three black pawns behaving like chess pieces (they advance vertically and "eat" diagonally). The winner is whoever manages to arrive on the opposite side or capture or block all the opponent's pawns. The first move is up to the white and the computer will play with the black. If we now consider all possible moves, we can construct diagrams with the positions for each move by eliminating, for convenience, all symmetrical positions.

To build this automaton, presented in several texts (Bolt, 1966), you need: 24 boxes of matches (empty), as many as there are possible configurations that black can be found; several colored tokens, paper, colored pencils, and glue. On each box of matches apply the positions relative to the situation after the white move with the colored arrows indicating the possible moves of black (Figure 16).

Figure 16. Hexapawn

The boxes contain tokens the same color as the arrows. When you make the first move, you take the box with the position reached on the chessboard. Initially, there are only two. You randomly extract a colored token and place it outside the box; then you make the relative move. This move will lead to another configuration and the white will be able to choose the move it considers best. The game continues like this until the end. If the robot wins, the tokens are put back, if it loses, the token that led to its defeat is removed. In this way you can say that, in a certain way, the robot learns to play correctly with a probabilistic algorithm. After 11 games it becomes unbeatable as it discards all the losing moves.

Another method could be to add more tokens with the color of the moves that didn't turn out to be (at least until then) losers. This form of "compensation" modifies the odds, favoring the most productive moves at the expense of the losing ones.

In Figure 17 the twenty-four configurations to be glued on as many matchboxes:

Road signs

The road signs are almost the same age as the streets themselves. Already at the time of the Roman Empire, along the newly built roads, each *milium* was placed on the edge of the roadway a cylindrical stone on which were engraved the distances to Rome. In more recent times, of course, traffic lights have been added to signals (Figure 18), which are used in road traffic at junctions, pedestrian crossings and in other situations where regular traffic flows are required.

Figure 17. Hexapawn schemes

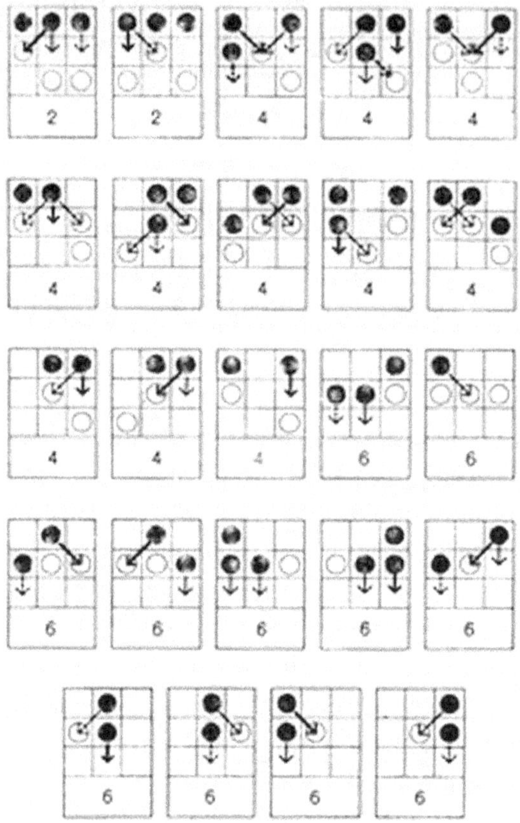

Figure 18. Different types of road signs

This kind of problem stems from an idea of Ana Serradó (2017).

What is the best possible route to move injured people from the car accident site to the hospital at a speed of 50 km/h as shown in Figure 19?

Figure 19. Road route

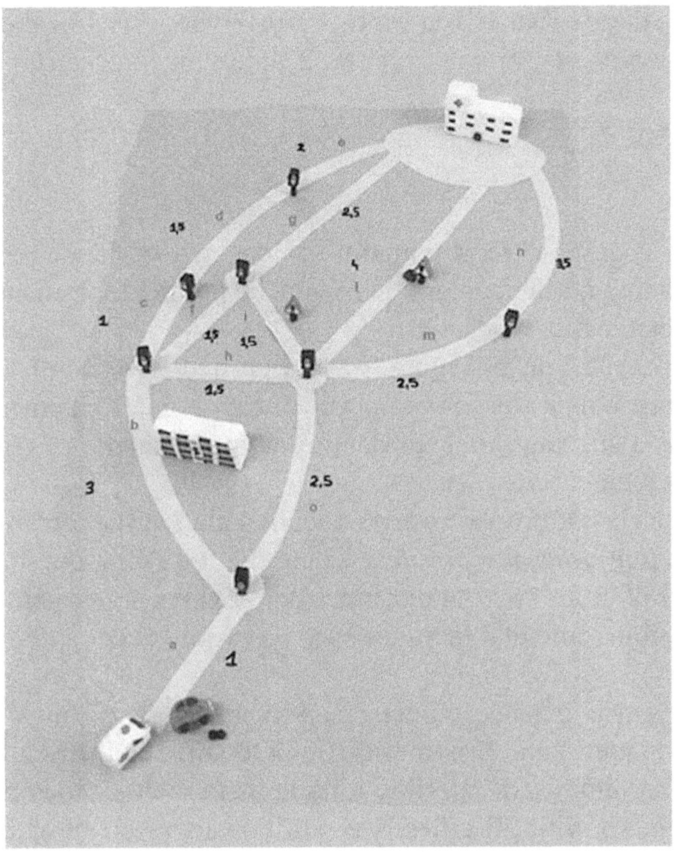

Analyze the map knowing the probability of a delay of 0.02 h:

1. due to a traffic light is 1/4
2. due to school (b) is 1/2
3. due to road works (i) is 1/3
4. due to partial obstruction of the carriageway (l) is 1
5. due to the presence of pedestrians crossing (c, d, i) is 1/2

The possible paths, as you can see by building the corresponding graph, are thirteen. Let us suppose for simplicity that the distances between the various nodes are the same and that therefore the only parameter to be taken into consideration is the one determined by traffic lights, works, etc.. Taking into account the indicated delay probabilities, we can determine that, adding the various travel times, the best route is a-o-l which has a time of 0.18 hours, quite close to the alternative route a-b-f-g (0.185 hours) while the worst route is the tortuous a-o-h-f-i-m-n in which 0.327 hours are foreseen for completion.

MAP

The map is a graphic representation of an area of land, reproduced to scale. Depending on the content there may be political, physical, historical, geological, etc. maps.

A map like the one in Figure 20 is a particularly useful tool to learn geography and to make use of geometric or trigonometric knowledge. It can also serve as a starting point for dealing with the theme of combinatorial probabilistic simulations.

One of the classic problems that concern a map of the territory is known as the *Travelling salesman problem* where the following question is asked: "Given a list of cities and the distances between each city pair, what is the shortest possible route that includes each city (only once) and returns to the city of origin?

The problem is considered extremely complex (it belongs to the class called NP-complete) and becomes difficult to solve as the number of cities increases. The number of possible itineraries to visit *n* cities grows as the factorial of *n*, so with 20 cities, we will consider you have 20! equal to 2432902008176640000 routes. There is a combinatorial explosion that makes the search for the optimal route through the enumerative method a desperate undertaking.

For this reason, heuristic solutions have been proposed which, although they are not able to find the optimal solution, can find an acceptably reasonable solution.

An interesting strategy is known as *Simulated annealing* and was born as a technique based on the similarity between the physical process of annealing solids and the resolution of combinatorial optimization problems.

How does the algorithm first defined by Kirkpatrick (1983) and others and then analyzed by several authors including Van Laarhoven & Aarts (1987)?

Figure 20. Map of Europe

After having identified an initial starting solution f(x), through the application of a random move (a random number is generated), the algorithm will identify a solution f(y) around the current solution. The first solution will be accepted if it is considered better than the current solution, otherwise you move in *y* with a probability established by a function that depends on a control parameter t > 0:

$$e^{\frac{f(y)-f(x)}{t}}$$

The procedure is repeated a number of times after which the parameter is lowered until the preset minimum is reached. The probability of accepting worse moves decreases as the value of the control parameter decreases because the algorithm only accepts better actions.

A simple case in which applying this method can be the following: a manager has to travel by private plane from London to several European cities and back to the city of departure. What is the shortest route? The longitude and latitude data for the cities concerned are as follows (Table 6):

Table 6. Coordinates of some European cities

	Latitude	Longitude
London	51.500	-0.100
Paris	48.900	2.300
Rome	41.900	12.500
Oslo	59.900	10.700
Moscow	55.700	37.600
Warsaw	52.200	21.000
Madrid	40.400	-3.700
Wien	48.200	16.400
Bucharest	44.400	26.100
Lisbon	39.00	9.10
Amsterdam	52.40	4.90
Copenhagen	55.70	12.50
Stockholm	59.40	18.10
Athens	37.90	23.70
Budapest	47.50	19.00
Tirana	41.30	19.80
Helsinki	60.20	24.90
Dublin	53.30	-6.20
Prague	50.10	14.40
Bern	46.90	7.50

In practice it starts with the first solution, for example from London you go to Paris, from there to Rome, then to Oslo and so on. At each subsequent step, the algorithm randomly selects a solution close to the current one, verifies it and then decides to move to it or stay with the current solution based on whether the new solution is better or worse than the current one. Knowing the locations of the cities it is possible to determine the relative distances. At this point, with the proposed method, different solutions can be simulated in order to verify what might be sufficiently acceptable. For example, a better solution than the initial one is: London, Dublin, Copenhagen, Vienna, etc. (Figure 21).

Figure 21. A possible solution

CONCLUSION

Games have always played an important role in learning mathematics in general, all the more so when it comes to probability. Several studies have revealed that game-based learning improves students' performance by increasing their motivation. According to Burguillo (2010): "The qualitative opinions provided by the students and the similarity of the surveys' results indicate that the use of friendly competitions provide a strong motivation that helps to increase the student performance".

One of the most relevant suggestions is to ask students to make predictions as it increases both their motivation and allows the teacher to have more control over the various phases of learning.

FUTURE RESEARCH DIRECTIONS

A great impulse to educational research could come from studies such as that on games with the use of augmented reality. The results of recent research (Sáez-López et al, 2019) showed that the experimental group achieved statistically significant improvements in performance, motivation, and level of collaboration. "The conclusion is that the dynamic activities managed in the intervention, which made use of augmented reality and localization, benefit teaching-learning processes, and encourage innovation and improvement through educational technology".

REFERENCES

Axelrod, R., & Hamilton, W. D. (1981). The evolution of cooperation. *Science, 211*(4489), 1390-1396.

Bartolomei, G., Giambalvo, O., & Macaluso, V. (2018). *Statistica o fortuna? Un modo per vincere giocando a dadi... Induzioni* (Vol. 57). Fabrizio Serra Editore.

Bolt, A. B., Harcourt, J. C., & Hunter, J. (1966). *We Built Our Own Computers (No. 1)*. CUP Archive.

Brutlag, D. (1994). Choice and Chance in Life: The Game of" Skunk. *Mathematics Teaching in the Middle School, 1*(1), 28–33.

Burguillo, J. C. (2010). Using game theory and competition-based learning to stimulate student motivation and performance. *Computers & Education, 55*(2), 566–575. doi:10.1016/j.compedu.2010.02.018

Dance, G., & Jackson, T. (2010, Oct. 7). Rock-Paper-Scissors: You vs. the Computer. *The New York Times.*

Delahaye, J. P. (1998). *Jeux mathématiques et mathématique des jeux.* Pour la Science.

Frosini, B. V. (1990). *Lezioni di statistica* (Vol. 1). Vita e pensiero.

Gardner, M. (1969). *The unexpected hanging: and other mathematical diversions.* Simon and Schuster.

Isaac, R. (2013). *The pleasures of probability*. Springer Science & Business Media.

Kirkpatrick, S., Gelatt, C. D., & Vecchi, M. P. (1983). Optimization by simulated annealing. *Science, 220*(4598), 671-680.

Leyton-Brown, K., & Shoham, Y. (2008). Essentials of game theory: A concise multidisciplinary introduction. *Synthesis Lectures on Artificial Intelligence and Machine Learning, 2*(1), 1-88.

Nahin, P. (2013). *Will You Be Alive 10 Years from Now? And Numerous Other Curious Questions in Probability*. Princeton University Press.

Sáez-López, J. M., Sevillano-García, M. L., & Pascual-Sevillano, M. (2019). Application of the Ubiquitous Game with Augmented Reality in Primary Education. *Comunicar: Media Education Research Journal, 27*(61), 66–76. doi:10.3916/C61-2019-06

Serradó, A. (2017). Empowering students' decision making in an everyday risk situation. *Quaderni di Ricerca in Didattica, 27*,111-114.

Van Laarhoven, P. J., & Aarts, E. H. (1987). Simulated annealing. In *Simulated annealing: Theory and applications* (pp. 7–15). Springer. doi:10.1007/978-94-015-7744-1_2

ADDITIONAL READING

Frayn, C. (2005, April). An Evolutionary Approach to Strategies for the Game of Monopoly. In CIG.

Koparan, T. (2019). Teaching Game and Simulation Based Probability. *International Journal of Assessment Tools in Education, 6*(2), 235–258. doi:10.21449/ijate.566563

Lu, P., & Wang, F. (2015). Heterogeneity of inferring reputation probability in cooperative behaviors for the spatial prisoners' dilemma game. *Physica A, 433*, 367–378. doi:10.1016/j.physa.2015.03.053

Miller, J., & Sanjurjo, A. (2019). A Bridge from Monty Hall to the Hot Hand: The Principle of Restricted Choice. *The Journal of Economic Perspectives, 33*(3), 144–162. doi:10.1257/jep.33.3.144

Stewart, I. (1996). Monopoly revisited. *Scientific American*, *275*(4), 116–119. doi:10.1038cientificamerican1096-116

KEY TERMS AND DEFINITIONS

Banach Problem: Inspired by a humorous reference to Banach's smoking habit.

Mixed Strategy: An assignment of a probability to each pure strategy.

Prisoner's Dilemma: A classical example of a game analyzed in game theory.

Pure Strategy: A complete definition of how a player will play a game.

Simulated Annealing: A probabilistic technique for approximating the global optimum.

Tit for TAT: A Strategy in game theory.

Travelling Salesman Problem: The shortest possible route that visits each city and returns to the origin.

Chapter 8
Probability and Simulations

ABSTRACT

This chapter is fundamental in educational terms as it deals with a topic that cannot be underestimated: the use of laboratory simulation. As long as we talk about probability in the classic sense, paper, pen, and maybe a pocket calculator are sufficient tools, but when we want to analyze a probabilistic model in depth, the use of a computer tool is essential. This support not only allows us to confirm hypotheses but, in some cases, it is indispensable. In the school it is unthinkable to work on large data sets (often not available) that would require not only appropriate software but also a different approach. The balance should be sufficiently significant but easy to handle dataset.

INTRODUCTION

The generation of random numbers is too important to be left to chance. ~ Robert R. Coveyou

Simulation is therefore a model of reality that allows to evaluate and predict the development of a series of events. The importance of simulation in the teaching of probability is an irrefutable fact. As Henry & Parzysz say (2011): "Of course the computer does not make use of the notion of probability when it performs a simulation: it just exhibits the effects of the equipartition principle on the random numbers that it generates (a principle which is part of its specifications). But even so, the use of computer simulations in the classroom, as pseudo-random generators, fosters a better understanding of

DOI: 10.4018/978-1-7998-3871-5.ch008

the notions of relative frequency, sample fluctuation, variability of empirical parameters and finally probability".

The simulation allows performing a random experiment with several tests that normally are not possible in class, of course this can be accepted provided that the simulation reflects the characteristics of the experiment to be referred to. Obviously, if the model is inadequate or even wrong the results are not reliable.

Simulation is also useful in another case: when you study a topic or a problem or an event and want to try to understand which theoretical model can be linked to it. Here is a simple example: you want to know what distribution can be underlying the toss of a fair coin, in other words what is the probability that launching 10 coins will occur 0, 1, 2, ..., 9, 10 heads. To answer this question we try to repeat the experiment virtually 100000 times and record the results. In Table 1 we can compare the empirical results with those of two known distributions.

Table 1. Comparison of distributions

	Empirical	Binomial	Poisson
0	0.0009	0.0010	0.0067
1	0.0101	0.0098	0.0337
2	0.0448	0.0439	0.0842
3	0.1171	0.1172	0.1404
4	0.2068	0.2051	0.1755
5	0.2457	0.2461	0.1755
6	0.2024	0.2051	0.1462
7	0.1177	0.1172	0.1044
8	0.0440	0.0439	0.0653
9	0.0099	0.0098	0.0363
10	0.0006	0.0010	0.0181

The graph (Figure 1) convincingly shows that the empirical distribution is correctly explained by the binomial.

In this case the theoretical answer was already known. However, many other examples can be formulated in which the simulation can provide useful indications on how to proceed to confirm or refute a hypothesis.

Figure 1. Comparison of distribution

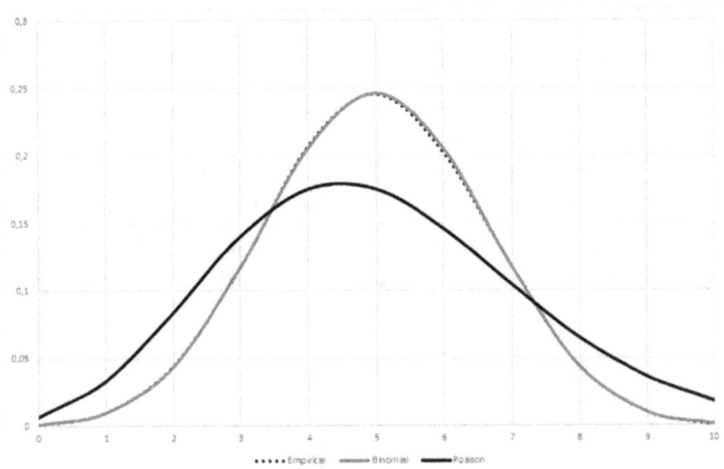

In conclusion we can say that the simulation provides useful indications when the random process is not easily definable. Moreover, if the simulation agrees with the theory in an acceptable way, we should feel comforted by the result.

The simulations are now used by social scientists, economists, meteorologists, demographers, etc. For example, in demography, one can take a cohort of individuals and simulate the mutations that have occurred over time as biological, environmental, psychological variables vary. In several simulations it is possible to obtain data, in reality not existing, but that could have been in the availability of the researchers if that particular model were true. Having more data archives available, it is possible to calculate statistical parameters as if numerous samples were available. We can therefore conceive the simulation as a virtual experiment.

Let us take as an example the price model of the French economist Léon Walras (1834-1910). The model considers a market where there are three players: producers, traders, and buyers. Their actions influence the *P* price of goods according to the relationship:

$$P_{t+1} = P_t + \mu\left(P_t^{-\alpha} - P_t^{\beta}\right)$$

In which the price at time t+1 depends on the price *Pt* at time *t* and three fundamental parameters:

1. the price elasticity constant $\alpha > 0$;
2. the elasticity constant of the production $0 < \beta < 1$;
3. a μ value, in turn linked to other factors, which influences the dynamics of the system.

In reality, the relationships underlying the model are more complex, but here they are omitted because we are more attracted by a different consideration. According to Walras, starting from arbitrarily random P_0 prices (he uses the term *hasard*) successive adjustments can be made until a stable solution is reached.

Cugno & Montrucchio, within a work on chaotic price behavior, have found that being $\mu_0 = \frac{1}{\alpha + \beta}$ e $\mu_1 = \frac{2}{\alpha + \beta}$, if:

$\mu < \mu_0$ then a stable solution is reached with Pt = 1;
$\mu < \mu_1$ the system is oscillating;
$\mu > \mu_1$ the system is unstable.

At this point, at each value of P_0, α, β and μ a particular result is determined. Of course the problem is that it is unthinkable to try the infinite series of values to be attributed. Walras is the father of the theory of equilibrium and it would be interesting to understand instead what is the probability of oscillating or unstable cases. A significant number of events (each with a random combination of parameters) can be simulated. As an example with 100000 simulations the average result was as follows:

μ_1 0.518
μ_0 1.393
μ_1 2.786
P_0 0.501

From the fact that $\mu < \mu_1$ it can be deduced that the relationship tends to oscillating price value. However, the solution should be taken with inventory benefit as you use random numbers generated by a computer. In a real economy, not all values are acceptable. In this context, the objective was not to verify established economic theories, but simply to understand how simulation can be used to address complex problems.

In the educational field, simulations sometimes take on the character of a game. In another situation, educational software is developed to reproduce experiences similar to the real thing. They are particularly useful in those fields where it is difficult or impossible to physically reproduce the actual conditions to be studied in the laboratory. According to Evans and Rosenthal: (2004) "For example, computers often use pseudorandom numbers to make games fun, accurate simulations, and searches efficient".

The solution of a probabilistic problem is not always easy, so it can be useful to approach the theoretical solution through computer simulation or using specific *apps*. For example, the BetOnMath experiment developed by Andrà (2015) and others of the Polytechnic di Milano has moved in this direction. One of the *apps* that stimulates students' attention the most is the one that simulates a type of lottery quite common in Italy, called "Gratta e Vinci" (Scratchcard). To participate you buy a card that is partially covered with a silver substance. By scratching it away, the words "You didn't win, try again" or the amount of the winnings appear. With the app, you can simulate the game by scratching any number of cards and checking how inadvisable it is to play this game. As an example, the following problem has been proposed in many classes: if you imagine investing the average monthly salary of an Italian (about 1500€) by buying 300 tickets, what is likely to happen? With great amazement, it occurred that at least a quarter of the salary ended up in the pockets of the State, in line with what was officially declared by the body that manages the game (26%) but unknown value to the students and probably also to most of the possible regular players. Once again it is shown that probability acquires a concrete meaning when confronted with real experiences.

At this point a short digression must be opened: to build a suitable random model you need tools to simulate the case.

The factors determining the acceptability of a method are essentially the following:

1. the numbers in the generated sequence must have the same probability of showing up;
2. the numbers must be statistically independent;
3. full reproducibility of the sequence is required;
4. the sequence must not repeat itself until after a great period.

One of the oldest methods for generating pseudorandom numbers between 0 and 5 capable of simulating the throwing of a dice is due to the monk Edvin (around 1250). You need to choose a four-digit number, square it and keep the middle four digits if possible. This is repeated several times taking the remainder of the division by six of the last number. For example, if we start from the number 8653, the following steps will generate the sequences 8744, 4575, 9306, 6061. If we calculate the remainder of the division of the number 6061 by 6 we get 1 and therefore it is as if we had rolled a die and obtained as a result two. The defect of the method resides that, in certain cases, the starting number is found after a few steps or 0 is obtained. It has been seen on several occasions that the toss of a coin, a dice, the drawing of a card from a deck or a numbered ball from an urn are events related to chance and therefore should, in theory, generate random numbers. However, these methods are not very practical, so we turn to deterministic methods that generate, under certain conditions, *Pseudo-Random numbers*.

For the generation of pseudo-random numbers with the help of a computer, arithmetic generators based on type formulas are chosen:

$$x_{n+1} = (\alpha \bullet x_n + \beta) \bmod M$$

The numbers α,β, are characteristic parameters of the model and *mod* indicates the rest of the Euclidean division of the first number for the second. To evaluate the goodness of a generator, the period of the generator and the independence of the outputs are considered. The maximum extension of the cycles is determined by the number *M,* to which is practically assigned a large value and that makes easy the division on a computer.

To understand the second aspect (independence) you can group the numbers generated in pairs so as to provide abscissae and ordinates of a square of unitary side, if the numbers are independent they will be evenly distributed on the square with frequency $1/M^2$. At this point you start with a random initial value, called the generator seed, for example taken from the clock by the computer and apply the algorithm.

For example, the generator $x_{n+1} = (x_n + 3) \bmod 10$ produces, starting from $x_0 = 0$, the sequence 0, 3, 6, 9, 2, 5, 8, 1, 4, 7 short and not independent (Figure 2).

A better set of parameters is for example a= 259, b = 0, M = 65536 with "generator seed" $x_0 = 725$ leading to the following uniform and independent distribution (Figure 3):

The generation of numbers as close as possible to an ideal of randomness is fundamental in the simulation of stochastic processes. Stochastic models (from the Greek στοχαστικός conjectural, speculative) take into account the variations of input variables and provide results in terms of "probability". Generally, stochastic models have a more complex structure than deterministic models and calculations are always performed with the help of the computer.

Figure 2. Non-independent generation

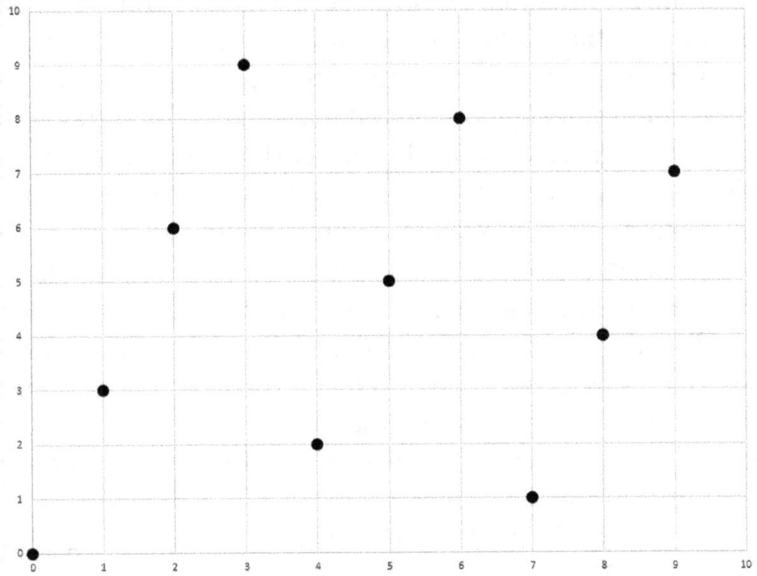

Figure 3. Independent generation

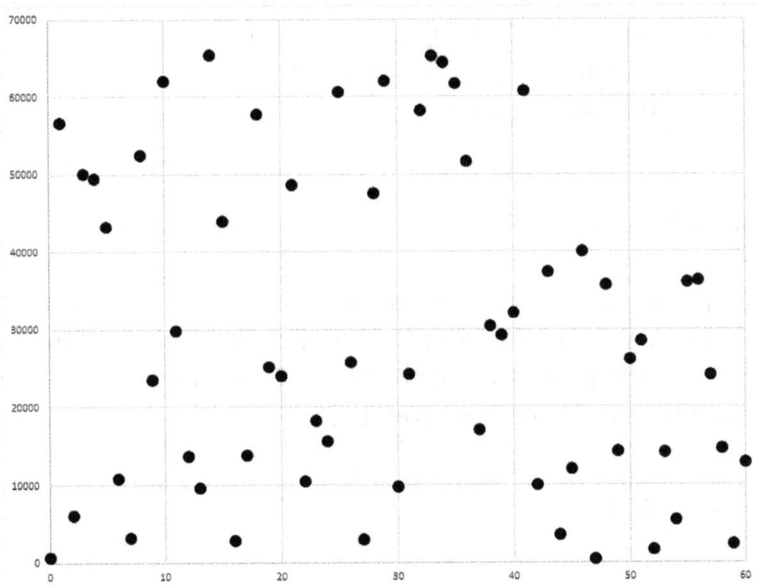

To clarify the differences between a deterministic model and a stochastic model, let us suppose that the performance of equities on the stock market can be inferred from the product of two variables of which we only know that the first can vary between 0.3 and 0.5 and that the second can vary between 0.7 and 0.9. A deterministic model could use as input the average values and therefore we can say that the probability can be $0.4 \bullet 0.8 = 0.32$. The stochastic approach instead takes into account the variability of the two input variables and the calculation of the incidence is made through a simulation with the Monte Carlo method. A program, taking into account the possible frequency distributions derived from historical series, can calculate a value for the two variables which are then multiplied among them. By repeating this process many times, not a single value but a frequency distribution is obtained.

With the stochastic model, one has the advantage of being able to predict - on a probabilistic basis - the entire range of possibilities of the phenomenon.

The formulated example is rather trivial but it is enough to refer to the vast scientific literature to realize the importance of the subject. As Obropta & Kardos (2007) say about urban stormwater quality models: "A deterministic model seeks to perfectly simulate the physical world through cause and effect relationships. A stochastic model uses statistical patterns of a particular phenomenon to simulate that same phenomenon". To differentiate the two model approaches, Nix (1994) suggested that, following the same input, a deterministic model always produces the same response, while a stochastic model always produces a different response.

PLAYING CARDS

A simple game is the following: we take a deck of 20 cards with 10 pairs of the same value. By turning two cards at a time and putting them back into the deck, what is the probability of getting at least one pair?

The theoretical probability is given by:

$$P = 10 \bullet \frac{C_{18.3}}{C_{20.5}} \approx 0.526$$

By performing 10000 simulations, values around 0.53 are obtained.

Let us examine a second problem, represented in Figure 4.

Figure 4. A, 2, 3

You start by laying six playing cards covered on the table: Ace, 2 and 3 of hearts; Ace, 2 and 3 of spades. Ace is worth one. Suppose the cards of hearts have positive values, while the cards of spades have negative values.

A proposes the following game to *B*: he will turn a card and then put it back in the deck, then *B* will turn a card. If the product of the two numbers that appear is positive, *B* wins a bet equal to the result given by the sum of the face value of the two cards, if the product is negative, *A* wins.

Suppose we play a number of games with three different situations. Looking at them, three questions arise:

1. is there a favorite?
2. *B* proposes to *A* to remove the 3 of spades and to restart the game with the five remaining cards, does the game change?
3. *B* proposes to *A* to eliminate the 3 of hearts and to restart the game with the five remaining cards, does the game change?

In response, the use of simulation is useful as only a large number of tests will or will not confirm the subjective assessment. An experiment with 10000 launches provided the following indications:

1. in the first case the game is balanced;
2. in the second and third case it is *B* who has an advantage

DICE

If you roll two regular dice several times you get a distribution that follows the theoretical distribution of the binomial (Figure 5).

Figure 5. Distribution of the throw of two dice

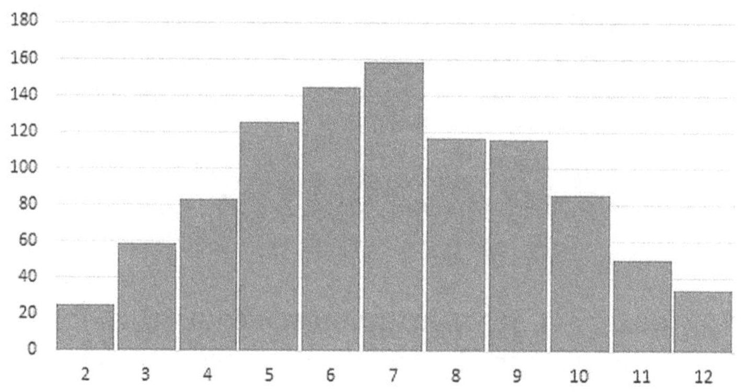

If instead you roll three dice and then discard the one with the highest score, apparently we continue to have the result of throwing two dice, but the distribution will have the same characteristics as the previous one? In Figure 6 the dice with the score of four should be discarded and then you would get as sum 3.

Repeating the experiment several times, with the elimination of the major result, the distribution shown in Figure 7 is obtained.

The distribution presents a positive asymmetry (long tail on the right), while the average of the first case is close to the theoretical 7, in the second case a value close to 5.5 is obtained.

Obviously, if you exclude the nut with the lower score, the distribution will present a negative asymmetry.

One of the simplest games you can play with dice is called HOG (Feldman & Morgan, 2003). The aim of the game is to get the highest possible score by following the rules below:

1. Each player decides how many dice to roll (from 1 to 12);
2. If none of the numbers he threw is 1, the score is the sum of the numbers he got;
3. If a 1 is rolled on one of the dice, the score is 0;
4. Whoever gets to 100 first wins.

Figure 6. Result of throwing three dice

Figure 7. Distribution of the throw of three dice with the elimination of the greater value

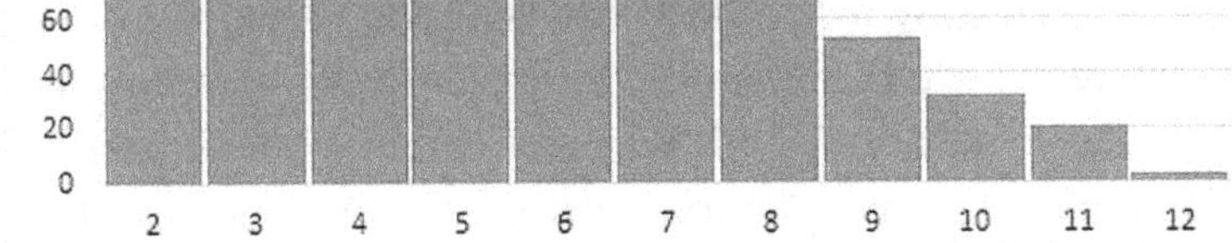

The game, in the simplicity of the rules, can be adapted to students of different ages as it can be articulated, in a more or less advanced way, on two different levels of complexity. This activity allows you to ask questions that require the use of two areas of mathematics. The first concerns combinatorics, the second concerns probability-based game strategies.

This game is above all a good example of an activity that induces to exploit computer science technology in order to create a simulation of the possible outcomes of the various strategies. A very simple computer program, written in BASIC, might look like this:

```
RANDOMIZE
INPUT "Number of rolls ", r
PRINT
PRINT "Number of Dice", "Average Score"
PRINT
FOR D = 1 TO 12
    t = 0
    FOR i = 1 TO r
        s = 0
        FOR j = 1 TO D
            V = INT(6 * RND(1)) + 1
            IF V = 1 THEN s = 0: GOTO 100
            s = s + V
        NEXT j
        100 t = t + s
    NEXT i
    PRINT USING "    ##                        ##.##"; D, t / r
NEXT D
```

With 100000 launches the following results were obtained:

TARGET

The back of the target consists of a circle (Figure 9) of which it is easy to determine the area by knowing the radius: $A=\pi r^2$.

An interesting alternative method for the calculation of the area is based on the already well-known Monte Carlo Method. Assuming we insert the target in a square whose dimensions (and therefore the area) we know, we throw a large number of darts. Then we calculate how many have fallen inside the target. By making the proportion between these and the darts thrown, we get an estimate of the target area. However, we must assume that the throws are random and independent.

By widening the discussion, you don't throw arrows on a target but with the computer, you can generate points on a circle. All of this is shown in Figure 10, where the square has area 4.

Figure 8. HOG simulation

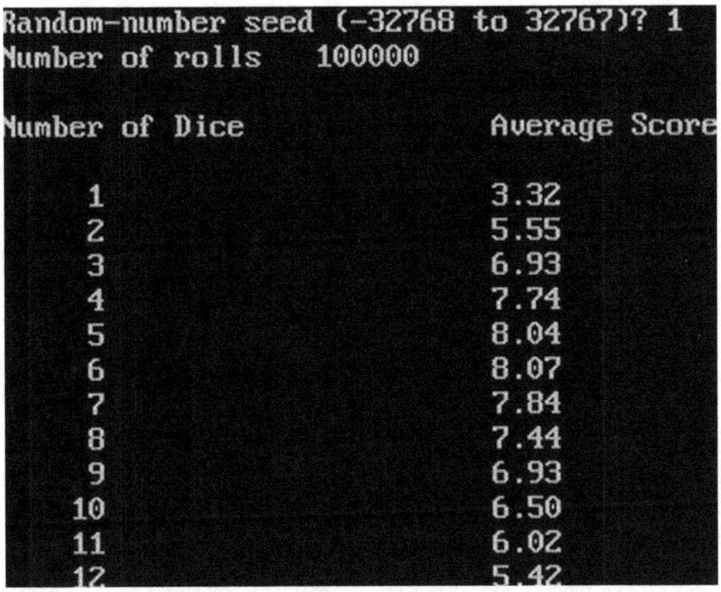

Figure 9. The rear of the target

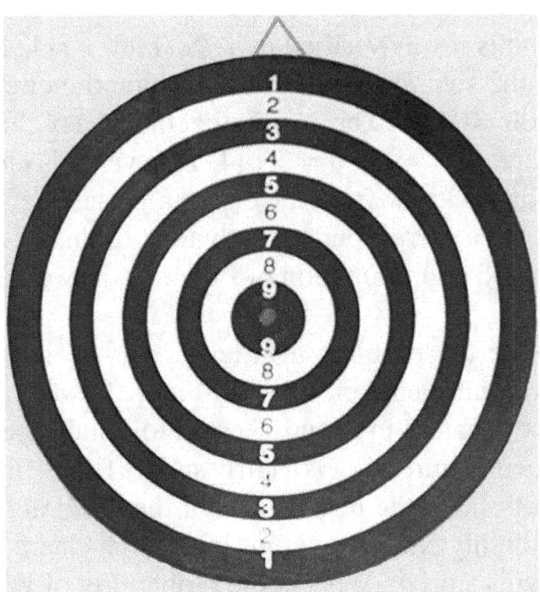

Figure 10. Monte Carlo Method estimation of the area

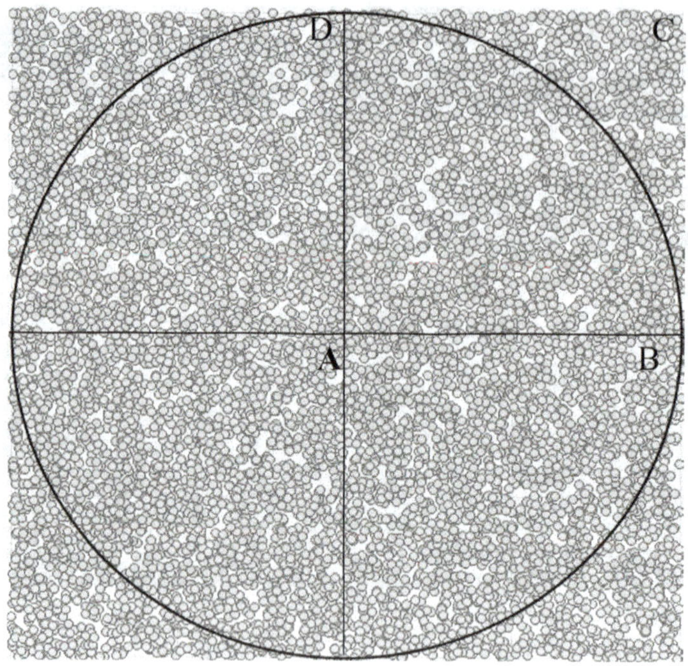

With 10000 points it was verified that 7860 fell inside circle *A* of radius 1 and the rest on the square on side 2 but outside the circle. Consequently P(A) = 7860/10000 = 0.786. This means that the area of the circle is 78.62% of that of the square, so 4 • 0.786 = 3.144. This value is close to the area of the circle which, as we know from the elementary geometry, is $\pi \cdot r^2$; being the unit radius the area corresponds exactly to π. In this case we can say that we have found a good approximation of Pi which, as we know, corresponds to 3.141592654...

Now let's change the problem a little bit. Given the top right portion (ABCD) of the original square on side 1 of Figure 9, we draw the arcs of the circumference with centers in *A* and *C* so as to obtain the circular spindle highlighted in gray in Figure 11. It is a fairly simple figure; the ancient Greeks gave it the name of one of the instruments of the pastoral civilization.

The player randomly chooses a point inside the square, if the point is in the grey part he wins a prize. What is the probability of winning?

We can approach the problem in two ways: consider the probability as the ratio between the grey part and the area of the square of the unitary side or take advantage of the simulation just seen.

Figure 11. Circular spindle

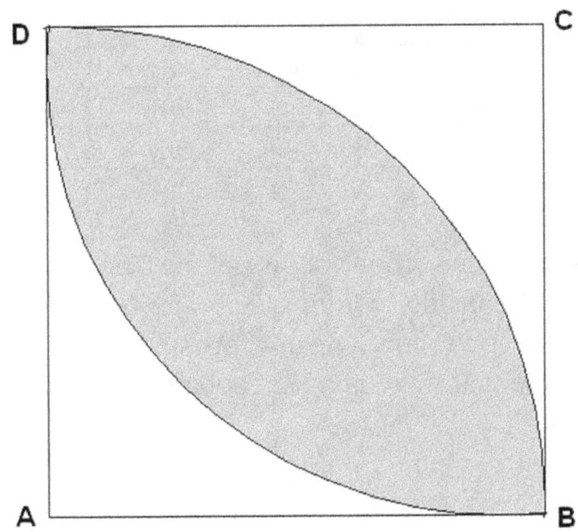

In the first case it is necessary to determine which area is grey. The circular sector ABD with the center in *A* is a quarter of the whole circle in Fig. 9: therefore $\pi \bullet r^2/4 = \pi \bullet 1^2/4 = \pi/4$ and consequently the remaining white portion, placed at the top in Fig. 10 is $1 - \pi/4$; similar reasoning can be done for the CBD sector with the center in *C*. Consequently, the two white portions are worth $2 \bullet (1 - \pi/4) = 2 - \pi/2$. The grey area is therefore obtained by subtracting the two white portions from the area of the square. Basically, $1 - (2 - \pi/2) = \pi/2 - 1 = 0.571$.

The problem can also be tackled by using the simulated throwing of darts already analyzed: in the ABCD dial presumably about 2500 of the total 10000 darts thrown have fallen; of these 7860/4 = 1965 have fallen in the inner part of the circle while the number of those fallen in the outer sector is obtained by simple subtraction: 2500 - 1965 = 535. The probability of the grey area should therefore be equal to $1 - (1965 - 2 \bullet 535)/1965 = 0.544$.

We can conclude with a problem in a Clifford Pickover (2002) book. The player has three darts at his disposal. On the second throw he gets a worse result than the first. Assuming the skill level is constant, what is the probability that the third attempt is worse than the first?

Apparently it seems impossible to provide a solution, but the reasoning is not complicated. Let's suppose that the shots can be distinguished according to the score achieved: High, Medium, Low.

There are six possible combinations, as can be seen from Table 2:

Table 2. Table with possible results

	1	2	3	4	5	6
1st shot	High	High	Medium	Medium	Low	Low
2nd shot	Medium	Low	High	Low	High	Medium
3rd shot	Low	Medium	Low	High	Medium	High

Since we know that the second shot is not the best and the first shot is not the worst, we do not consider columns 3, 5 and 6. Of the three remaining columns, only the first two show a result consistent with the premise and this allows us to conclude that there is a 2/3 probability that the third attempt is worse than the first.

CHESSBOARD

An empty chessboard (Figure 12) lends itself to many questions.

Figure 12. Empty chessboard

On a chessboard you can present many probabilistic games, one of them is the Franc-Carreau (Fair-square in English), practiced since the Middle Ages. This game consists of flipping a coin on the chessboard and checking its final position: entirely inside a square (Franc-Carreau) or straddling one of the edges or. The problem was dealt with by Georges-Louis Leclerc, Count of Buffon (1733). These are his words: “Here's a problem that's kept me busy these past few days, and which may be to Mr. de Moivre's taste. You may not know what we call in French the game of Franc-carreau. In a room paved with tiles, one throws a shield into the air. If it falls on a single tile, it is said to fall straight down, and the one who threw it wins. If it falls on two or more tiles, the one who threw it loses. This is a problem to be solved and has no difficulty in finding the probability of winning or losing, the tiles and the shield being given”.

This is a problem that requires geometric analysis. Calling the side of box *S*, a coin of radius *r* just touches the edge of the square of the unit if its center is on an inner square of side (L - 2 r) as shown in Figure 13. The probability that the center of the coin is inside the inner square is $(L - 2r)^2$. Since the square has an area of L^2, the probability sought is

$$P = \frac{(S-r)^2}{S^2}$$

We know that the probability of an event can be obtained approximately by performing a large number of times an experiment and calculating the frequency of the results of this event. For example with a 20 cm side of the box and a coin with a diameter of 5 cm the expected value is 0.5625. Simulation performed with 10000 flips gave a result of 0.565. In this case the results of the simulation tend to coincide with those expected.

FOOTBALL STICKERS

Many children made a collection of figurines, sold in special envelopes (Figure 14), which were then attached to an album.

The problem is that, while at first filling in the boxes is easy, over time, with each purchase, many duplicates accumulate. Faced with his son's insistence, a mathematical father has found a formula to calculate how many cards he will have to buy on average in order for his son to complete the album. If you indicate with n the number of boxes to fill in he will have to buy a number of stickers (s):

$$s = n \cdot \left(1 + \frac{1}{2} + \frac{1}{3} + \ldots + \frac{1}{n}\right)$$

Figure 13. Franc-Carreau conditions

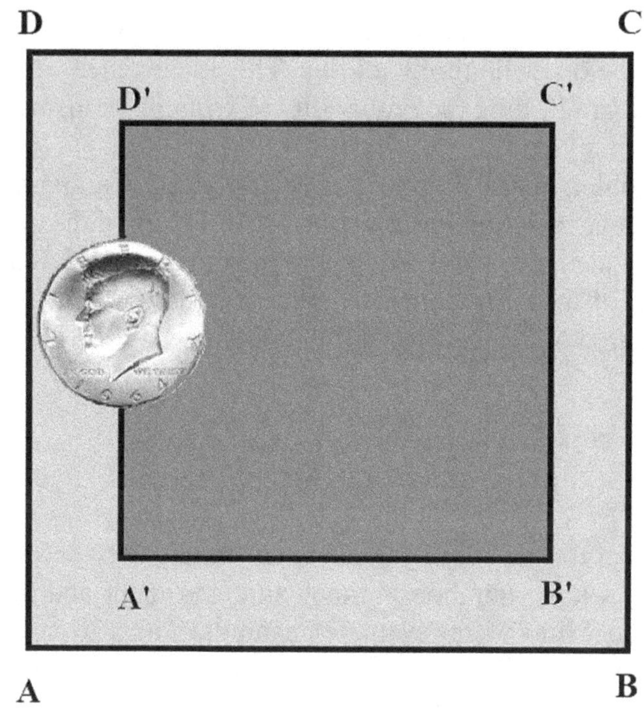

The father, to convince the son that the number s is given by the previous formula, could use a simulation. In fact the solving formula is not easy to discover and the Monte Carlo method is an excellent tool to approach the solution. Using a software (in this case R), with n = 50 and 10000 tests you get the distribution graph, as you can see in Figure 15. The result is distributed among many values, but it is evident that the number of necessary figurines will be quite high. We underline the fact that the formula is valid only if the figurines are put in the envelopes in a random way. In reality there have been many cases where the manufacturer has made some stickers rarer in order to sell more envelopes.

Figure 14. Envelopes with figurines

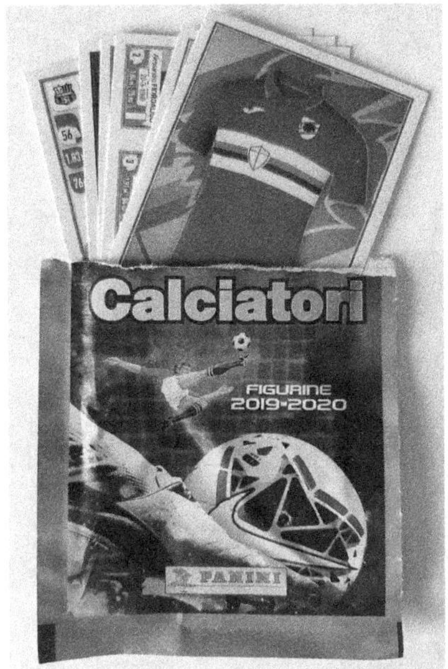

Figure 15. Distribution of stickers

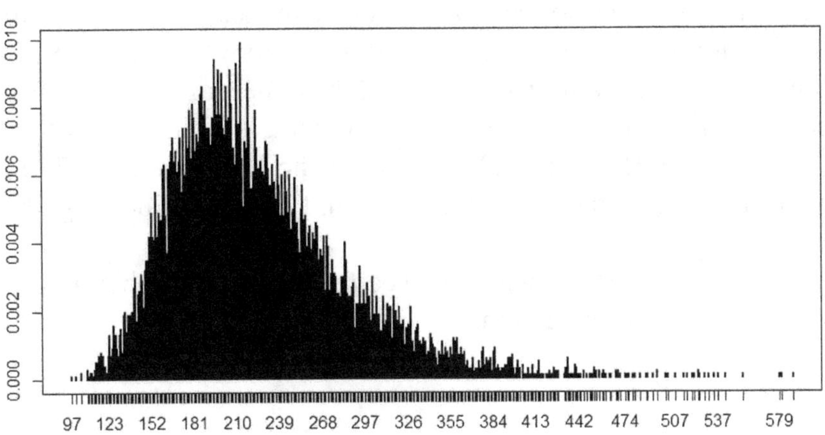

BUS

The problem of waiting at a bus stop (Figure 16) is a classic. Let's look at a particular version.

Pascal lives in a small town that has a particular characteristic: from the bus station in the center he can take a vehicle to go south to the sea or north to the lake. Not knowing what to choose Pascal decides to leave the house completely randomly and take the first bus leaving. Southbound buses depart at exact time (10:00, 11:00, 12:00, etc.) while northbound buses depart with a ten-minute time lag (10:10, 11:10, 12:10, etc.).

Figure 16. Bus

To his amazement, he notices that he went to the sea more while he thought that the distribution of the voyages would be more equitable. Why is his reasoning (despite the name evoking one of the fathers of probability) wrong?

Since Pascal reaches the bus station at random, it is more likely to happen in the long interval (50 minutes) than in the short interval (10 minutes). In fact, about four times out of five he will find himself collecting shells on the beach.

This is a classic example where you can simulate the problem in Excel, a simple macro like this is enough:

```
Sub Macro()
n = InputBox("# tests")
For i = 1 To n
    Calculate
        If Range("F2") > Range("B2") Then south = south + 1
Else north = north + 1
    Next i
Range("H2") = south
Range("H3") = north
End Sub
```

Setting n = 10000, it turns out that Pascal will travel south probably 82% of the time.

A complex problem of operational research, which has been behind almost 50 years of research, is known as the school bus route problem (SBRP). The most recent research has begun to consider the impact of uncertainty, both from the point of view of the uncertainty of demand and the uncertainty of travel time. This approach has led to significant cost savings for a school bus transport system. The objectives of the text do not allow to address the issue, for a summary of the various research trends please refer to the article by Ellegood (2020) and its extensive bibliography.

METER

The meter (Figure 17) allows us to make length measurements and is therefore used to verify the height of people, especially growing children.

Figure 17. Camel-shaped meter

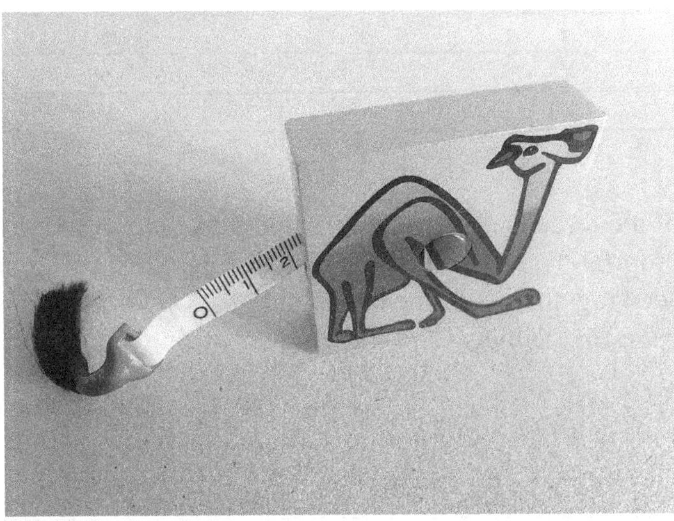

The debate on the factors that can influence growth is still open; in principle there is agreement that there are many.

From the measurement of the heights of 1000 real subjects it has been possible to obtain summary data consisting of:

1. minimum height: 146 cm;
2. maximum height: 200 cm;
3. average: 170 cm;
4. median: 169 cm.

Remember that the *median* is the value that separates the upper half from the lower half of a probability distribution; in this case the average and the median are similar. Now, simplifying a lot, we can consider three elements that can influence growth: feeding, sport and health, heredity. It is curious to wonder if these factors can be simulated by rolling three dice, obviously giving a "weight" to each result so that it can be adapted to the scale of the real data. The weights chosen are shown in Table 3.

Table 3. Association between dice result and weight given

dice	height factor
1	48
2	53
3	55
4	58
5	61
6	66

To each of the three factors was attributed the result of the launch which, weighed appropriately, constituted the result of the three components which were then added together to define the height of 1000 virtual individuals. For example, for the first subject the following situation could occur (Table 4):

Table 4. Example of height estimation

	Dice		Weighted Dice			
feeding	sport & health	heredity	feeding	sport & health	heredity	sum
2	6	3	53	66	55	174

At this point the two different results are compared (Table 5):

Table 5. Comparison of results

	observed	estimated
minimum	146	144
maximum	200	198
average	170	170
median	169	170

The result of the simulation is fascinating, but it must be stressed that it is only a pretext, without scientific validity, useful to play a bit with the concept of randomness.

PHONE

Figure 18 shows a very traditional cordless phone that we can use to introduce a classic game.

Figure 18. Phone

A traditional board game, called the telephone chain, consists of placing participants in a row. The prompter whispers a word in the ear of the first who then has to whisper it in the ear of the second, and so on. At the end of the chain the initial word is sometimes different. Another game can be connected to this game, proposed by Paul Nahin (2013). Let us imagine that we have *n* people who tell the truth with probability *p* or lie with probability (1 - p). At the first person in the queue it is said 1 and the transmission continues, of course liars will change as heard in its opposite -1. What is the probability that the message arrives correctly at the end of the chain?

Obviously, to get a 1, the liars must be even in number since the lies are eliminated in pairs. This situation can be simulated by flipping a fair coin.

If *n* is even there will be an even number of heads (truth), if *n* is odd the heads must also be odd. It can be demonstrated that, whatever the value of *n*, the probability that the result is 1 is given by:

$$P(1) = \frac{1}{2} + 1/2 \bullet (2p-1)^n$$

Suppose now to experience this procedure with 25 people and p = 0.96.

$$P(1) = \frac{1}{2} + \frac{1}{2} \bullet (2 \bullet 0.96 - 1)^{25} = 0.562$$

If you make a simulation you get a value remarkably similar to the theoretical one.

SOCCER BALL

The soccer ball (Figure 19) is the symbol of a sport, probably the most popular in the world, at least for fans. Every game starts with the touch of this object. But what happens afterwards? The obvious answer is that you play a game, in probabilistic terms you could say that the first kick to the ball starts a random event lasting 90 minutes or more.

As the game Risk simulates a war between states you can conceive a game that simulates a football match. The problem is that it is a situation in which many factors come into play: not only the relative strength of the two teams but also a complex of random events.

Figure 19. Old soccer ball

A good simulation takes into account a real situation and tries to build a suitable but, if possible, not too complicated model. An interesting model has been suggested by Alan Parr (Retrieved from https://nrich.maths.org/1184) and from that we will try to develop a similar model.

You can start with a characterizing element and then gradually add more and more details in order to improve its realism. For example, a first step could be to predict goals, i.e. how many times the ball will enter the net.

In the Italian Serie A football championship, taking into account the year 2018-19, shots on goal were, on average, 13.9 while goals were 1.3. This means that about 10.7 shots are needed to score a goal. The data for the other important European Championships are: Premier League (United Kingdom) 11.6 and 1.3 which means 8.9 shots per goal; La Liga (Spain) shows 12.2 shots and 1.5 goals with a ratio of 8.1; for the Bundesliga (Germany) the ratio is 9.6 as the shots are on average 13.5 and the goals 1.4. If you take these four figures it results, approximately, that the ball enters the goal every nine shots.

These numbers can be the basis for playing a game, just roll an octahedral dice nine times (the choice of this dice instead of the classic one derives from the fact that the results are better suited to the data found in the various leagues). Each time the roll gives the number 8 you score a goal. With 20000 simulated matches the results are those shown in Table 6:

Are these figures realistic enough?

Table 6. Simulation results in terms of goals

goal	%
0	30.34%
1	38.32%
2	22.05%
3	7.41%
4	1.61%
5	0.27%
6	0.01%
7	0.00%

If we compare them with the average of the values of the various countries (in which the different shot / goal ratio has been taken into account) we notice that perhaps we can be satisfied and proceed with the game, in fact these are the percentages found in 2018-19 (Table 7):

Table 7. Comparison with four football leagues

goal	Series A	Premier League	La Liga	Bundesliga	Average	Simulation
0	32.89%	28.95%	32.89%	25.82%	30.14%	30.34%
1	35.00%	36.32%	40.26%	35.62%	36.80%	38.32%
2	22.11%	21.05%	17.89%	23.53%	21.15%	22.05%
3	6.84%	9.21%	6.84%	12.09%	8.75%	7.41%
4	2.63%	3.42%	1.32%	2.29%	2.42%	1.61%
5	0.53%	0.79%	0.79%	0.33%	0.61%	0.27%
6	0.00%	0.26%	0.00%	0.00%	0.07%	0.01%
7	0.00%	0.00%	0.00%	0.33%	0.08%	0.00%

Basically we can think of two contenders as two football teams, each player rolls nine times the dice and this random event will decide whether the match ends even or with the victory of one of the two. For example if the two sequences are 1, 5, 2, 5, 6, 4, 3 and 7, 7, 1, 4, 7, 1, 2, 4, 6 both teams do not score and the match ends 0 - 0. The sequences 1, 5, 8, 5, 2, 5, 6, 4, 8 and 7, 7, 1, 4, 8, 1, 2, 4, 6 instead lead the first player to win the match 2-1 and the goals scored are, for statistics, 3.

This model is obviously only basic, you can improve it by introducing some other features that reflect the real matches. This requires the dice roll to match the probability of the events taken into account as closely as possible. It is well known that teams do not all have the same strength (it is no coincidence that bookmakers take this into account when defining quotas, as has been highlighted in the third chapter). To make the whole thing more credible it can be established that the strongest team can benefit from an additional shot: for example if an 8 comes out then they can attribute themselves an additional goal. A further variation may take into account an element present in real matches, penalty kick. At the end of a round a normal six-sided dice can be used: if 6 is rolled out the team gets a penalty kick but if, by rolling an octagonal dice, 7 or 8 is rolled out, the penalty is wrong or parried. This may change the final result.

As you can see, simple dice can work to simulate a random event such as a football match: a search in soccer statistics sites and careful use of simple objects such as dice are the basis on which to work.

PLANNER

A planner like the one in Figure 20 is nothing more than a sort of calendar on which the commitments are marked. In this case, as a decorative element, but perhaps also as a sports reminder there is a Spinner. This object is a variation of the old wheel of fortune, a widespread symbol in medieval iconography, which used it to depict cyclicity: alternation of day and night, phases of the moon, motion of the planets. The wheel of fortune is one of the Tarot cards used for divination purposes.

Observing this figure provides an opportunity to take up some of the concepts we have seen before and lead us back to the field of probabilities.

The problem we are analyzing actually deviates a little from the object shown. It was proposed in 1996 by NAEP (National Assessment of Educational Progress) and then studied by Shaughnessy & Ciancetta (2002). During the school year 1999-2000, the two mathematicians proposed to 652 students, grade 6-12, in 28 math classes the problem related to the rotation of two spinners, shown in Figure 21.

The problem is interesting because it deals with the relationship between Spinner and *sample space*.

Figure 20. Planner with Spinner

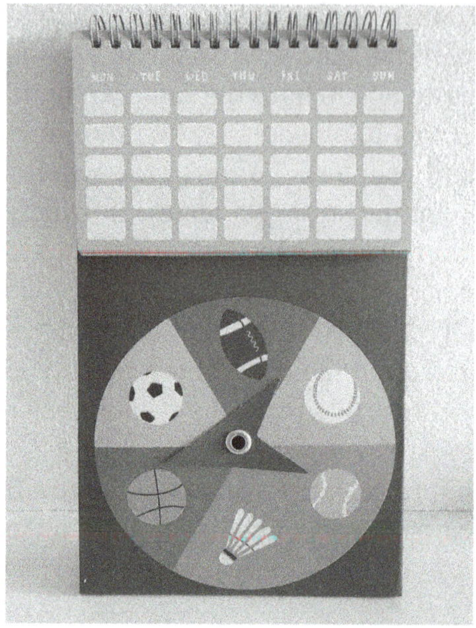

Figure 21. Spinner task

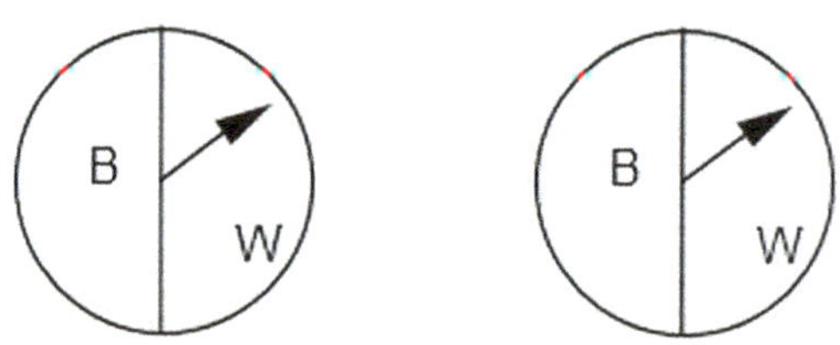

The two fair spinners above are part of a carnival game. A player wins a prize only when *both* arrows land on black after each spinner has been spun once.

Jeff thinks he has a 50-50 chance of winning. Do you agree?

A Yes B No Justify your answer.

Many students, especially from lower classes, replied that there was a 50% probability because the two Spinners have an equal amount of B and W. The correct answers given by the students of the upper classes were motivated by two different settings: there were those who considered the space of events (BB, BW, WB, WW) and then based on the classical probability to establish that the probability of victory only occurs in one out of four cases. Others were convinced by a series of empirical evidence (10 Spinner throws) and therefore referred to the frequentist probability.

POLYPHEMUS

The Odyssey is considered, together with the Iliad, one of the masterpieces of the ancient epic. The ancients attributed these poems to a poet named Homer, a figure who perhaps really lived or, more probably, legendary. A careful reading of these texts highlights several references that can stimulate probabilistic reasoning. One of these refers to the famous encounter between Ulysses and the cyclops Polyphemus. Homer tells that Ulysses, during his long journey, landed on the Land of the Cyclops. Driven by curiosity, he reached the cave of Polyphemus (Figure 22), where he and his companions were captured. After blinding the giant, Ulysses managed to escape from the cave with a ruse. Realizing that the Greeks had escaped, Polyphemus moved to a promontory, from where, blindly, he began to throw rocks into the sea, in an attempt to sink the ship that was taking off.

Figure 22. Reproduction of Polyphemus

Suppose we represent marine space as a geometrically lattice system. Polyphemus is in the coordinates (0;0). Ulysses' boat can be placed at any of the coordinate points (x; y) with x and y between 1 and 6. The blind cyclops launches 3 boulders that can fall at one of the points on the grid where the boat can be found. What is the probability that Polyphemus will strike the king of Ithaca? And how does the probability of hitting at least one ship change if they are 2 or 3?

Let us start with the simplest solution to the problem: what is the probability that by launching a Polyphemus rock it will hit Ulysses' ship? The event can be represented as in Figure 23.

Figure 23. Polyphemus launches the boulders towards the ship

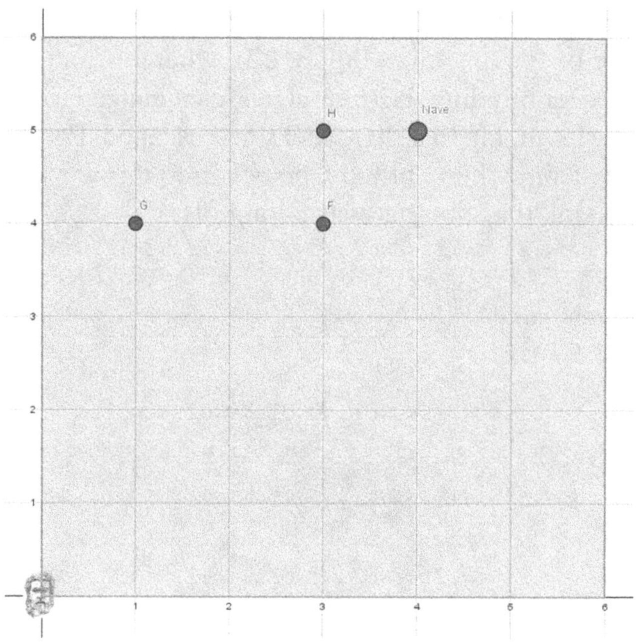

Obviously, the probability will be 1/36, but we can also formulate the problem differently by asking ourselves what is the probability of not hitting the ship and then calculating the opposite probability, that is:

$$1-\left(\frac{35}{36}\right)^{1}$$

As you can easily verify, the two values are the same. The second version, however, is easier for later calculations. For example, if the boulders are 2 the formula becomes:

$$1-\left(\frac{35}{36}\right)^{2}$$

Whereas if there are 2 ships and 3 boulders we can write:

$$1-\left(\frac{34}{36}\right)^{3}$$

In general we can say that, indicating with N the ships and M the boulders, you will have:

$$1-\left(\frac{36-N}{36}\right)^{M}$$

We report the data in Table 8:

Another way to deal with the problem is to simulate the event a fairly large number of times. For example, 1000000 simulations we obtained (Table 9):

TOUCAN

Over the years we have witnessed the development of epidemics within a human population, but also within the animal population. In the human field, we remember the case of smallpox in England and Wales in the years between 1837 and 1839, the Spanish flu that struck between 1918 and 1920, the Asian flu carried by the virus called H2N2, the H3N2 or Hong Kong flu that caused the flu pandemic of 1968, the SARS of 2003, the H1N1 virus of 2009, the Ebola epidemic particularly intense from 2013, the Zika virus of 2015 and the more recent Covid-19 of 2020.

Table 8. Theoretical probabilities

Typology	Formula	P(hit)
1N 1M	$1-\left(\frac{35}{36}\right)^1$	0.028
1N 2M	$1-\left(\frac{35}{36}\right)^2$	0.055
1N 3M	$1-\left(\frac{35}{36}\right)^3$	0.081
2N 1M	$1-\left(\frac{34}{36}\right)^1$	0.056
2N 2M	$1-\left(\frac{34}{36}\right)^2$	0.108
2N 3M	$1-\left(\frac{34}{36}\right)^3$	0.158
3N 1M	$1-\left(\frac{33}{36}\right)^1$	0.083
3N 2M	$1-\left(\frac{33}{36}\right)^2$	0.160
3N 3M	$1-\left(\frac{33}{36}\right)^3$	0.230

In the case of animals, it is more correct to speak of an epizootic disease and important phenomena have occurred, with less uproar, over time, we only mention Venezuelan equine encephalitis in 1995 and the much more terrible avian flu or H5N1, which started at the end of 2003 in the south-east and involved more than 150 million birds.

The development of epidemics can be mathematically modelled through the use of simulations. One of the most famous basic models is the 1920's model known as *Reed-Frost model* due to Lowell Reed (1886-1966) and Wade Frost (1880-1938) which was later subject to extensions, refinements, and modifications. Simplifying for educational needs: this model provides

Table 9. Empirical probabilities

Typology	P(hit)
1N 1M	0.028
1N 2M	0.055
1N 3M	0.081
2N 1M	0.055
2N 2M	0.106
2N 3M	0.155
3N 1M	0.081
3N 2M	0.155
3N 3M	0.224

for a subdivision of the population into three categories, Susceptible (S), Infected (I), Resistant (R) so we also talk in terms of the SIR model. In order to adopt it, several conditions are necessary: every infected person is expected to become ill, heal and become immune; the incubation period is fixed; the conditions remain constant. In the model there is a probabilistic term, the probability q of not having efficient contact, derived as opposed to the probability p that infection will occur if one subject is receptive and the other infected at time t.

Since we said that this is a simplified model we will apply it to a small population of toucans (Figure 24), particularly sensitive to *Yersinia pseudotuberculosis*.

Figure 24. Toucan

The Reed-Frost model is as follows:

$$I_{t+1} = S_t\left(1 - q^{I_t}\right)$$

Suppose we have, in a given area, a population of 100 birds and that one of them is infected (I), then we will have 99 possible susceptibility (S); let's say that the probability of an efficient contact is 2%, so q = 0.98. At this point the model generates the trajectories of the three categories (Figure 25).

Figure 25 Infection trend with p=0.02

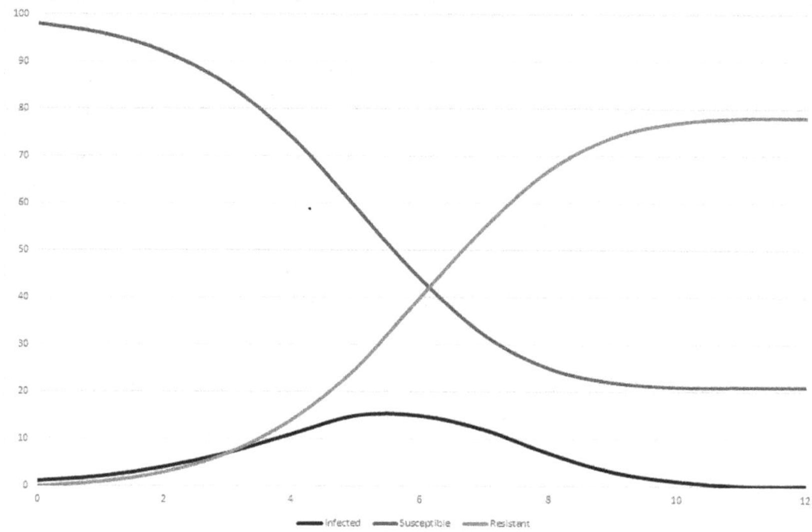

As can be seen, the infection has a peak and then tends to fade as resistant toucans grow. However, it is sufficient to change the risk of infection, for example to 5% to get a different graph (Figure 26).

SNAKES AND LADDERS

Snakes and Ladders is a traditional board game, developed in England, but which refers to a much older Indian game that describes both the importance of chance and the simultaneous presence of good (ladders) and evil (snakes) as reported by Ivars Peterson (1998) which however adopts the American term

Figure 26. Infection trend with p=0.05

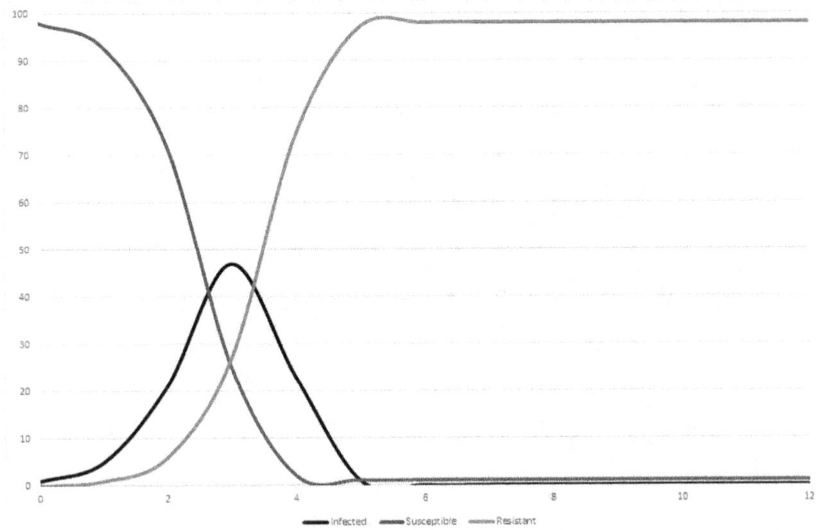

Chutes and Ladders. It is a path game rather similar to the game of the goose. The traditional board represents a path consisting of rows and columns. The path is made more complex by a certain number of "ladders" and "snakes" that cross the board, joining two boxes of different rows (Figure 27).

Similar to what happens in the game of the goose, players advance as many squares as the result of rolling a dice. A marker that arrives at a square "at the foot" of a ladder is moved to the square at the top of the ladder; conversely, a marker that arrives at a square with the mouth of a snake "goes back" to the tail. The game ends when a player reaches or passes the last square.

The size of the table and the presence of snakes and ladders make the study complex and difficult to assess how many moves on average are needed to reach the end of the path. The simulation facilitates the task. The movement along a path can be seen as a *Markov chain* where the rows indicate the starting box and the columns indicate the arrival box.

The most commonly used way to represent A status and p probabilities is the scheme shown in Table 10. Technically it is a square matrix.

This matrix is called a *transitional matrix* and each line, in which the sum gives 1, represents a probability distribution. The initial situation, which expresses the initial configuration, can be defined with a vector V, i.e. a matrix with only one row and n columns. It is possible to analyze the evolution of the process by making the product of the vector V for the matrix P.

Figure 27. Snakes and Ladders

Table 10. Representation of status and probabilities

		Final status				
Initial status		A_1	...	A_j	...	A_n
	A_1	p_{11}	...	p_{1j}	...	p_{1n}
	...	...	...	...	...	...
	A_t	p_{i1}	...	p_{ij}	...	p_{in}
	...	...	...	...	...	...
	A_n	p_{n1}	...	p_{nj}	...	p_{nn}

The matrix indicates the probability of transition, in other words, the probability of passing from status A to status B by rolling a dice. For example, to go from square 1 to square 5 the probability is 1/6. This is a generalization of the pattern of repeated tests. By multiplying *V* by *P* you can obtain successive matrices that indicate the probability of going from a certain square to another by rolling the dice. In Figure 28 the first two matrices.

Figure 28. First two transition matrices

1 move	0	1	2	3	4	5	6	7	8	9	10
0	0.00	0.17	0.17	0.17	0.17	0.17	0.17	0.00	0.00	0.00	0.00
1	0.00	0.00	0.17	0.17	0.17	0.17	0.17	0.17	0.00	0.00	0.00
2	0.00	0.00	0.00	0.17	0.17	0.17	0.17	0.17	0.17	0.00	0.00
3	0.00	0.00	0.00	0.00	0.17	0.17	0.17	0.17	0.17	0.17	0.00
4	0.00	0.00	0.00	0.00	0.00	0.17	0.17	0.17	0.17	0.17	0.17
5	0.00	0.00	0.00	0.00	0.00	0.00	0.17	0.17	0.17	0.17	0.33
6	0.00	0.00	0.00	0.00	0.00	0.00	0.00	0.17	0.17	0.17	0.50
7	0.00	0.00	0.00	0.00	0.00	0.00	0.00	0.00	0.17	0.17	0.67
8	0.00	0.00	0.00	0.00	0.00	0.00	0.00	0.00	0.00	0.17	0.83
9	0.00	0.00	0.00	0.00	0.00	0.00	0.00	0.00	0.00	0.00	1.00
10	0.00	0.00	0.00	0.00	0.00	0.00	0.00	0.00	0.00	0.00	1.00

2 move	0	1	2	3	4	5	6	7	8	9	10
0	0.00	0.00	0.03	0.06	0.08	0.11	0.14	0.17	0.14	0.11	0.17
1	0.00	0.00	0.00	0.03	0.06	0.08	0.11	0.14	0.17	0.14	0.28
2	0.00	0.00	0.00	0.00	0.03	0.06	0.08	0.11	0.14	0.17	0.42
3	0.00	0.00	0.00	0.00	0.00	0.03	0.06	0.08	0.11	0.14	0.58
4	0.00	0.00	0.00	0.00	0.00	0.00	0.03	0.06	0.08	0.11	0.72
5	0.00	0.00	0.00	0.00	0.00	0.00	0.00	0.03	0.06	0.08	0.83
6	0.00	0.00	0.00	0.00	0.00	0.00	0.00	0.00	0.03	0.06	0.92
7	0.00	0.00	0.00	0.00	0.00	0.00	0.00	0.00	0.00	0.03	0.97
8	0.00	0.00	0.00	0.00	0.00	0.00	0.00	0.00	0.00	0.00	1.00
9	0.00	0.00	0.00	0.00	0.00	0.00	0.00	0.00	0.00	0.00	1.00
10	0.00	0.00	0.00	0.00	0.00	0.00	0.00	0.00	0.00	0.00	1.00

Of course, the process is complicated by the presence of solutions that may favor advancement or relegation. For example, the player who has reached square 42 in the diagram in Figure 29, with the roll of the dice can find himself in one of the following squares: 77, 44, 45, 5, 47, 11. For this reason a game can end in a short time or it needs many rolls. A simulation repeated several times can only provide an average value. By way of example, I simulated 10 games that ended with an average of 44 throws, with a minimum of 11 and a maximum of 142 throws. Various authors have expressed themselves on the matter, in some cases referring to slightly different versions of the game, however all agree that the variance is definitely high.

These chains, named after the Russian mathematician A.A Markov (1856-1922), are particularly useful for dealing not only with situations described in this game but above all with more significant situations such as: market evolution, genetic problems, changes in the composition of a population, etc.

In the case of a Markovian variable the dependence on the previous result is given by:

$$P\left(X_{n+1}=j|X_n=i\right)=\begin{cases} p & per\ j=i+1 \\ 1-p & per\ j=i-1 \\ & otherwise \end{cases}$$

Figure 29. Possible results starting from box 42

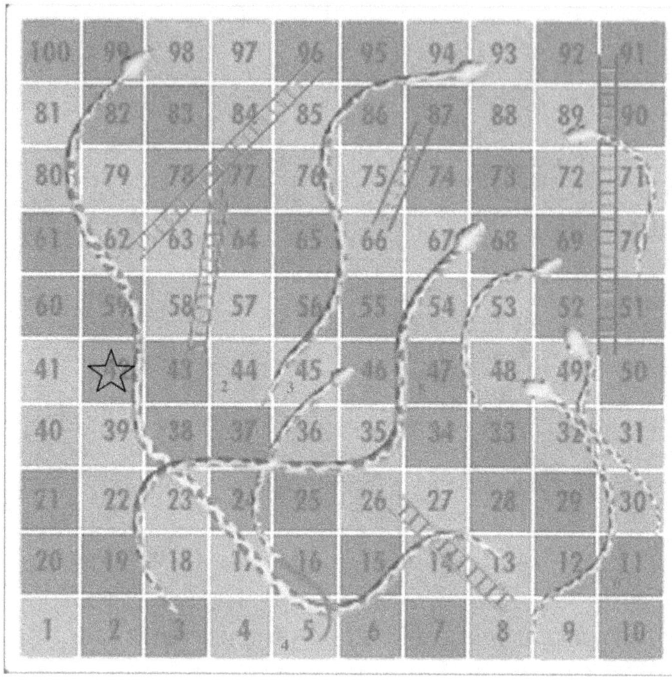

The results from the *Aleatory variable*. X_{n+1} are called "system states". We will have therefore that a system will evolve from one state to another with probability *p*. In this case we are faced with a typical aleatory behavior with temporal order.

A simple example can be used to clarify the concept.

Every day I can decide whether to buy fresh bread or tap into the frozen bread supply. Although I am a bit lazy, I decide not to eat frozen bread two days in a row, and if I have eaten fresh bread the day before, I decide what to do by flipping a coin. There are two possible results, so the system states are A_1 = "eating frozen bread" and A_2 = "eating fresh bread". The times are the days. The transition probabilities are expressed in the transition matrix:

$$P = \begin{pmatrix} 0 & 1 \\ 0.5 & 0.5 \end{pmatrix}$$

which can also be traced back to an oriented graph (Figure 30).

Figure 30. Transition graph

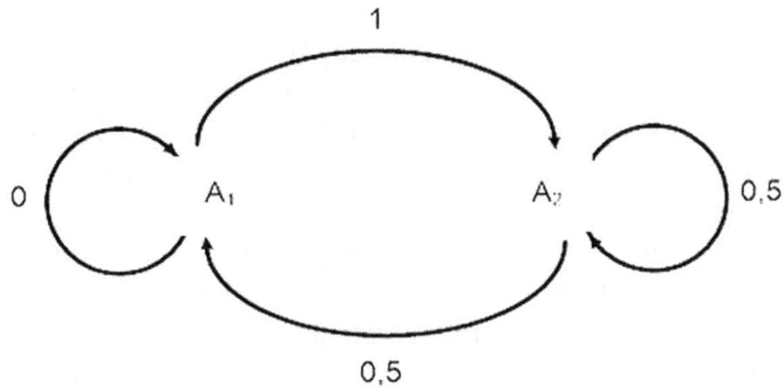

Assuming that the initial situation is that today I ate frozen bread, then $V_0 = (1\ 0)$ the next state vector is given by the product of this vector for the probability matrix. In conclusion:

$$V_1 = V_0 p = \begin{pmatrix}1 & 0\end{pmatrix}\begin{pmatrix}0 & 1\\ 0.5 & 0.5\end{pmatrix} = \begin{pmatrix}0 & 1\end{pmatrix}$$

$$V_2 = V_1 p = \begin{pmatrix}0 & 1\end{pmatrix}\begin{pmatrix}0 & 1\\ 0.5 & 0.5\end{pmatrix} = \begin{pmatrix}0.5 & 0.5\end{pmatrix}$$

$$V_3 = V_2 p = \begin{pmatrix}0.5 & 0.5\end{pmatrix}\begin{pmatrix}0 & 1\\ 0.5 & 0.5\end{pmatrix} = \begin{pmatrix}0.25 & 0.75\end{pmatrix}$$

$$V_4 = V_3 p = \begin{pmatrix}0.25 & 0.75\end{pmatrix}\begin{pmatrix}0 & 1\\ 0.5 & 0.5\end{pmatrix} = \begin{pmatrix}0.375 & 0.625\end{pmatrix}$$

Continuing the calculation, if you follow the evolution of the experiment you get that the values converge to $\begin{pmatrix}\frac{1}{3} & \frac{2}{3}\end{pmatrix}$.

CONCLUSION

Some games present difficulties with regard to school use, e.g. Snakes and Ladders has an excessive running time to be properly exploited. In this case, reduced versions can be used which retain the basic elements of the problem. Later on, a simulation based on the original structure can be shown. When we set the process of solving a problem, the original problem must be correctly transformed into the simulated model. A prerequisite is that the model is equivalent, from a probabilistic point of view, to the original problem.

The theme of simulation is fundamental in the teaching of probability. The possibility of repeating an experiment an extremely high number of times in a short time is particularly important. This possibility of repetition refers to a frequent approach and makes good, for example, the sense of the law of large numbers. The interweaving of problems and simulation models represents a fundamental node.

FUTURE RESEARCH DIRECTIONS

The simulation process is not simple, in fact it requires at least three conditions: 1) the modelling must make the complexity of the problem while maintaining a sufficient level of simplicity; 2) the use of the computer tool must not constitute a "black box" for the student; 3) the teacher must possess the necessary skills and therefore it is essential to think about what we could define the "learning curve" of the tool adopted.

REFERENCES

Andrà, C., Parolini, N., & Verani, M. (2015). Using gambling simulators to foster awareness about gambling risks: A focus on emotions. *Digital Experiences in Mathematics Education*, *1*(1), 59–78. doi:10.100740751-015-0005-1

Buffon, G. L. Count of. (1733). Solution de problèmes sur le jeu du Franc-carreau, in Essai d'arithmétique morale, Histoire Naturelle, générale et particulière. *Imprimerie Royale*, 46-148.

Ellegood, W. A., Solomon, S., North, J., & Campbell, J. F. (2020). School bus routing problem: Contemporary trends and research directions. *Omega*, *95*, 102056. doi:10.1016/j.omega.2019.03.014

Evans, M. J., & Rosenthal, J. S. (2004). *Probability and statistics: The science of uncertainty*. Macmillan.

Feldman, L., & Morgan, F. (2003). The pedagogy and probability of the dice game hog. *Journal of Statistics Education: An International Journal on the Teaching and Learning of Statistics*, *11*(2). Advance online publication. do i:10.1080/10691898.2003.11910708

Henry, M., & Parzysz, B. (2011). Simulating random experiments with computers in the classroom: indispensable but not so simple. Actes de Congress of the European Society for Research in Mathematics Education (CERME7).

Nahin, P. (2013). *Will You Be Alive 10 Years from Now?: And Numerous Other Curious Questions in Probability*. Princeton University Press.

Nix, S. J. (1994). *Urban stormwater modeling and simulation*. CRC Press.

Obropta, C. C., & Kardos, J. S. (2007). Review of Urban Stormwater Quality Models: Deterministic, Stochastic, and Hybrid Approaches 1. *Journal of the American Water Resources Association*, *43*(6), 1508–1523. doi:10.1111/ j.1752-1688.2007.00124.x

Peterson, I. (1998). *The jungles of randomness: A mathematical safari*. Penguin.

Pickover, C. A. (2002). *The mathematics of Oz: mental gymnastics from beyond the edge*. Cambridge University Press.

Shaughnessy, J. M., & Ciancetta, M. (2002). Students' understanding of variability in a probability environment. In *Proceedings of the Sixth International Conference on Teaching Statistics: Developing a Statistically Literate Society*. International Statistical Institute.

ADDITIONAL READING

Althoen, S. C., King, L., & Schilling, K. (1993). How long is a game of snakes and ladders? *The Mathematical Gazette*, *77*(478), 71–76. doi:10.2307/3619261

Pfannkuch, M., & Ziedins, I. (2014). A modelling perspective on probability. In *Probabilistic Thinking* (pp. 101–116). Springer. doi:10.1007/978-94-007-7155-0_5

KEY TERMS AND DEFINITIONS

Markov Chain: A model in which the probability of each event depends only on the state attained in the previous event.

Median: The value separating the higher half from the lower half of a probability distribution.

Pseudo Random Numbers: Numbers generated by a deterministic algorithm that produces a sequence with statistical properties similar to those of a sequence of numbers generated by a random process.

Redd-Frost Model: A mathematical model of epidemics.

Sample Space: An experiment or random trial is the set of all possible outcomes.

Chapter 9
Probability and Statistics

ABSTRACT

Statistics and probability both fall within the broader scope of the theory of random phenomena. The first deals with providing probability distributions adaptable to the various real random phenomena, and the second deals very often with a random sample to describe its properties or infer to the underlying probabilistic model and the estimation of its parameters. The chapter tries to show this connection by reporting examples that are more or less known but that are characterized by being unconventional. Other objects could have been taken into consideration but those chosen are characterized by a closer link with the calculation of probabilities.

INTRODUCTION

Statistics are like bikinis. What they reveal is suggestive, but what they conceal is vital. ~ Aaron Levenstein

Statistics is an autonomous discipline, whose history has something fascinating (Hald, 1998), but with a strong link to probability (Rohatgi & Saleh, 2011). Both probability calculation and statistics are mathematical tools developed to study case-dependent phenomena whose outcome cannot, therefore, be predicted with certainty. Probability provides a quantitative estimate of the ease with which an event may or may not occur. Statistics describe populations and their variability or serve to make inferences from a sample. The link between the two disciplines is strong, especially if we consider the frequentist

DOI: 10.4018/978-1-7998-3871-5.ch009

approach to probability. The examples are innumerable and therefore we will limit ourselves to only a few before introducing some objects that better clarify the subject.

For example, we could ask ourselves whether the influence of certain religious practices could, with a certain degree of probability, influence the course of the disease.

We take our cue from a 2006 research (Benson et al, 2006) on the therapeutic effects of prayer. Many patients undergoing coronary bypass surgery and susceptible to complications were divided into three groups equally divided according to religious affiliation. Group A could receive or not receive the prayer activity and received it, group B was placed in the same conditions as the previous one but did not receive the prayer, group C received, as promised, the prayer (Table 1).

Table 1. Results of the experimentation

Group	A	B	C
N	604	597	601
Complications	315	304	352
C/N ratio	0.52	0.51	0.59

The relationship can be understood as the likelihood of complications based on the treatment of the three groups. Interesting are the conclusions: Prayer has had no effect on healing without complications, but above all there are more complications among those who receive the expected prayer.

We are not interested here in evaluating the usefulness or otherwise of prayer, but only in pointing out that randomized controlled trials (RCTs) are experimental studies that make it possible to evaluate the effectiveness of a specific treatment in a given population. Randomization makes it likely that other variables not considered will be evenly distributed across groups. In this case the differences observed between the three groups can be attributed to the main variable (prayer) and lead to a probabilistic evaluation.

The second field of investigation concerns land transport, the question is: what is the probability of being involved in an accident?

Let us look at data from some European countries. In the Netherlands, in 2018, the number of deaths in accidents involving cyclists or caused by cyclists (206) exceeded the number of deaths in cars (201). The high number of fatal cycling accidents is certainly linked to the increasingly extensive use of bicycles (especially electrically assisted bicycles). In Italy, cycling seems to be the most dangerous choice of mobility in absolute terms: the risk rate of a fatal accident is 2.18%, higher than buses (0.48%), or trucks (0.67%), or cars (0.78%), or mopeds (1.06%) and motorcycles (1.96%). In France, a 2016 study shows that mortality among motorists is substantially stable, while among cyclists there is a strong growth of +12%.

All these estimates are derived from statistical collections, these data are essential as they allow us to reason in terms of probability. Unfortunately, the data are not easily comparable because of the different methods of collection and the different periods to which the surveys refer.

Another field in which statistics provide opportunities for probabilistic analysis concerns data on the age and gender distribution of a population. These data are represented with a graph called "population pyramid". It is composed of two histograms arranged symmetrically (to differentiate between males and females) around the vertical axis representing the ages. In abscissa is indicated the amount of population for each age class (Figure 1).

One of the most recent studies on the subject is due to John Weeks (2020) while the authorship of the demographic pyramid is uncertain, the first to present data in this mode were the astronomer Edmond Halley (1656-1742) and the statistician John Graunt (1620-1674), who also calculated the first mortality table. Graunt built his tables by observing deaths by age, gender, and cause of death in London. These tables show the probability of death between birthdays.

The mortality table is mainly used in the insurance sector, as it provides the technical basis for the calculation of insurance premiums. By showing the numerical consistency of past and current age groups, it provides indications to predict, for example, how many older people there will be in a certain number of years. In this sense, population pyramids can be used to predict the future.

BOOK

The book is something special, as Christopher Morley (1890-1957) said: "When you sell to a man a book, you don't sell him 12 ounces of paper and ink and glue - you sell him a whole new life".

We will use this object as a starting point to analyze the relationship between probability and statistics. Among the books, we choose a small masterpiece, *The Gold-Bug* (Figure 2).

Figure 1. Italian age pyramid (2019)

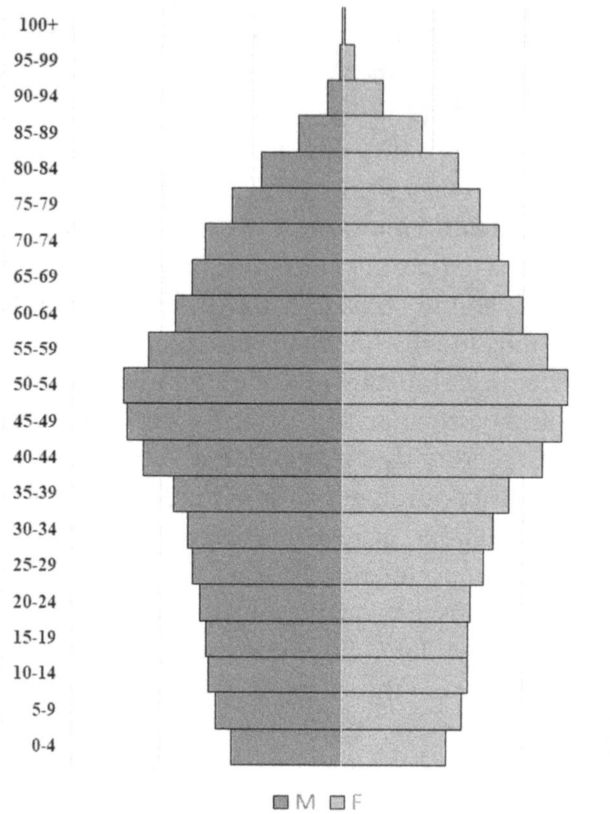

Imagine dropping a grain of rice on a page of this book. Is it more likely that the grain covers the "e" or the "j" typeface? Of course the answer seems obvious, in the English language the first letter occurs about 140 times more frequently than the other. In the Italian language this ratio rises even to 640! Each language differs from the others according to the probability of finding a certain character within a text. On the whole these characteristics constitute a real fingerprint of the language. For example, English boasts a high frequency of consonants while in Italian the presence of vowels is stronger (Figure 3).

This consideration is the basis for language studies and statistical cryptanalysis. Let us remember that cryptography is the science of protecting information by making it incomprehensible so that it can only be understood by the recipient.

Figure 2. Cover of The Gold-Bug

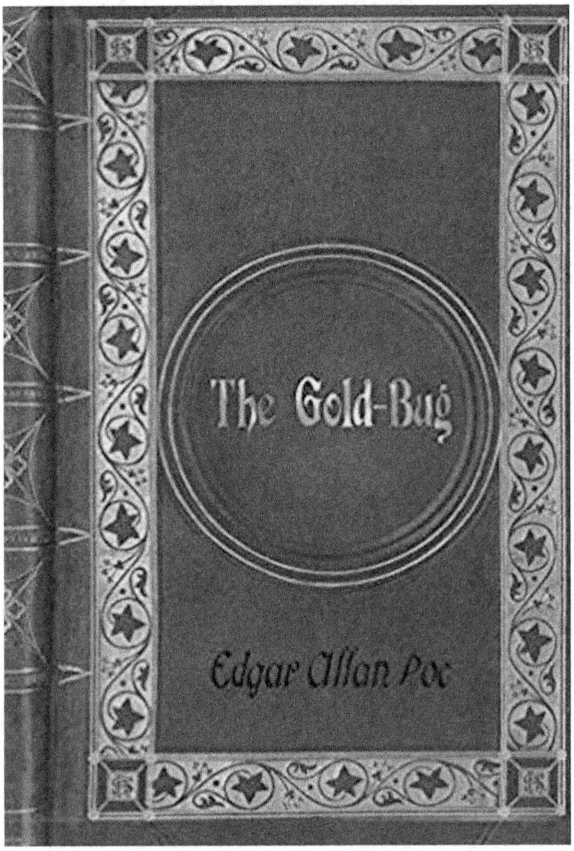

Figure 3. Comparison between English and Italian

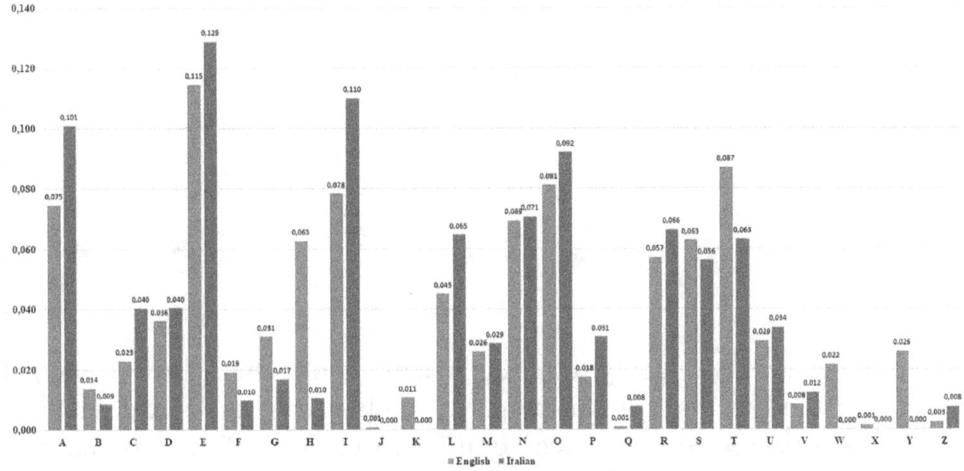

As far as the linguistic aspects are concerned, one cannot fail to mention *Zipf's law*, due to George Kingsley Zipf (1902-1950), an American linguist who studied the occurrences in different languages (zipf, 1932).

The frequency of any word is, empirically, inversely proportional to its rank in the frequency table. Thus the most frequent word occurs about twice as often as the second most frequent word, three times as often as the third most frequent word, etc.

The law is simple but effective. By setting k as the maximum frequency value and as variable parameter a, the frequency F is given by:

$$F = k \bullet rank^{-a}$$

For example, analyzing the ten most frequent words in the text already mentioned in the previous chapter of Shaughnessy and Ciancetta we can find a fair agreement with the law (Figure 4).

Figure 4. Analysis with Zipf model

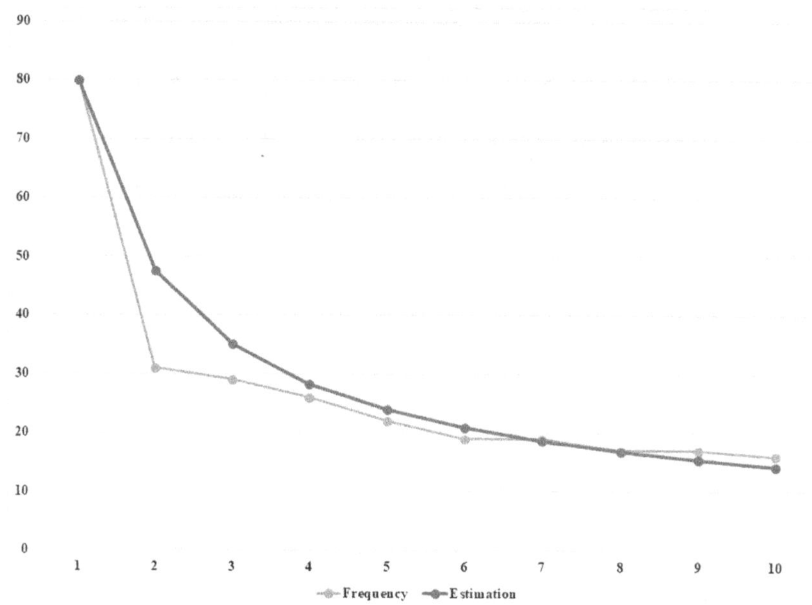

Zipf's law seems to apply to all languages even if the reason is not yet well understood, perhaps it can be partly explained by statistical analysis of randomly generated texts. According to Li (1992): "The facts that the frequency of occurrence of a word is almost an inverse power law function of its rank and the exponent of this inverse power law is very close to 1 are largely due to the transformation from the word's length to its rank"

But let's get back to the subject of cryptography. Edgard Allan Poe (2016) was also involved in the aforementioned story *The Gold-Bug*. The story, for those who did not know it, can be summarized as a treasure hunt guided by a coded message. Due to a series of fortuitous circumstances, the protagonist Legrand discovers that a parchment contains a cipher message written with sympathetic ink. When heated slightly, it shows this aspect (Figure 5):

Figure 5. The coded text of the parchment

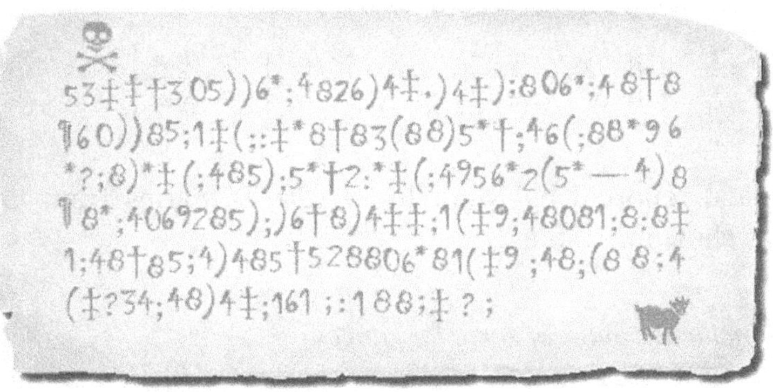

For Legrand ''the solution is by no means so difficult as you might be led to imagine from the first hasty inspection of the characters". The protagonist continues: ''Now, in English, the letter which most frequently occurs is e. Afterward, the succession runs thus: a o i d h n r s t u y c f g l m w b k p q x z. E predominates so remarkably, that an individual sentence of any length is rarely seen, in which it is not the prevailing character". With a series of reasoning, partly statistical and partly linguistic, he comes to decipher the text: ''A good glass in the bishop's hostel in the devil's seat forty-one degrees and thirteen minutes northeast and by north main branch seventh limb east side shoot from the left eye of the death's-head a bee-line from the tree through the shot fifty feet out.'''

A good example of a link between literature, probability, and statistics, possible thanks to the fact that the message has been encrypted using a single substitution code, "8" instead of "e", ";" instead of "t", "4" instead of "h" and so on. This method is largely outdated because simple substitution cryptography is too easy to make clear and therefore the example has a mainly historical value. This does not detract from the fact that other more evolved forms are widespread among the military, banks, companies that sell online, etc..

Another aspect related to cryptography refers to the coincidence index, widely treated by the cryptographer William Friedman (1987): it measures the probability that, taken at random from a text, two letters are equal. The probability of selecting a certain letter of which there are *n* occurrences in a text made up of *N* characters is obviously p = n/N; the probability that the next one is the same is p = (n-1)/(N-1). Applying the law of the product in case of independent events we can obtain the index for all the letters of the alphabet, placing *k* as a constant that depends on the number of letters of which the alphabet is composed (26 for the Anglo-Saxon one):

$$IC = k\bullet\left(\frac{n_a}{N}\bullet\frac{n_a-1}{N-1}\right)\left(\frac{n_b}{N}\bullet\frac{n_b-1}{N-1}\right)\left(\frac{n_c}{N}\bullet\frac{n_c-1}{N-1}\right)\cdots\left(\frac{n_z}{N}\bullet\frac{n_z-1}{N-1}\right)$$

This index, reported by Mark Frary (2017), varies from language to language as shown in Table 2.

Table 2. Coincidence index of some languages

Language	IC
English	1.73
Russian	1.76
Italian	1.94
French	2.02

The knowledge of this index would seem to be of little interest, in fact it has great importance to reveal the number of alphabets used and fundamental for the decryption of a message.

LOVE DROPS

This object (Figure 6) would seem very distant from the topic we are dealing with; in reality the mathematical implications are absolutely intriguing. First of all it must be said that Love Me Drops is an herbal product, classified as an aphrodisiac and stimulant; a help for all those who seek love.

Figure 6. Love drops

But is there a secret formula for love?

Perhaps a chemical or a magic filter can help us, but only mathematics can provide us with tools to better understand how to deal with this feeling.

Love dynamics are definitely complex and apparently impossible to deal with mathematical models. Several examples contradict this thesis. Sergio Rinaldi (1988), starting from the work of Frederic J. Jones (2014), from the University of Wales, has proposed and analyzed a minimal model of love dynamics between a poet and a woman. The model is adapted to the case of a famous 14th century Italian poet, Petrarch; he sang Laura's praises in his *Canzoniere*. There is a cyclical behavior of the two lovers, with alternating periods of love exaltation and periods of despair. The probability associated with Petrarch's passion can be seen in the author's graph (Figure 7) .

The case just examined refers to a relationship that lasts over time and with a single woman, but the objective here is different, it is to find the ideal partner (or at least the most acceptable one).

Figure 7. Graph of the love between Petrarch and Laura

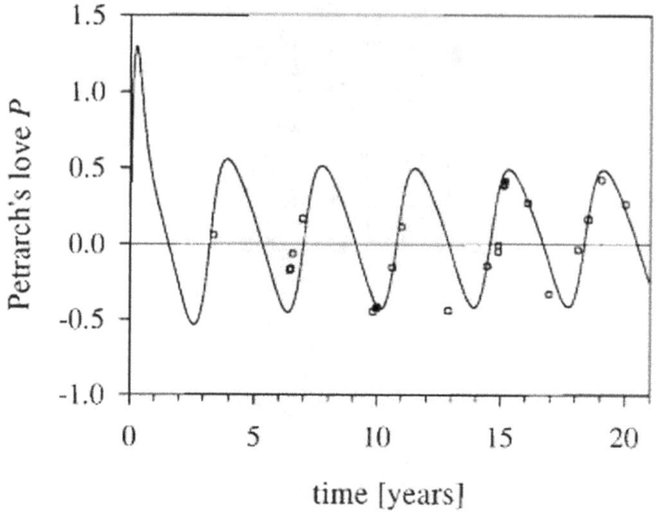

The most interesting suggestion comes to us from mathematician Hannah Fry (2015), in one of her books introduces the “37% rule”. But what is it? What is the link between probability and statistics?

We can think of the possibility of finding the right person based on the “optimal arrest theory”. An algorithm defines the limit before which to discard all the “experienced” partners and beyond which to start considering the idea of settling in. In other words, suppose we can classify the possible partners with a grade, mathematics tells us that the most efficient strategy is to discard 37% of the first experiences and then choose the partner who will first exceed the maximum score achieved so far. This is the best method to maximize the chances of finding the perfect partner, i.e. it is the best strategy to increase the probability of not making a mistake that could be irreparable, but obviously does not ensure a 100% success rate. There may also be other strategies on a set of *n* partners.

Some may object that elements that cannot be labeled as rationally conscious choices may enter into the choice, as particular alchemies that unleash our senses may be skillfully manipulated. The decision on the vote to be given depends on many factors, not least the way one presents oneself publicly. Can a seductive dress and a pair of 12-heeled shoes deflect the decision-maker and not let him/her take into account other aspects such as intelligence, reliability, caring skills, etc.? The answer probably doesn’t exist. It may be fun, however, to examine a geometric and statistical research of the Institute of Physics in London led by Dr. Paul Stevenson in which, in the formula for calculating the heel height suitable for each woman, the probability peeps out.

The formula for calculating the ideal heel height is as follows:

$$h = Q\left(12 + \frac{3s}{8}\right)$$

where;

h is the maximum heel height (in cm)
s is the size of the shoe (UK women's size).
Q is a sociological factor and has a value between 0 and 1 defined as follows:

$$Q = \frac{p(y+9)L}{(t+1)(A+1)(y+10)(L+20)}$$

Let's see the meanings of the variables:

1. The variable p is the probability that the heels make men dizzy. 1 if all men look back with admiration, 0 if men look the other way;
2. The y indicates experience in using heels (in years);
3. The L indicates the cost of the shoes (in British pounds);
4. The t indicates the age of the shoe (in years);
5. The A indicates how many glasses of wine are expected when the shoes are worn: 0, 1, 2 and so on.

Beyond the judgement that can be made of the equations, the presence of factor p, that is, a subjectively assessed probability, is relevant.

Remaining within this theme, one can imagine a love game in which the girl puts to the suitor the following condition: the relationship can begin if the boy manages to form a "bracelet". The girl holds six pieces of twine in her hand, the ends of which come out of her fist, and the suitor has to tie them. When the girl opens her fist, if a ring has formed, then she will accept the court. Difficult task? Apparently it seems that the girl has not given the boy a lot of chances, but that's not the case.

There are fifteen possible ways of knotting the heads that come out at the top. If you choose one garment, it can be knotted with one of the other five, the second free garment can be tied with only three and finally there are only two garments left to tie; 5 • 3 • 1 = 15. At this point it is a matter of knotting the other ends, if you choose one garment it is obvious that it is not convenient to use the same twine that was used for the first knot, this reduces the possibilities to 4 • 2 • 1 = 8. The probability of succeeding in the test is therefore:

$$P = \frac{8}{15} \cong 0.53$$

The boy has practically more than 50% chance of being accepted, which is quite counterintuitive. If he fails, the young man could ask for a proof of appeal, in this case the probability increases a lot, in fact the opposite probability is 0.47 and two refusals give a probability 0.47 • 0.47 = 0.22; then the chances for the boy increases to 78%.

The problem is easy to generalize. For two strings the probability is 1, for four it is 2/3, for six it is 2/3 • 4/5, for eight 2/3 • 4/5 • 6/7 and so on.

This game is an adaptation of the game of grass strands in vogue among girls in rural areas of the Soviet Union, presented by the famous mathematician and scientific popularizer Martin Gardner (1971).

MINI SOCCER

Probability, intended as an attempt to identify structures in random events, plays an important role in the analysis of football. A football game has, to a certain extent, aspects of predictability, but it remains essentially a random phenomenon as well as a Mini Soccer game (Figure 8).

The analysis of this sport can be carried out at different levels, the most widespread being the statistical approach. For example, you can easily answer the question whether playing at home increases the chances of winning (1 indicates the home win, X the draw and 2 the victory of the visiting team).

If we look at the Italian league championship, the years from 2011 to 2019, the advantage of playing at home is evident. In any case, it can be seen that, except for some fluctuations, there are no significant variations over the years, as shown in Table 3.

Figure 8. Mini Soccer

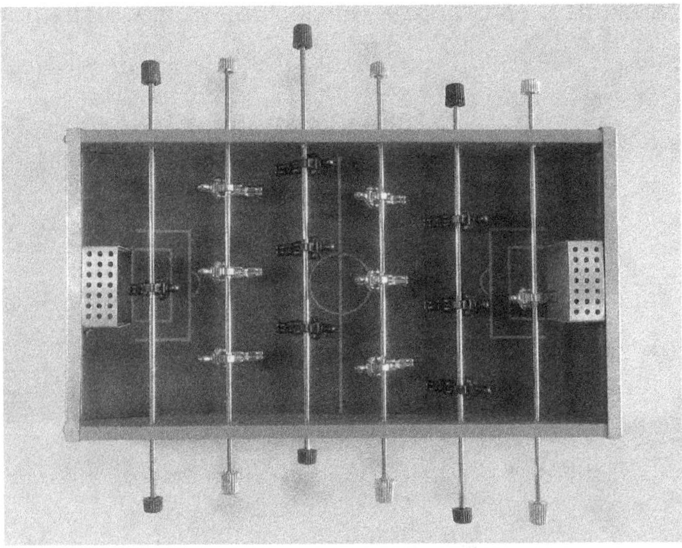

Table 3. Historical series results Italian series A

	11-12	12-13	13-14	14-15	15-16	16-17	17-18	18-19	Average
1	46%	47%	47%	40%	46%	48%	43%	43%	**45%**
X	29%	25%	24%	32%	25%	21%	22%	29	**26%**
2	25%	28%	29%	28%	29%	31%	35%	28%	**29%**

But the discussion can be broadened.

Another issue concerns the predictions that can be made regarding the outcome of a match, considering the history of the teams examined and also comparing the results obtained with the same teams already faced. The value that expresses the condition of the teams, and therefore the quantification of the relative strength of the contenders, is synthetically defined by VF (Value of Form). In other words, this numerical value can be considered as an indicator of a team's chances of winning.

The football match is a random event but it is possible to determine a precise pattern. There are several classification methods for pattern recognition, which can be divided into two different types: Non-Sequential Classification Methods that include machine learning algorithms that do not consider the time factor (e.g. Bayesian models) and Sequential Classification Methods (e.g. Markov models).

Football is an incredible mine of cues: for example, how do the chances of scoring a goal differ according to the position from which the shot is taken (Figure 9)?

Figure 9. Goals by position

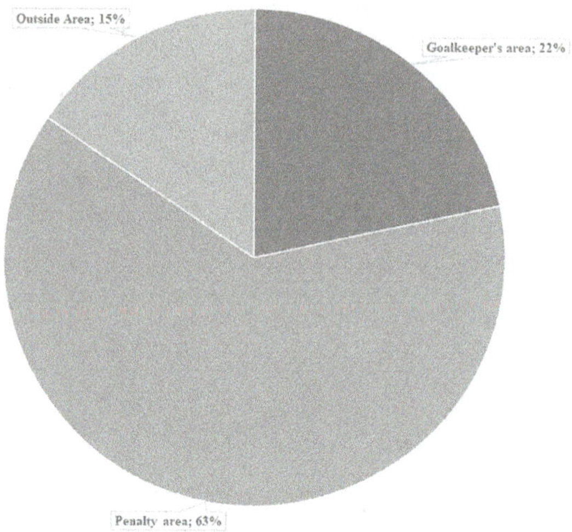

A particularly studied aspect is the relationship between sports results (economists would speak of productivity) and the total wages of a football team. Various analyses confirm that the teams that spend the most to pay their players get better results. Starting from this consideration, we want to ask ourselves a further problem: is there a relationship between sports results and the distribution of payroll within the team? In other words: is it more likely that the best results come from a team in which players are paid very differently or is the opposite more likely?

To tackle this issue, it is first of all necessary to collect data on players' salaries and results achieved at the end of the season in terms of points. Then we must define an indicator capable of measuring the greater or lesser wage concentration. In the case of income the concentration is maximum when only one individual receives all the income of the team while in the case of equal distribution all income is divided into equal parts between players.

The ultimate goal of the study is to synthesize the information on the phenomenon through a single numerical value, called concentration index.

The choice of the indicator to use fell on *Gini's index*, a number between 0 and 1. The Gini's index is a measure used mainly to measure the inequality of a wealth distribution. Low values of the coefficient indicate a homogeneous distribution. Most developed European nations show Gini coefficients between 0.24 and 0.36 for the population's income,

For Italian football teams this index is between 0.19 and 0.36 for the year 2015-16, between 0.16 and 0.26 for the following year, between 0.24 and 0.40 for the year 2017-18. For the year 2018-19 the index ranges from 0.22 for Cagliari to 0.42 for Juventus.

The understanding of the phenomenon is clearer by using a graphical representation, easy to interpret, called the Lorenz curve.

First of all, the data must be in order, the abscissae will show the cumulative frequencies F_i of the income recipients, while the cumulative quantities Q_i related to the income shares will be shown on the abscissae. In the case of income, this will mean carrying over the income shares and income shares, respectively. The break of concentration is obtained by joining on the Cartesian plane the points of coordinates (F_i, Q_i). The bisector indicates the equitable distribution.

The graphs of the two teams that in 2018-19 have the lowest (Figure 10) and highest (Figure 11) index value are very indicative.

Figure 10. Lorenz curve of Cagliari

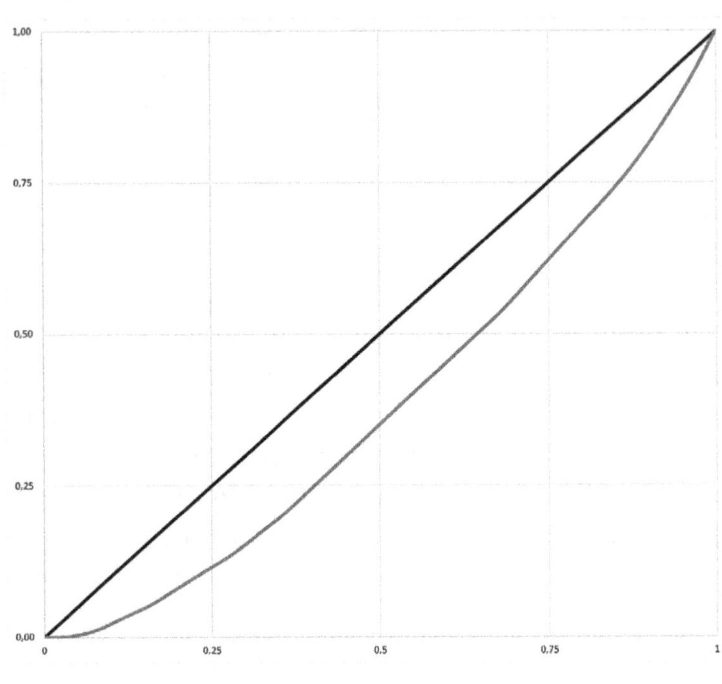

Figure 11. Lorenz Curve of Juventus

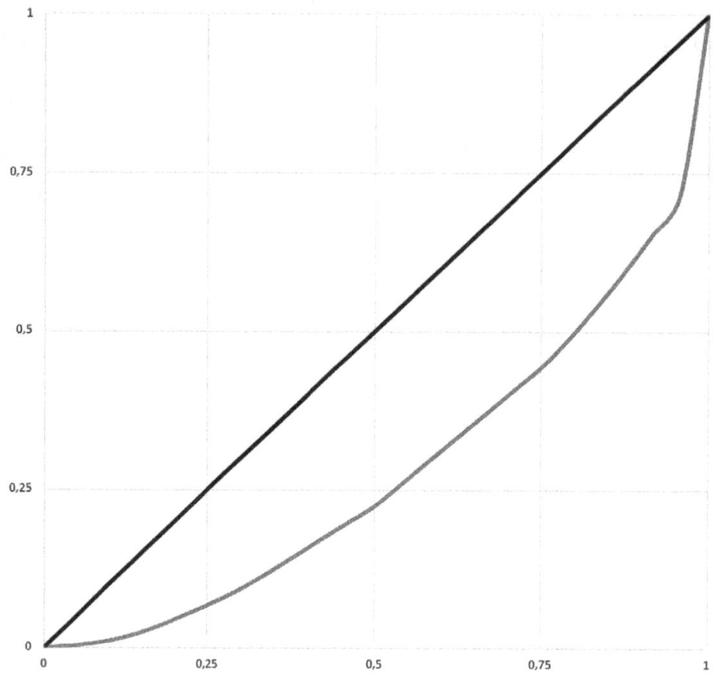

The graphs allow us to answer the initial question: better results are more likely to be achieved by teams with higher wage imbalances. In fact, Juventus has won all the championships since 2011-12 while teams like Cagliari have often found themselves at the bottom of the league table or relegated. Of course there are exceptions that leave space for a "necessary" unpredictability.

BASKETBALL

Another sport that is easy to relate to probability and statistics is basketball (Figure 12). And we are not talking about odds and betting, but other aspects.

One of them is related to the probability of making a basket. Among the many attempts to provide the definitive answer to the problem we can cite the proposal of Professor Larry Silverberg (2011) on the infallible method to go to the basket: "These shots were launched from 7 ft above the ground, with 3 Hz of back spin, and assumed a standard deviation in launch speed that corresponds to a 70% direct shot from the free throw line". This technique should provide a 20% advantage over normal modes.

Figure 12. Table Basketball

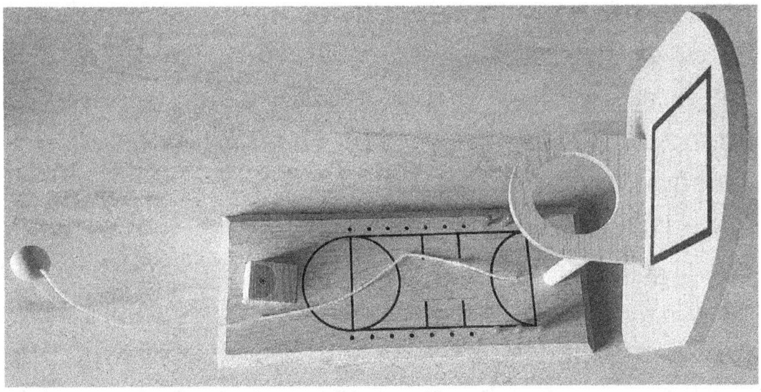

The most incredible aspect is the possibility to determine, thanks to software, in real time, what can happen in a game. To get an idea of the potential we suggest seeing the TEDTalk 2015 by Professor Rajiv Maheswaran, with the commentary of an old NBA final. The game is recreated in a computer simulation: for each phase, the odds are automatically determined. Here is his commentary on the game and considerations on the simulation: “Miami was down by three, there was 20 seconds left. They were about to lose the championship. A gentleman named LeBron James came up and he took a three to tie. He missed. His teammate Chris Bosh got a rebound, passed it to another teammate named Ray Allen. He sank a three. It went into overtime. They won the game. They won the championship. It was one of the most exciting games in basketball. And our ability to know the shot probability for every player at every second, and the likelihood of them getting a rebound at every second can illuminate this moment in a way that we never could before”.

Those interested in simulations of the possible distributions of this game can refer to the work “Basketball” developed by Theodosia Prodromou in her doctoral thesis.

One of the most used phrases in basketball, but not only that, is “He’s on fire”. A basketball player starts to hit the basket and it seems that he is in a state of grace that helps to increase the loot. The issue has been examined by several scholars and the dominant opinion is that it is an illusion. This thesis has been questioned by Miller & Sanjurjo (20180 They argue that it is a phenomenon that really exists: they believe that after a certain number of shots have been made, the probability of making even more shots is greater than the probability of making mistakes. Thanks to the data obtained from SportVU’s optical tracking we can see that players who have managed to make a good performance, scoring more consecutive shots than their standards, tend to try to go to the basket again, this determines a small but significant Hot Hand effect. As the researchers Bocskocsky, Ezekowitz & Stein (2014) say, somewhat sarcastically, “At the very least, our findings cast doubt on the overwhelming consensus that the Hot Hand is a fallacy. Perhaps the next time a professor addresses the Harvard Men’s basketball team, the Hot Hand will not be so quickly dismissed”.

Another aspect to be taken into consideration concerns the evaluation of the results of the basket shots. Suppose a coach has to evaluate the performance of two players: *A* collected 10 baskets and 10 attempts failed, *B* collected 8 baskets and 12 failures. So *A* has a higher chance of hitting the basket than *B*: 10/20 > 8/20; the coach should prefer *A*.

This analysis is superficial, however, does not take into account that the baskets have a different value. Depending on the position at the time of shooting, you may have three-point or two-point scoring. If you split the data (the values in Table 4 are fictitious but can be considered reliable for our purposes) the situation may change.

Table 4. Shooting data in basket

	A			B		
	Three-point field goal	**Two points field goal**	**Total**	**Three-point field goal**	**Two points field goal**	**Total**
Goal	1	9	10	5	3	8
No goal	3	7	10	10	2	12
Total	4	16	20	15	5	20

In shots from outside the area the probability that *A* makes a three-point basket is 25% (1/4), that of *B* rises to 33% (5/15). In shots from under the basket *A* reaches 56% (9/16) while *B* ensures a 60% (3/5) of useful shots. It seems that *B* is more productive and therefore the first judgment must be reversed!

The example shows how the probability assessment changes based on additional information not initially present in the dataset. A phenomenon not taken into account changes the structure of the data in our possession and may reverse the previous estimate. This is an example of the *Simpson Paradox*, named after the statistician Edward Hugh Simpson (1922-2019).

A verification of Simpson's paradox often cited, took place in the 1970s at the University of Berkeley when 44% of boys and 35% of girls who had applied were admitted to degree courses. The data, reported by Bickel, Hammel & O'Connell (1975), were as follows (Table 5):

Table 5. Data entered at Berkeley 1973

	Admit	Deny
Men	3738 (44.27%)	4704 (55.73%)
Women	1494 (34.57%)	2827 (65.43%)

This apparent discrimination caused the protests of feminist associations that sued the University, accusing it of sexist discrimination. However, it was shown that the cumulative result depended on Simpson's paradox and that, if one looked at the situations of individual departments, admissions were substantially balanced. Starting from the real data and analyzing the departments where both boys and girls had applied, it turned out that there had been no form of discrimination. In the paradox, the hidden variable was the preference for departmental enrolments.

Returning to basketball, we can conclude the argument with a combinatorial problem. The NBA draft is an annual event of the National Basketball Association in which the thirty teams can choose new players. The mechanism is reminiscent of a lottery, from fourteen numbered balls four balls are drawn at random to determine the winner. The order of the numbers drawn is not significant so these are combinations so you will have $C_{14.4} = 1001$. There is actually one unassigned combination: 11-12-13-14. The 1000 four-digit combinations are assigned to the 14 teams in the lottery. In case the extracted combination has already been assigned, the mechanism is repeated. From 2019 the percentages change and become: for the first three teams 140 combinations with a 14% probability of being able to make the first choice; for the others, the values decrease up to the 5 combinations of the last one to which a 0.5% probability is assigned.

TABLE OF LOGARITHMS

The calculation of logarithms is an inverse operation with respect to exponentiation: $a^x = b$. The logarithm of a positive real number is defined, in a positive base different from 1, the exponent to which this base must be raised to obtain the given number. The tool that allowed the practical use of logarithms was the logarithm tables (Figure 13). The first of them was completed by Henry Briggs (1561-1630) in 1617.

Figure 13. Portion of a logarithm table

N	Mant.	N	Mant.	N	Mant.	N	Mant.	N	Mant.	N	Mant.	N	Mant.
		50	69 897	**100**	00 000	**150**	17 609	**200**	30 103	**250**	39 794	**300**	47 712
1	00 000	**51**	70 757	**101**	00 432	**151**	17 898	**201**	320	**251**	39 967	**301**	47 857
2	30 103	**52**	71 600	**102**	00 860	**152**	18 184	**202**	535	**252**	40 140	**302**	48 001
3	47 712	**53**	72 428	**103**	01 284	**153**	18 469	**203**	750	**253**	312	**303**	144
4	60 206	**54**	73 239	**104**	01 703	**154**	18 752	**204**	30 963	**254**	483	**304**	287
5	69 897	**55**	74 036	**105**	02 119	**155**	19 033	**205**	31 175	**255**	40 654	**305**	48 430
6	77 815	**56**	74 819	**106**	02 531	**156**	19 312	**206**	387	**256**	824	**306**	572
7	84 510	**57**	75 587	**107**	02 938	**157**	19 590	**207**	597	**257**	40 993	**307**	714
8	90 309	**58**	76 343	**108**	03 342	**158**	19 866	**208**	31 806	**258**	41 162	**308**	855
9	95 424	**59**	77 085	**109**	03 743	**159**	20 140	**209**	32 015	**259**	330	**309**	48 996
10	00 000	**60**	77 815	**110**	04 139	**160**	20 412	**210**	32 222	**260**	41 497	**310**	49 136
11	04 139	**61**	78 533	**111**	04 532	**161**	20 683	**211**	428	**261**	664	**311**	276
12	07 918	**62**	79 239	**112**	04 922	**162**	20 952	**212**	634	**262**	830	**312**	415
13	11 394	**63**	79 934	**113**	05 308	**163**	21 219	**213**	32 838	**263**	41 996	**313**	554
14	14 613	**64**	80 618	**114**	05 690	**164**	21 484	**214**	33 041	**264**	42 160	**314**	693
15	17 609	**65**	81 291	**115**	06 070	**165**	21 748	**215**	33 244	**265**	42 325	**315**	49 831
16	20 412	**66**	81 954	**116**	06 446	**166**	22 011	**216**	445	**266**	488	**316**	49 969
17	23 045	**67**	82 607	**117**	06 819	**167**	22 272	**217**	646	**267**	651	**317**	50 106
18	25 527	**68**	83 251	**118**	07 188	**168**	22 531	**218**	33 846	**268**	813	**318**	243
19	27 875	**69**	83 885	**119**	07 555	**169**	22 789	**219**	34 044	**269**	42 975	**319**	379

Here we are not interested in the aspect related to the calculation using tables (which has produced generations of frustrated students) as fortunately there are now pocket calculators. Instead we are interested in a historical aspect that could also have been dealt with in the chapter dedicated to distributions, we insert it here because it lends itself to many statistical considerations.

In the late 1800s, mathematician and astronomer Simon Newcomb (1835-1909) was using a logarithm table to make some calculations. He noted that the pages closer to the beginning were more worn, while those towards the end were practically new.

The observation fell into oblivion until, in 1938, the American physicist and engineer Frank Benford (1883-1948) sought an answer. He then began to collect an impressive amount of data: population, area of nations, length of rivers, etc. He studied the frequency of the first digit of these numbers and discovered that the first digits had no uniform distribution, there were more 1 than 2, more 2 than 3 and so on up to 9. Certainly, not all sets of numbers follow what will later be called *Benford's law,* for example, if you consider the height in meters of human beings then only the first significant digits 1 and 2 appear.

Why do numbers collected from so many origins follow Benford's law? An answer was provided by the statistician Theodore Preston Hill, known for demonstrating an important theorem regarding Benford's random variable, but his analysis is complex and transcends the objectives of this book. Here we are interested in considering what this law actually means. In terms of probability distribution it can be said that for each number n between 1 and 9, it is given by:

$$P(n) = \log\left(1 + \frac{1}{n}\right)$$

If you try to build a Multiplication table large enough (e.g.: 50 x 50) you can check the distribution of the first digits (Figure 14):

Figure 14. Comparison between the table and Benford's law

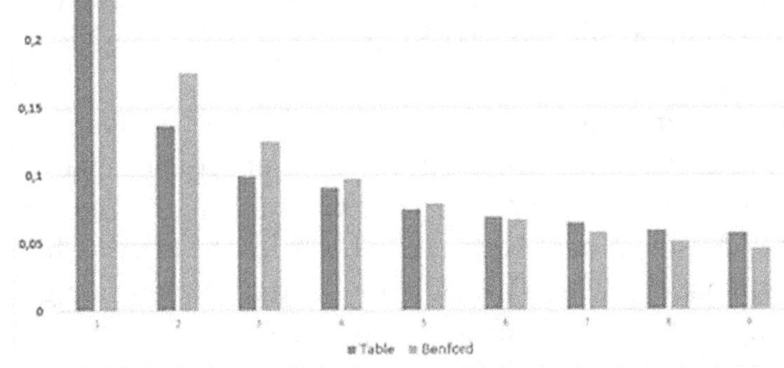

Distribution is far from uniform because it favors smaller numbers at the expense of larger ones.

One area where Benford's law can apply is that of corporate financial statements. Since the 1970s, the possibility of using this law to detect possible forgeries was discovered. If you want to manipulate the books or issue false invoices, you must avoid generating numbers at random. In stories about Benford's law, we often hear about Mark Nigrini for his work on using Benford's law as an audit and accounting tool to detect anomalies in company data. He allegedly exposed fraudsters because their statements did not follow the law.

However, some caution is necessary because only a randomly chosen set of numbers obeys this law, while a set of "real" data in which limits have been imposed may not follow this law. If we consider the number of inhabitants of the states of the US federation, the report appears very clearly, if we consider the postal codes this does not happen because they follow an administrative logic.

Always remaining in the field of logarithms, we like to remember a statement by Poincaré: "What is the probability that the fifth decimal of a logarithm taken at random from a table is a 9? There is no hesitation in answering that this probability is 1/10". Does this statement apply to any digit and any position? Let's leave the reader with the pleasure of discovery.

CHOCOLATE

Chocolate is a food derived from the seeds of the cocoa plant. The word derives from the term of Amerindian origin *nahuatl cocholatl*. Chocolate is produced in the most varied forms; the most common is the bar (Figure 15).

Starting from this tablet, various games can be generated, the most greedy of all was invented by the American mathematician David Gale (1921-2008), but it was the aforementioned Martin Gardner (1971) to give it the name: "Chomp".

The game is interesting both for its algorithmic and combinatorial aspect. In summary: the two players, in turn, choose a chocolate square and eat all the squares above and to the right of the chosen square. The square at the bottom left is poisoned and the player who is forced to eat it has lost. The possible combinations with a bar of *r* rows and *c* columns are given by the combinations $C_{r+c,c}$ which, in the case of the image, are

$C_{2+3,3} = \frac{5!}{2!3!} = \frac{5 \times 4 \times 3 \times 2 \times 1}{2 \times 1 \times 3 \times 2 \times 1} = 10$. With larger tablets the game becomes more complicated and fun (and at risk of indigestion!).

Figure 15. Chocolate bar

But the purpose of the book is to talk about probability and then the chocolate will be examined from another point of view. Although it may seem bizarre, there have been many scholars who have asked the question: does chocolate consumption increase the chances of winning the Nobel Prize?

Everything stems from having found a certain degree of correlation (mutual relationship between two variables) between the two events and this, according to some, is due to the beneficial effect of flavonoids contained in cocoa that could have a positive impact on cognitive functioning. Several notes can be made to this interpretation, not least the risk of confusing correlation with causality (reciprocal relationship between two variables where one causes the other). A very well-founded objection is that flavonoids are present in many other nutrients and that the correlation, in these cases, has not been found. Without entering into more complex and articulated research, it is quite easy to take data on per capita chocolate consumption and the number of Nobel Prize winners (every 10 million inhabitants) for about twenty countries from the Internet. Out of curiosity, the correlation coefficient, a number that indicates the strength of the link between the two variables, is 0.71 on a scale from 0 to 1.

Indeed, the arrangement of points on the graph would seem to indicate a certain relationship: Switzerland is a large consumer of chocolate and has a high ratio of Nobel Prize winners to population, while Spain is in the opposite position (Figure 16).

Figure 16. Scatterplot of the relationship between chocolate consumption and Nobel Prize winners

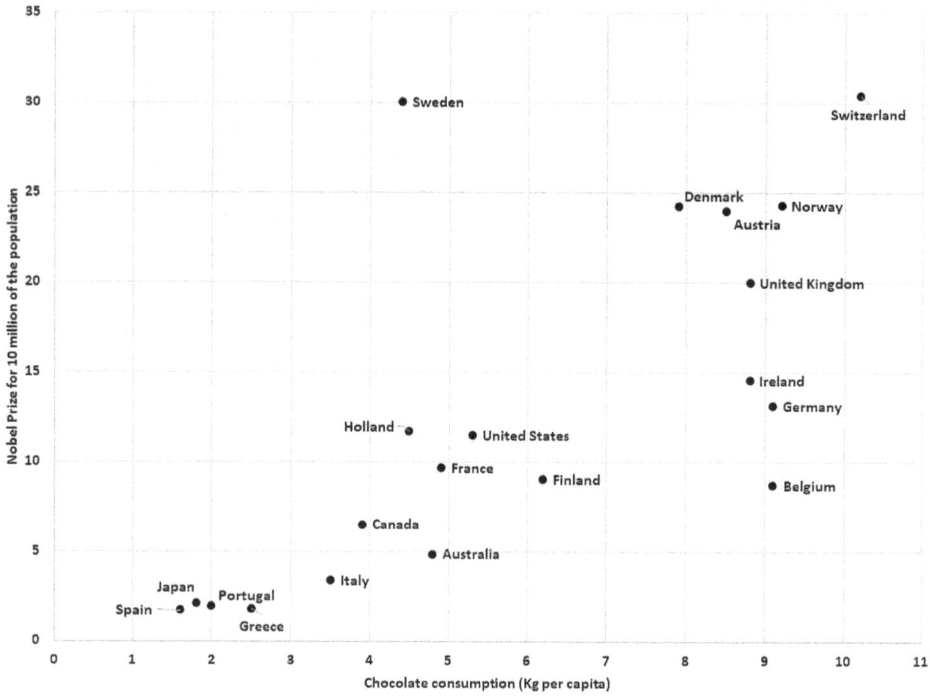

However, we must repeat ourselves: correlation is not synonymous with a causal relationship. In this case, we can strongly assume that it is a spurious correlation. Other variables could affect the model. One is particularly interesting: the age of the Nobel Prize winner. The analysis made by some economists Jones (2014) shows that the average age of the winner of the Prize for Physics, Chemistry and Medicine is 45-50 years. To use their words: "Great scientific output typically peaks in middle age". Then, however, to return to our theme, we should demonstrate that chocolate consumption grows with age.

BUTTONS

To conclude the chapter we can use a small object, usually flat and round shaped, used to close the clothes: the button (Figure 17).

Figure 17. Buttons

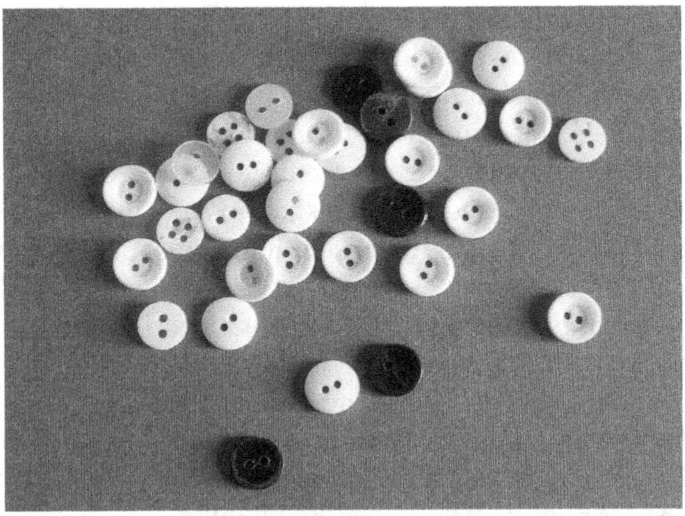

A small group of buttons can be used to estimate a characteristic of a population, introducing an extremely important concept: *probabilistic sampling*.

Suppose you want to count the fish in a pond: do you have to fish them all, count them and put them back in the water again? No, we can settle for an estimate. We could, for example, proceed as follows:

1. we catch a certain number M of fish; each fish is marked in some way and released;
2. let a certain amount of time pass so that the fish are scattered all over the lake and we make a second catch;
3. we count the total n of the fish caught this time and the number s of those among them that are marked.

With this data, we are able to estimate, through a calculation, the total N of fish present in the pond. Let us assume that the ratio between the number of fishes marked s and the total number of fishes n of the second catch is equal to the corresponding ratio between the *M* fishes caught the first time (and marked) and the fishes in the pond, i.e. s/n = M/N. From this we can determine the estimate N ≈ M n/s.

We can now try to simulate the procedure, through a non-transparent container in which there are a certain number of white buttons (the number of fish to be estimated). To start, we extract a certain number of white buttons and replace them with the same number of black buttons and put them back in the container. After mixing well, we take a handful of buttons out of the bag and count the s/n ratio between black and white buttons. If, for example, the second time the fish/button drawn are 10% of the total, it is reasonable to assume that the total number of fish/button drawn is 10% of the total number of fish in the pond. To get a more accurate estimate we can repeat this operation other times and calculate the average of the forecasts.

One thing to keep in mind is that when simulating "fishing", the elements only need to occur once. It is as if we have a list numbered from 1 to N. This means that it is randomized sampling without repetition. If in a spreadsheet such as Excel we use the function = RANDBETWEEN (1,N) it is likely to find a duplication of values (sampling with repetition). There are several possibilities to get around this obstacle, including using a macro. However, there is a simpler way as can be seen in Figure 18 where the calculation is shown and the formulas to be used are highlighted.

Of course this capture-recapture process can be applied to countless situations. For example Kit Yates (2019) uses it to count snails in its own garden and Pere Grima (2011) to estimate taxis in a city. More concretely, sampling is applied, for example, to electoral surveys.

CONCLUSION

As we have said, the link between probability and statistics is relevant, even if according to some scholars, statistics have their autonomy and should be closer to the economic and social sciences, even if with the evident presence of a mathematical apparatus. Or perhaps a definition collected from a post could be accepted: in a joking way one could say that probability "studies" the future while statistics "studies" the past ".

Figure 18. Sampling without repetitions

	A	B	C	D	E	F	G	H	I
1									
2	1	0.573588233767847	13	13		1	=RAND()	=RANK(G2,$G$2:$G$31)	=INDEX($A$2:$A$31,H2)
3	2	0.159789607983092	24	24		2	=RAND()	=RANK(G2,$G$2:$G$31)	=INDEX($A$2:$A$31,H2)
4	3	0.866073934074209	5	5		3	=RAND()	=RANK(G2,$G$2:$G$31)	=INDEX($A$2:$A$31,H2)
5	4	0.389422710858255	20	20		4	=RAND()	=RANK(G2,$G$2:$G$31)	=INDEX($A$2:$A$31,H2)
6	5	0.122674062594769	26	26		5	=RAND()	=RANK(G2,$G$2:$G$31)	=INDEX($A$2:$A$31,H2)
7	6	0.586621146306521	12	12		6	=RAND()	=RANK(G2,$G$2:$G$31)	=INDEX($A$2:$A$31,H2)
8	7	0.879567705163534	4	4		7	=RAND()	=RANK(G2,$G$2:$G$31)	=INDEX($A$2:$A$31,H2)
9	8	0.321993133707665	22	22		8	=RAND()	=RANK(G2,$G$2:$G$31)	=INDEX($A$2:$A$31,H2)
10	9	0.0213990517483001	29	29		9	=RAND()	=RANK(G2,$G$2:$G$31)	=INDEX($A$2:$A$31,H2)
11	10	0.548261830869832	15			10	=RAND()	=RANK(G2,$G$2:$G$31)	
12	11	0.590049861628461	11			11	=RAND()	=RANK(G2,$G$2:$G$31)	
13	12	0.404776059158545	18			12	=RAND()	=RANK(G2,$G$2:$G$31)	
14	13	0.65474200939386	9			13	=RAND()	=RANK(G2,$G$2:$G$31)	
15	14	0.809406111760792	6			14	=RAND()	=RANK(G2,$G$2:$G$31)	
16	15	0.800277635741043	7			15	=RAND()	=RANK(G2,$G$2:$G$31)	
17	16	0.403959152049647	19			16	=RAND()	=RANK(G2,$G$2:$G$31)	
18	17	0.087824607868332	27			17	=RAND()	=RANK(G2,$G$2:$G$31)	
19	18	0.615933075069957	10			18	=RAND()	=RANK(G2,$G$2:$G$31)	
20	19	0.943947795219945	1			19	=RAND()	=RANK(G2,$G$2:$G$31)	
21	20	0.427510270101112	17			20	=RAND()	=RANK(G2,$G$2:$G$31)	
22	21	0.0316736418464211	28			21	=RAND()	=RANK(G2,$G$2:$G$31)	
23	22	0.913872479208594	2			22	=RAND()	=RANK(G2,$G$2:$G$31)	
24	23	0.234343638673159	23			23	=RAND()	=RANK(G2,$G$2:$G$31)	
25	24	0.894395019622427	3			24	=RAND()	=RANK(G2,$G$2:$G$31)	
26	25	0.468912958684865	16			25	=RAND()	=RANK(G2,$G$2:$G$31)	
27	26	0.015436559942055	30			26	=RAND()	=RANK(G2,$G$2:$G$31)	
28	27	0.354042107072332	21			27	=RAND()	=RANK(G2,$G$2:$G$31)	
29	28	0.552872441647453	14			28	=RAND()	=RANK(G2,$G$2:$G$31)	
30	29	0.15869796229811	25			29	=RAND()	=RANK(G2,$G$2:$G$31)	
31	30	0.714189283947582	8			30	=RAND()	=RANK(G2,$G$2:$G$31)	

FUTURE RESEARCH DIRECTIONS

There are several aspects on which perhaps it would be good to focus the research, here we would like to outline only one of them: the link between probability, risk, and decision: the field is vast as it affects all disciplines. The common feature is the need for reflection on what characterizes a correct reasoning and an effective decision. An object, not present in the text, is a small container of CO2, this gas certainly contributes to the increase in the average temperature of the planet. A great debate is open on the validity of the various models that estimate the impact of CO2. This is an area that involves probabilistic estimates, risk assessment and, consequently, roles and tasks of policy makers. The collection of data and information on the topic, the (re)construction of simple models, the possibility of setting up role-playing games are just some of the possibilities.

Another field of research concerns seismic hazard risks. In this case, we are talking about the probability of an earthquake in a certain geographical location in a certain period. An interesting project is related to the use of artificial neural networks (ANN, Artificial Neural Network). For a summary, see Jena (2019).

REFERENCES

Benson, H., Dusek, J. A., Sherwood, J. B., Lam, P., Bethea, C. F., Carpenter, W., ... Drumel, D. (2006). Study of the Therapeutic Effects of Intercessory Prayer (STEP) in cardiac bypass patients: A multicenter randomized trial of uncertainty and certainty of receiving intercessory prayer. *American Heart Journal, 151*(4), 934–942. doi:10.1016/j.ahj.2005.05.028 PMID:16569567

Bickel, P. J., Hammel, E. A., & O'Connell, J. W. (1975). Sex bias in graduate admissions: Data from Berkeley. *Science, 187*(4175), 398–404. doi:10.1126cience.187.4175.398 PMID:17835295

Bocskocsky, A., Ezekowitz, J., & Stein, C. (2014). *Heat check: New evidence on the hot hand in basketball.* Available at SSRN 2481494

Frary, M. (2017). *Decipher – The greatest codes ever invented and how to break them.* Elwin Street Productions.

Friedman, W. F. (1987). *The index of coincidence and its applications in cryptanalysis.* Aegean Park Press.

Fry, H. (2015). *The mathematics of love: Patterns, proofs, and the search for the ultimate equation.* Simon and Schuster.

Gardner, M. (1971). *Martin Gardner's Sixth Book of Mathematical Games from "Scientific American".* Freeman & Co.

Grima, P. (2011). *La certeza absoluta y otras ficciones: Los secretos de la estadística.* R.B.A.

Hald, A. (1998). *A History of Mathematical Statistics from 1750 to 1930.* Wiley.

Jena, R., Pradhan, B., Beydoun, G., Sofyan, H., & Affan, M. (2019). Integrated model for earthquake risk assessment using neural network and analytic hierarchy process: Aceh province, Indonesia. *Geoscience Frontiers.*

Jones, B., Reedy, E. J., & Weinberg, B. A. (2014). *Age and scientific genius (No. w19866).* National Bureau of Economic Research. doi:10.3386/w19866

Li, W. (1992). Random texts exhibit Zipf's-law-like word frequency distribution. *IEEE Transactions on Information Theory*, *38*(6), 1842–1845. doi:10.1109/18.165464

Maurage, P., Heeren, A., & Pesenti, M. (2013). Does chocolate consumption really boost Nobel award chances? The peril of over-interpreting correlations in health studies. *The Journal of Nutrition, 143*(6), 931–933. doi:10.3945/jn.113.174813 PMID:23616517

Messerli, F. (2012). Chocolate Consumption, Cognitive Function, and Nobel Laureates. *The New England Journal of Medicine, 367*(16), 1562–1564. doi:10.1056/NEJMon1211064 PMID:23050509

Miller, J. B., & Sanjurjo, A. (2018). Surprised by the hot hand fallacy? A truth in the law of small numbers. *Econometrica, 86*(6), 2019–2047. doi:10.3982/ECTA14943

Poe, E. A. (2016). *The Gold-Bug*. CreateSpace Independent Publishing Platform.

Prodromou, T. (2008). *Connecting thinking about distribution* (Doctoral dissertation). University of Warwick.

Rinaldi, S. (1998). Laura and Petrarch: An intriguing case of cyclical love dynamics. *SIAM Journal on Applied Mathematics, 58*(4), 1205–1221. doi:10.1137/S003613999630592X

Rohatgi, V. K., & Saleh, A. M. E. (2011). *An introduction to probability and statistics* (Vol. 910). John Wiley & Sons.

Silverberg, L. M., Tran, C. M., & Adams, T. M. (2011). Optimal Targets for the Bank Shot in Men's Basketball. *Journal of Quantitative Analysis in Sports, 7*(1). Advance online publication. doi:10.2202/1559-0410.1299

Weeks, J. R. (2020). *Population: An introduction to concepts and issues*. Cengage Learning.

Yates, K. (2019). The Maths of Life and Death. Academic Press.

Zipf, G. K. (1932). Selected studies of the principle of relative frequency in language. Academic Press.

ADDITIONAL READING

Spiegel, M. R., Schiller, J. J., Srinivasan, R. A., & LeVan, M. (2001). *Probability and statistics* (Vol. 2). Mcgraw-hill.

van Dijke-Droogers, M., Drijvers, P., & Bakker, A. (2020). Repeated sampling with a black box to make informal statistical inference accessible. *Mathematical Thinking and Learning*, *22*(2), 116–138. doi:10.1080/10986065.2019.1617025

KEY TERMS AND DEFINITIONS

Benford's Law: Describes the probability distribution with which the first digit of numbers appears in many examples of data collections.

Gini's Index: A measure of statistical dispersion and is the most commonly used measurement of inequality.

Probabilistic Sampling: Process of random selection of a number of individuals for a study.

Simpson Paradox: A situation in which a relationship between two phenomena appears to have changed due to hidden variables.

Zipf's Law: An empirical law describing the frequency of an event as a function of its position in the decreasing order with respect to the frequency of that event.

Chapter 10
Probability and Chaos

ABSTRACT

Chaos theory is a relatively recent and often misunderstood field of study because it is wrongly considered by laymen as tied to random events. Chaotic systems have properties similar in many ways to those of stochastic processes, for example punctual unpredictability. What is interesting is the fact that potentially controllable situations such as deterministic ones are instead difficult to interpret. It is quite natural to consider some of these situations in a text that deals with probabilities. Even though we continue to stress, these are different fields. To make their way through the many possible examples, fairly simple objects have been preferred.

INTRODUCTION

I say unto you: a man must have chaos yet within him to be able to give birth to a dancing star. ~ Friedrich Nietzsche

One of the fundamental questions that man asks himself concerns the origin of the universe. The answers can be sought in science, religions, myths. The starting point for all these reflections is always chaos. Chaos is presented in Genesis (the first book of the Bible) as a state of imperfection from which, by divine work, the Universe is shaped. For the Greeks it was *Chaos* who ruled a world and only his descendants gave shape to reality. Man, however, asked himself other profound questions to which he answered by building new myths. As the writer William Somerset Maugham (1944) said: "The

DOI: 10.4018/978-1-7998-3871-5.ch010

faculty for myth is innate in the human race. .... It is the protest of romance against the commonplace of life".

We open a small parenthesis that lends itself well to compare chaos, randomness, and order. The ancient populations sought answers by observing the movement of the stars and the shapes that could be recognized. In fact, people have always grouped the stars that appear close together and have therefore recognized (or, better, believed they could identify) the constellations: random clusters of stars that could be considered as formations with a structure. Each culture has built its own figures by giving them different names and stories according to its own traditions. We can say that different cultures have created different constellations. It is curious, however, that some constellations are common in civilizations far apart in a spatial and temporal sense. The 88 constellations are divided into three groups: the 12 constellations of the Zodiac, which are located along the ecliptic and are therefore traversed by the Sun in its apparent motion on the celestial vault during the year; 36 constellations listed by Claudius Ptolemy (about 100-175) in his Almagest, now rising to 38; the remaining constellations have been identified since 1600 in the empty spaces. Some symbols of animals such as Taurus, Scorpio, Leo indicated the quadrants of the sky corresponding to the equinoxes and solstices, as they were located in antiquity. With the precession of the equinoxes the function passed to the following constellations, Aries, Libra, Cancer (Figure 1). In many iconographies the signs of the zodiac are associated with the months of the year, often linked to representations of the agricultural work characteristic of each month.

One question arises spontaneously, does the presence of agglomerations justify the "divine" theory of constellations or do the stars have a random distribution?

If we try to "generate random stars" on a plane (Figure 2) we are amazed by the apparent presence of regularity and we can understand how the ancients built a cosmogony. Observing the starry sky the human being often tries to identify rules and patterns and in any randomly generated figure you can try to identify relationships.

In three-dimensional space the stars that form the same constellation can be separated even by enormous distances; vice versa, two or more stars that on the celestial sphere appear perhaps extremely far apart, in three-dimensional space others that belong to the same constellation can be relatively close. The example given here is not only a curiosity, but it allows us to underline the fact that random points seem to have an internal rule and present themselves with quite counterintuitive groupings and empty spaces.

Figure 1. Constellation

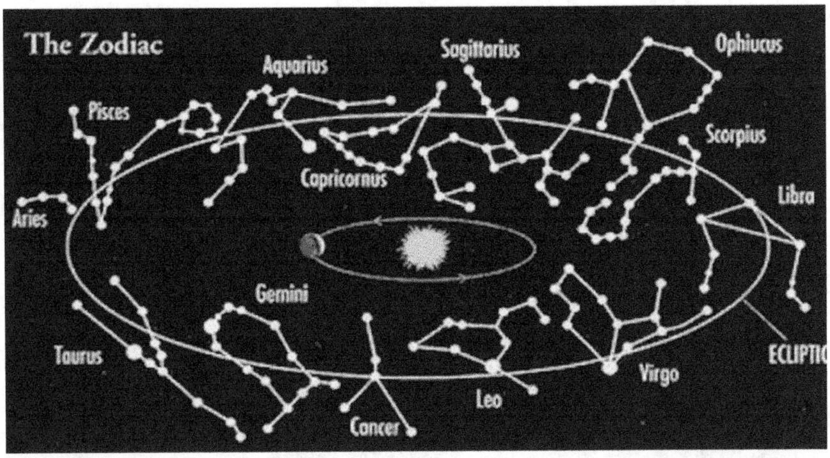

Figure 2. Starry sky simulation

In this regard, it is worth mentioning the mathematician and philosopher Frank Plumpton Ramsey (1903-1930). Among other things, he studied the conditions in which order must appear within disorder, we can say that he is the father of the theory that supposes the presence of regularity between randomly generated objects. Let us consider a sequence of numbers as in the case of the game of billiards called 9-pool. In this case there are nine colored balls numbered from 1 to 9 plus a white ball, to which we assign the value 0. We can say that the balls constitute an increasing sequence. Let's now consider shuffling the balls randomly. Well, also in this case there is a form of regularity; it is possible to find a subsequence. Using a pool cue, players must hit the white ball to shuffle the nine colored pool balls in ascending numerical order. You can also determine the characteristics of this event, which has a large number of possible cases: 10! = 32628800. It can be shown that the subsequence of n2+1 numbers will have a length of at least n+1. In this case n is worth 3 and therefore we can expect to find, within the ten balls, a subsequence of at least four balls. For example, if the random mixture has placed the balls in the order 5, 4, 9, 6, 7, 0, 2, 1, 3, 8 we can immediately see that the balls 4, 6, 7, 8 are arranged in ascending order (Figure 3).

Figure 3. Balls arranged in random order

After this digression, we return to the real theme of the chapter: chaos as it is understood in the modern sense.

The chaotic world surrounds us and we often do not realize it, there are many examples: weather phenomena, heart rhythm, smoke swirls, toys with lights flashing without regularity. And more: the vibrations that caused the collapse of the Tacoma Narrows Bridge, the dynamics of animal populations, the behavior of non-linear electrical circuits, the behavior of price changes, etc.. Many phenomena are unpredictable and even attempts to apply the probability calculation are futile. In this case several scholars have introduced a new concept, that of chaos (Gleick, 2008). The foundations of chaos theory date back to the early twentieth century, a great mathematician like Henry Poincaré (1854-1912) expressed the fundamental consideration: "A very small cause which escapes our notice determines a considerable effect that we cannot fail to see, and then we say that the effect is due to chance".

In the study of complex systems it is said that a system tends to chaos when it has completely unpredictable and irregular behavior. Let's say that a phenomenon shows a deterministic chaos regime when one has a sensitive dependence on initial conditions or an initial uncertainty that grows over time in order to determine the long-term unpredictability of its evolution.

It has been shown that in humans the interval between one heartbeat and another is chaotic and that the regular heartbeat is a predictor of impending cardiac arrest. There are many simulations on the net that model disparate situations and we refer to them.

Just as chaos is generated, it is possible to suppress it with appropriate operations that effect on the system parameters. An article by Ott, Grebogi & Yorke (1990), from the University of Maryland, introduces the fact that "the system behavior could be changed by switching the temporal programming of the small parameter perturbations to stabilize different orbits". The result of this possibility is of great importance, for example it is adopted to stabilize an excessively irregular heartbeat.

This introduction can be completed by the words of Ian Stewart (1997): "That's two more buzzwords – 'stochastic' and 'deterministic'. Laplacian determinism is already familiar to us. 'Stochastic' means 'random'. To understand the phenomenon of chaos we shall need to discuss their meanings further, because in its present form the definition is a paradox. Deterministic behavior is ruled exact and unbreakable law. Stochastic behavior is the opposite: lawless and irregular, governed by chance. So chaos is 'lawless behavior governed entirely by law'".

BALLOON

We have already said that the heartbeat has chaotic elements, but the examples could be multiplied. If we observe, in autumn, the fall of the leaves we can see that the trajectories are highly unpredictable. A dripping tap is a classic example of a chaotic phenomenon. Let us consider a tap on which a drop is hanging. The force of weight and the forces of adhesion and cohesion act on it. At a certain point, the prevailing weight force will cause the drop to fall down. As the flow rate increases, the drop detaches when the overlying liquid column is still oscillating due to the detachment of the previous drop. There is, in this case, an "interference" between the droplet formation frequency and the oscillation frequency of the liquid column. Small variations in flow lead to a transition from linear to chaotic (Cahalan, 1990).

A further example is provided by a balloon (Figure 4).

Figure 4. Balloon-Globe

If we inflate a balloon and then let the air out freely we see a chaotic zigzag. Imperceptible changes in the initial phase can amplify into huge changes in flight direction. In fact, the deflating balloon follows Newton's Third Law but determining the probability that the balloon is at a certain point would seem to be a superhuman feat, to say the least. Perhaps the only person able to solve the problem empirically is a boy used to video games. As Kevin Kelly (2009) wittily says: "Yet a teenager reared on Nintendo could learn how to catch the balloon. Not infallibly, but better than chance. After a couple dozen tries, the teenage brain begins to mold a theory-an intuition, an induction-based on the data. After a thousand balloon takeoffs, his brain has modeled some aspect of the rubber's flight".

Perhaps not all hopes of dominating chaos are lost!

BILLIARD

The game is played in a rectangular field (generally in 2x1 proportion) surrounded by banks and whose surface is covered with a characteristic green cloth, perfectly smooth (Figure 5). There are six holes: four in the corners and two in the center of the long banks, only international billiards have no holes. It is played using spherical balls called marbles. The marbles are moved by using a rod called pool cue. In billiards, depending on the presence or absence of holes, the number of balls and/or pins, the rules and the points count, many games can be played in four main categories: Italian billiards, American billiards, Russian billiards, and British billiards.

Figure 5. Billiard

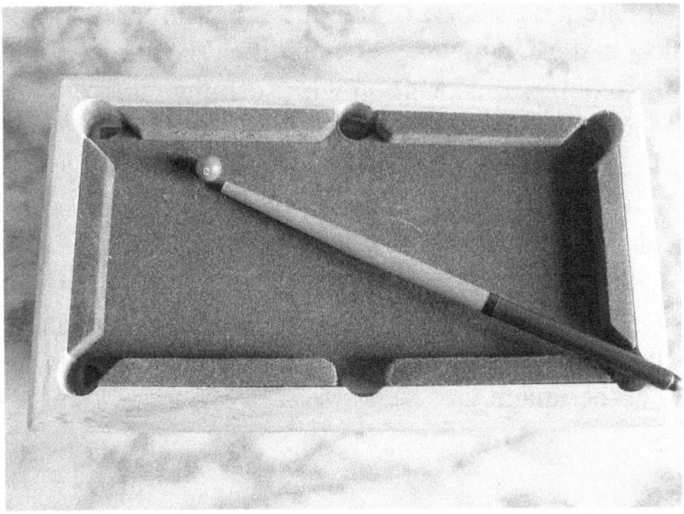

That billiards is related to probability results from a posthumous work by Thomas Bayes published posthumously in 1763. He wrote: “consider a billiard table in which, from the short bank, a ball is thrown, blindfolded, and then stops at a certain (unknown) point on the piano. A series of marbles are then thrown at random by another subject, if they stop on the left they are considered ‘success’, if they stop on the right they are catalogued as ‘failure’” (Figure 6).

Figure 6. Bayes Billiards

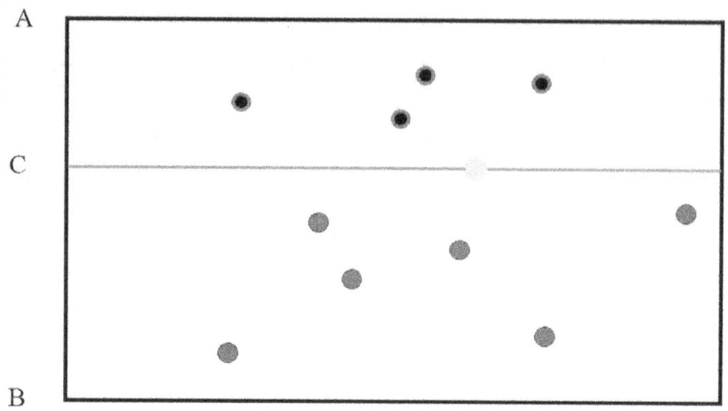

Having no information other than the frequency of successes and failures, how can I estimate the position of the initial ball in terms of distance from the long banks? In the end, given a sufficiently large number of marbles, you should be able to define this position with sufficient approximation.

In today's games, the player tries to hit a bouncing ball. To be successful, you need to have a minimum of geometrical knowledge. When a ball bounces off the edge of a normal pool table, the trajectory before and after the bounce is symmetrical to the normal line at the edge, i.e. the angle of incidence is equal to the angle of reflection (Figure 7). To make a precise shot, we can help ourselves with a simple rule: let's suppose to place a mirror parallel to the edge of the table, so as to see the reflected image of the ball to be hit. Now just aim at the image.

Figure 7. Trajectory

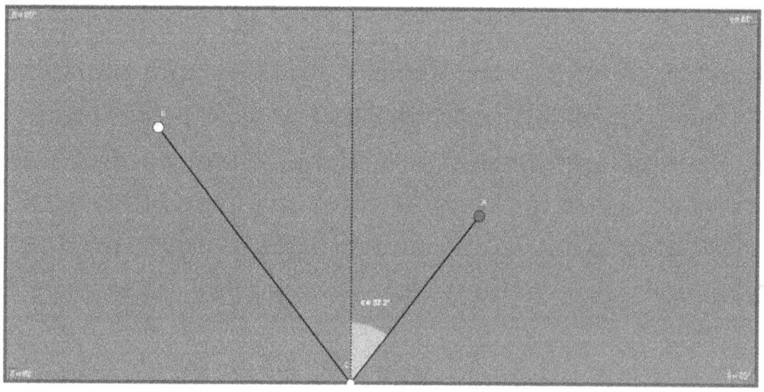

Of course, if the angle at which the first ball is hit changes slightly, after a certain number of rebounds the final position is different from the one assumed before starting. In any case it is possible to determine the probability that the ball stops at a certain point.

There are also elliptical billiards. In this case, the ball rebounds are more difficult to predict, because the ball's trajectory strongly depends on the exact position of the points where it rebounds. The study of the different trajectories gives an idea of chaos (Figure 8), the evaluation of probability becomes more difficult.

Figure 8. Elliptical billiards

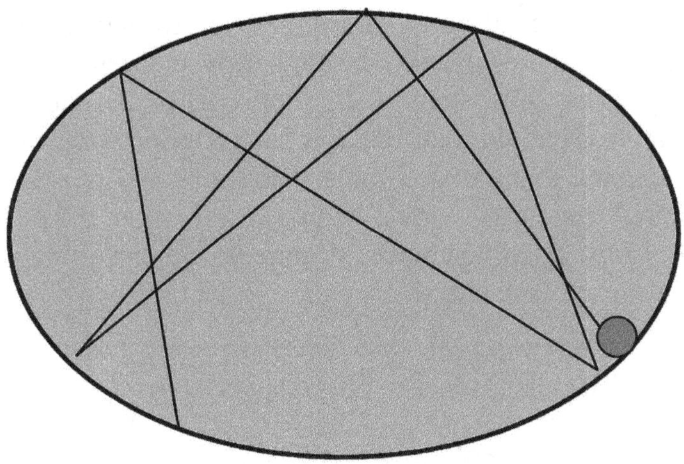

The observation of the game of billiards has stimulated several conjectures. For example: is it possible to estimate the probability that a billiard ball bouncing (in the absence of friction and gravity) inside an empty cubic die can return to its starting point after touching all the faces? The answer was found by Polish mathematician Hugo Steinhaus (1887-1972). He demonstrated that there is at least one path that leads the ball back to the starting point. The details can be found in an article published in Science by Barry Cipra (1997). One may wonder if the same thing can happen, in three-dimensional space, for the Platonic dice, seen in the first chapter, such as the tetrahedron, the octahedron, the dodecahedron and the icosahedron. In 1997 Matthew Hudelson (1997), using a computer, verified that this is possible. We can, therefore, say that $p = 1$.

DECISION MAKER

Let's consider a metal sphere suspended by a rigid rod (pendulum) and place some magnets under the base. We place the sphere in the position of coordinates (x_0, y_0) and let the system evolve under the action of the magnetic field generated by the magnets. The object described is a Random Oscillating Magnetic Pendulum (ROMP). When the pendulum is released it begins to oscillate irregularly to stop eventually attracted by one of the magnets at its base. Its motion is extremely sensitive to initial conditions: a small variation in the starting point causes large changes in trajectory so that it becomes impossible to predict where the pendulum will go.

In its movement, the pendulum is subject to the force of gravity, friction due to air and mechanical parts, and forces due to magnets in the plane. In theory, knowing these forces, we could express the motion in terms of differential equations. In practice, even if there is a law that allows us to obtain the position of the pendulum at time *t*, it has a random behavior due to the presence of magnets that act as "strange" attractors. The system will follow extremely complicated trajectories. After a certain time, the pendulum will stop near one of the possible choices (Figure 9). The resulting motion is an example of "deterministic chaos".

Figure 9. Decision Maker

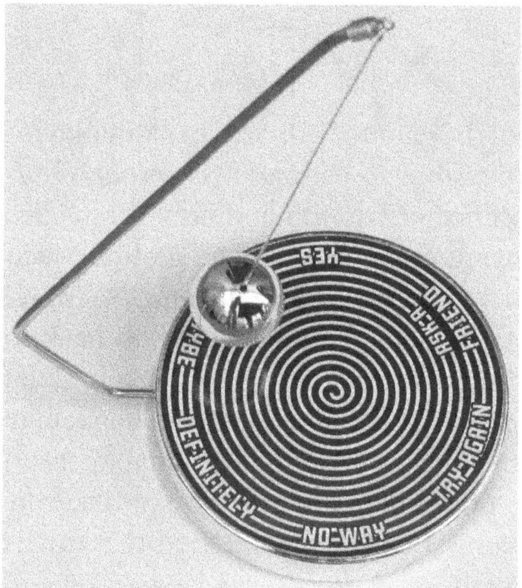

Using the Decision Maker we can color the trajectories with the color corresponding to the magnet near which the trajectory will stop. Trajectories that start close can end on different magnets: a small variation in the initial conditions of the system can lead to profoundly different consequences. Figure 10 shows the results that can be visualized thanks to the different colors of the areas on which the pendulum has stopped depending on the starting point. Each color indicates the final state of the pendulum, in other words it indicates the probability that the pendulum will stop on that particular portion of the plane. Starting from any point then we can no longer predict where it will stop, these are in fact points of instability in the system. When the pendulum during its motion is in a point of bifurcation it is as if it were in a point of unstable equilibrium. It is therefore impossible to determine its evolution a priori.

Figure 10. Ball arrival zones

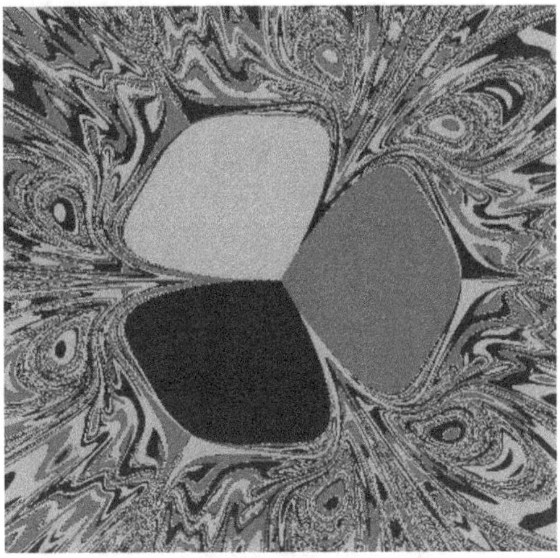

PIRATE

Ivar Ekeland (2006), in his capacity as a popularizer, dealt, among other topics, with chaos. One of the examples given refers to a toy, a rag acrobat that rotates irregularly around a fixed bar. Counting the turns the acrobat makes in each direction, the author notes that these are alternating movements (first clockwise and then counterclockwise) of type 5 2 2 2 3 2 3 1 4 1 2 3 4 5 2 4. In this sequence there is no regularity, digit 2 appears more frequently but this does not happen when new attempts are made.

The first consideration is that the mechanical instrument behaves differently each time and no protocol can predict what the next figure will be from the previous ones. This consideration is expressed in an amusing way by Ekeland (2006): “Let’s say it in other ways, a la Cyrano. Libertarian: the only rule is that there is no rule. Metaphysics: the past does not determine the present. Practical: if you wanted to store ten million of these figures in your computer, you could only save them one by one: there is no program that saves time or space”.

But what’s behind this device? A hydraulic wheel. Water (or sand) falls on the central part of the wheel, its weight sets the wheel in motion. After a period of regularity, as the flow increases, the rotation becomes chaotic as the containers are partially filled and can rise to the other side before emptying. The weight of the water (or sand) will slow down and reverse the direction of travel. The mechanism should be deterministic and the turns should be predictable, in reality this is not the case because the starting conditions do not exactly reproduce. Chaos theory tries to understand why small initial differences can be at the root of considerable discrepancies. Not having that kind of equipment available, you can make up for it with a small trapezoidal toy that is easy to find, in which a pirate performs evolutions when approaching and releasing the lower extremes (Figure 11).

Figure 11. Pirate

The interesting aspect is that the evolutions of the figure depend on the type of pressure-release and are not totally predictable, even if some positions occur more frequently. Once again we are faced with a chaotic model.

Similar features can be found in one of the best-known models, the double pendulum. It is a physical system consisting of two pendulums attached to each other's ends (Figure 12).

Figure 12. Double pendulum

Its dynamic behavior is sensitive to small variations in initial conditions and the path is highly unpredictable and disorderly, which is not the case for a simple pendulum. A normal pendulum does not have a chaotic oscillation. Thanks to a light source we can make chaos visible (Figure 13).

HEXBUG

The Hexbug Nano are automated robots that, thanks to the energy of a battery, move on a special track like real insects. These small electronic creatures manage to turn themselves around after turning over. Nano Hexbugs are also able to change direction when they encounter an obstacle in their path (Figure 14).

Figure 13. Double Pendulum long exposure (File:DPLE.jpg: George Ioannidis)

Figure 14. Hexbug runway

Initially the insect, equipped with 12 legs, moves in a straight line; however, the encounter with an obstacle causes a change of direction that is not predictable: this movement is known as *Random walk*.

The type of movement is reminiscent of Brownian motion, i.e. the disorderly motion of particles present in fluids or fluid suspensions. It was discovered by the Scottish botanist Robert Brown (1773-1858), who in 1827 observed with a microscope pollen particles suspended in water: their rectilinear motion suffered sudden rejects in different directions. The type of motion is reminiscent of the random walk of a drunk man (Figure 15).

The theory of random walking was first coined by the French mathematician and statistician Louis Bachelier (1900). He believed that the price movements of shares were, like the steps taken by a drunk, unpredictable. Over time, applications grew. For example, in population genetics, random walks describe the statistical properties of genetic drift; in mathematical ecology, random walks are used to describe the movements of individual animals; in computer science, they are used to estimate the size of the Web.

Figure 15. Walk of the drunk

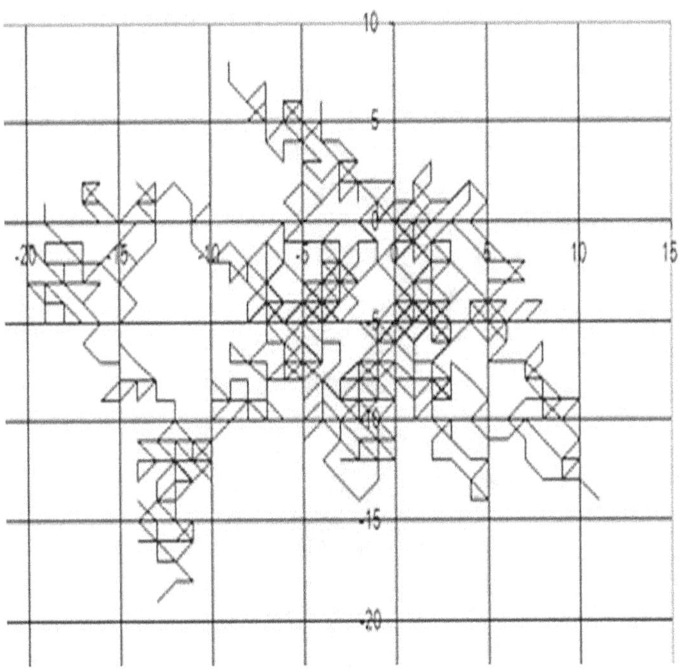

BREADSTICK

The breadstick is a product of Italian gastronomy and in particular of Turin. Its composition is similar to that of bread, but its elongated and thin shape is characteristic (Figure 16).

This food lends itself well to an analysis called "baker's processing". It takes its name from a kneading operation applied by bakers: the dough is cut in half; the two halves are stacked on top of each other and the operation is repeated several times. The baker thinks that this series of operations allows the dough to mix evenly and, in some ways, predictably. This is not true. Let's imagine placing a poppy seed in the dough to make a breadstick (supposed to be of unit length) at a random point and applying the transformation: you stretch the breadstick doubling its length and then fold it back on itself. There are many questions to be answered. Where does the seed end up after a series of stretching and folding? What happens if, in a second experiment, the seed is placed at an exceedingly small distance from the first point? Is it possible to estimate the probability that the seed ends up at a certain point in the breadstick?

Figure 16. Grissini

First of all, it must be said that the law of folding is:

$$S(x)=\begin{cases}2x & 0\le x<1/2\\ 2-2x & \frac{1}{2}\le x<1\end{cases}$$

It can be examined how two trials, which have a small initial gap of one millionth, lead, over time, to different paths (Figure 17).

Figure 17. The path of poppy seeds in the breadstick

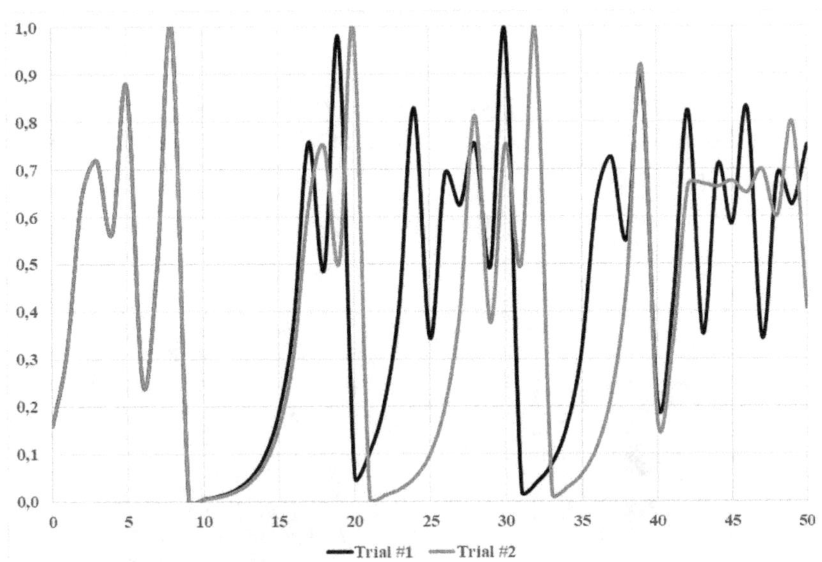

This operation is often cited in chaos theory. The theme comes up again when we manipulate images on the computer. In this case the transformation occurs in the discrete because the image is formed by a certain number of pixels. Each point is exchanged with another and therefore after a few steps the image is no longer recognizable. In the continuous case of the seed in the breadstick mixture, it is very unlikely that there will be a return to the starting point. The interesting aspect of the discrete case is that, after a certain number of iterations, we return to the starting image. For example, using an Applet programmed by Delahaye and Mathieu (1997) we can see that the image of the Earth in format 256 • 256 pixels returns after only 17 repetitions (Figure 18).

Figure 18. Some frames of the sequence

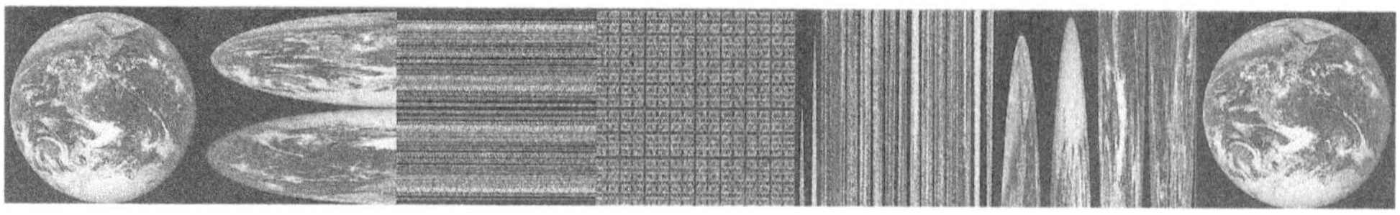

However, the time needed to return to the starting point remains uncertain, which can be exceptionally long. It can only be said that the probability of return is 1.

ANTS

Ants (Figure 19) are Hymenoptera insects with highly evolved and widely studied social behavior. In this case, however, they serve only as an example of any animal species.

Figure 19. Ants

To describe the development and behavior of a species, the simplest model is the *Logistic map*, a non-linear dynamic equation:

$$x_{n+1} = r \cdot x_n (1 - x_n)$$

Where x means a number between 0 and 1 representing the ratio of the existing population to the maximum possible population in a year and r the reproduction rate. What happens in the anthill as time goes by? Everything depends on the value assumed by *r*.

When this parameter varies, different behaviors are observed:

1. *r* less than 1: the population is extinguished;
2. *r* between 1 and 2: the population will tend to stabilize on a value;
3. *r* between 2 and about 3.6: the population fluctuates around increasing values multiples of two;
4. higher values: it triggers a chaotic trend.

For example for $r = 3.9$ and $x_0 = 0.9$ you will have the trend of Figure 20:

Figure 20. Population graph

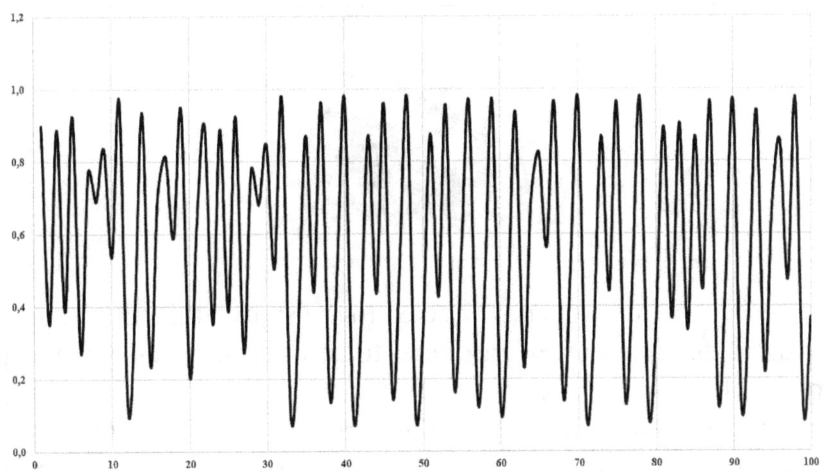

BEER

It is well known that beer (Figure 21) is a product with very ancient origins. Even before the 5th century B.C., Mesopotamian peoples were already producers of beers made from barley and wheat. The Greeks also appreciated beer and consumed a lot of it. In Italy, the equivalent of the first pub probably dates back to 83 AD when the governor of Britain brought master brewers from today's Gloucester to Rome. The medieval development is mainly due to the monasteries. The oldest monastic brewery is that of an abbey near Munich (Germany), dated 724 AD.

Figure 21. Beer

In this case the interest is not directed to the study of the drink, but to the mathematical implications, in particular to an interesting case known as the El Farol problem.

This was the name of a club in Santa Fe, New Mexico, where an Irish band played on Thursday nights. Brian Arthur, an economist at the Santa Fe Institute loved to go and listen to it, but he didn't like to be there when the place was too crowded. From there he took the cue to formulate a problem that called into question a complex system model. The problem was then taken up by John Casti (1996) and was the subject of a school experiment by the author of this book. Let's examine what is the common thread of experimentation. Let's assume that the people concerned know the number of patrons who have attended the venue in the past few weeks, possible patrons try to predict how many people will visit the venue in the following week. Let's suppose that each person decides independently to go to the venue only if his prediction is that the patrons will be less than x. Since people rely on different predictions, some of them will go to the venue, while others will stay at home having predicted that it will be too crowded. The new number of customers will be known the next day and everyone will update their forecast. Week after week, the number of admissions appears to be the product of a random phenomenon, although no random factor suggests what the actual number of customers will be. The number of those who come to the venue each week is, however, a purely deterministic function: it is dictated solely by individual predictions which, in turn, are functions of the number of patrons in the preceding weeks.

To verify El Farol's model, in a 2001 article, I hypothesized a population of 14 potential customers willing to go to the bar if they had less than 10 attendances: each of them based their decision on a specific statistical criterion, for example:

1. arithmetic mean of the presences of the whole period;
2. arithmetic mean of the last month's attendance;
3. arithmetic mean of the last two months' attendance;
4. harmonic mean of the number of persons present over the entire period;
5. geometric mean of the presences of the whole period;
6. mode of attendance over the entire period;
7. mode of attendance for the last month;
8. mode of attendance for the last two months;
9. median attendance over the entire period;
10. attendance forecast based on the Ordinary Least Squares;
11. attendance forecast based on the previous week;
12. attendance forecast based on the previous two weeks;
13. attendance forecast based on one month earlier;
14. attendance forecast based on data from two months earlier.

By defining in a random way the number of people present in the premises in the first three months (corresponding to 12 weeks) it was possible to start with a historical series, from which, elaborate the following forecasts. For example: 2, 12, 7, 6, 1, 9, 5, 1, 6, 12, 6.

Simulating the phenomenon for the following days, the following graph of presences was obtained (Figure 22):

There are three important considerations:

1. The oscillations stabilize around the average value of 9, respecting enough the value assumed by Arthur (0.6 x);

2. Although the criteria adopted are entirely deterministic, the graph shows a chaotic trend;
3. A different initial series leads to different results.

Figure 22. Simulation trend

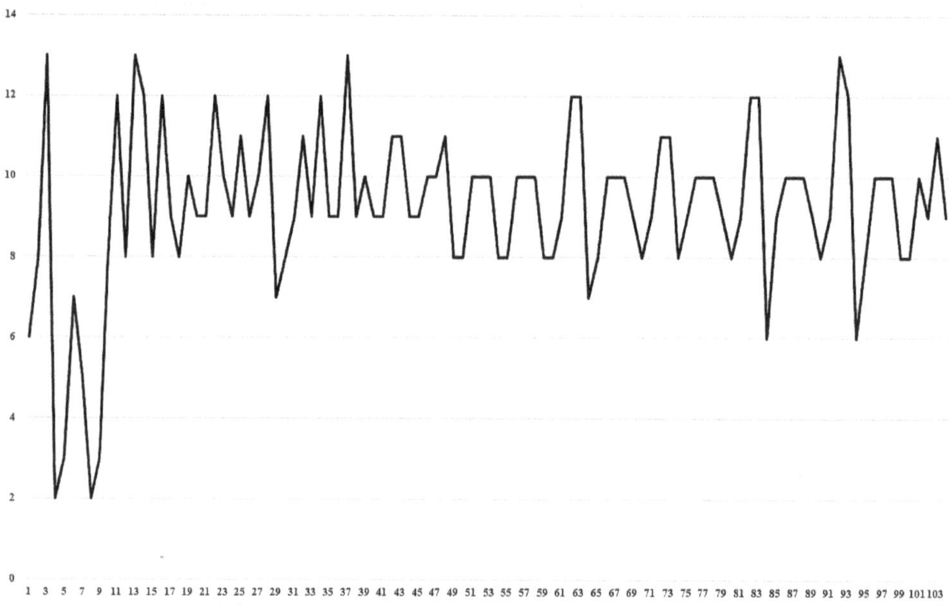

CONCLUSION

This chapter, although it can be considered a deviation from the probability argument, we believe represents a significant passage to define the discrimination between chance and chaos. It will then be up to the teacher to evaluate the opportunity to present this topic to broaden the student's cultural and cognitive horizon.

To conclude, we could make the words of Ian Stewart (2011) our own: "The discovery of chaotic dynamics implies that deterministic systems may not be predictable in any meaningful sense. The best-known source of unpredictability is sensitivity to initial conditions (popularly known as the butter ñ y effect), in which small errors or disturbances grow exponentially ".

FUTURE RESEARCH DIRECTIONS

Even a trivial spring can be a useful cue to point out that chaos is not just an argument with little practical application. Is it essential for industry to determine whether a batch of wire is suitable for making springs? At the micrometric level, a wire can be more or less thick, if you wrap it around a spool you can make experimental observations that produce a list of numbers that represent the value of the quantity observed at regular time intervals. The spatial sequence is a time series from which technicians can derive the attractor. Let us remember that the attractor is the space that contains all the possible states of a system and within which its evolution can be represented. In this case it resembles an elliptical spot. If this spot is compact, then the wire is good and therefore can be used for the production of springs.

The example of the spring has been deliberately excluded from the list of "chaotic" objects because the argument can be overly complex as can be seen from the text in the additional readings.

REFERENCES

Bachelier, L. (1900). *Théorie de la Spéculation*. Gauthier-Villars. doi:10.24033/asens.476

Cahalan, R. F., Leidecker, H., & Cahalan, G. D. (1990). Chaotic Rhythms of a Dripping Faucet: A simple drop detector transforms the computer into a temporal microscope, revealing a variety of rhythms in the common leaky tap. *Computers in Physics*, *4*(4), 368–382. doi:10.1063/1.4822928

Casti, J. L. (1996). Seeing the light at El Farol: A look at the most important problem in complex systems theory. *Complexity*, *1*(5), 7–10. doi:10.1002/cplx.6130010503

Cipra, B. (1997). How to play Platonic billiards. *Science*, *275*(5303), 1070–1070. doi:10.1126cience.275.5303.1070

Delahaye, J. P., & Mathieu, P. (1997). Images brouillées, images retrouvées. *Pour la Science*, *242*, 102–106.

Drivet, A. (2001). Decisioni in situazioni caotiche a El Farol? *Induzioni, 22*.

Ekeland, I. (2006). *Le Chaos*. Pommier.

Gleick, J. (2008). *Chaos: Making a New Science*. Penguin.

Hudelson, M. (1997). *Periodic Omnihedral Billiards in Regular Polyhedra and Polytopes*. Academic Press.

Kelly, K. (2009). Out of control: The new biology of machines, social systems, and the economic world. Hachette UK.

Maugham, W. S. (1944). *The Moon and Sixpence: A Novel by Somerset Maugham*. Penguin Books.

Ott, E., Grebogi, C., & Yorke, J. A. (1990). Controlling chaos. *Physical Review Letters*, *64*(11), 1196–1199. doi:10.1103/PhysRevLett.64.1196 PMID:10041332

Stewart, I. (1997). Does God play dice? The new mathematics of chaos. Penguin UK.

Stewart, I. (2011). Sources of uncertainty in deterministic dynamics: an informal overview. *Philosophical Transactions of the Royal Society A: Mathematical, Physical and Engineering Sciences, 369*(1956), 4705-4729.

ADDITIONAL READING

La Fata, L. (2010). Il Caos Deterministico. PhD thesis.

Shinbrot, T., Grebogi, C., Wisdom, J., & Yorke, J. A. (1992). Chaos in a double pendulum. *American Journal of Physics*, *60*(6), 491–499. doi:10.1119/1.16860

Wood, D. A. (2006). Making better springs using aspects of chaos theory. *Proceedings of the Institution of Mechanical Engineers. Part C, Journal of Mechanical Engineering Science*, *220*(3), 253–259. doi:10.1243/095440606X79136

KEY TERMS AND DEFINITIONS

Chaos: Temporal evolution with sensitive dependence on initial conditions.

Logistic Map: Model representing the dynamic evolution of populations.

Random Walk: Describes a path that consists of a succession of random steps.

Conclusion

INTRODUCTION

This branch of mathematics [Probability] is the only one, I believe, in which good writers frequently get results which are entirely erroneous. ~ Charles Pierce

The teaching of probabilistic concepts has a rather long history and includes both theoretical analysis and empirical research. Before the 1970s the classical version of probability was dominant and the use of combinatorial calculus was widespread. The frequentist approach became space later, thanks also to the possibility of using IT tools. The subjectivist position, on the other hand, was not particularly included in the didactics as well as the axiomatic approach (not dealt with in the text), which was considered too formal (Batanero et al., 2016). More recently, a further step has been taken towards the modelling of random situations, a sort of "applied probability". Batanero & Díaz (2007) have proposed an interesting scheme, including some positions not covered in the text, shown in Figure 1.

In parallel to this story, there has been continuous research, measurable over decades, to understand both the difficulties related to probability and the most appropriate tools for teaching this discipline.

The main problems are: the omnipresent random or systematic error, the difficulty of overcoming conceptual obstacles that are hidden everywhere, the lack of attractiveness of certain content or methods.

Moreover, it cannot be denied that probabilistic reasoning is quite different from other types of reasoning based on causal thinking, logical thinking or "magic thinking". A typical example of magical thinking is the following: two manufacturers of equivalent products both organize a prize competition. The first offers a cruise and the second a trolley. Even if the potential buyer is aware that the chances of winning the trolley are greater than the chances

Figure 1. Historical summary diagram

Table 1. Elements Characterizing Different Historical Meanings of Probability

Meaning of Probability	Problem Field	Tools	Representations	Properties	Some Related Concepts
Intuitive	• Games of chance • Divination	• Random generators	• Colloquial expressions	• Unpredictable • Opinion, belief	• Luck • Fate
Classical	• Games of chance	• Combinatorics • Proportions • A priori analysis of the structure	• Pascal Triangle • Enumerations • Combinatorial formulas	• Quotient of favorable cases divided by number of all cases possible • Equiprobable simple events	• Expectation • Fairness
Frequentist	• Repeatible experiments	• A posteriori recording of statistical data • Fitting mathematical curves • Mathematical analysis	• Statistical tables/graphs • Simulations • Density curves • Random number tables • Distribution tables	• Limit of relative frequencies in long runs • Objective character, based on empirical facts	• Relative frequency • Universe • Random variables • Probability Distribution
Propensity	• Physical situations, including single cases	• A priori analysis of the experimental set-up		• Physical disposition or tendency • Objective character • Applicable to single cases • Related to the experimental conditions	• Propensity • Probabilisitic causal tendency • Virtual frequency
Logical	• Widening inferences	• A priori analysis of the space of possibilities • Propositional logic • Inductive logic	• Formal language • Notation of conditional probability •	• Objective degree of belief • Relationship between two statements • Possibilities can be given unequal weights • Generalizes implication • Revisable with experience	• Evidence • Hypothesis • Degree of implication
Subjective	• Uncertain events, even non-repeatable	• Bayes's theorem • Conditional probability decision theory •	• Notation of conditional probability • Graphs of prior and posterior distributions	• Subjective character • Revisable with experience	• Likelihood • Exchangeability • Utility
Axiomatic	• Random experiments	• Set theory and algebra • Algebra • Analysis	• Boolean diagrams and set notations	• Measurable function	• Sample space • Borel sets

of winning the cruise, the desire to win the most valuable prize can drive the purchase of the first product, ignoring the poor chance of winning. The magic operated by thought comes from the illusion of being able to change reality only by virtue of one's own desire. Even when the probability is known, so-called wishful thinking prevails.

An interesting starting point is the so-called perceptual theory (Zadeh, 2001). The human mind seems to be predisposed to ignore the rules of statistics and the probability calculus, which involves many errors and it would be said that there are cognitive illusions comparable to perceptual illusions. This is particularly true when we are faced with inaccurate probabilities or, as in the case presented below, there are ambiguities in the definition of the problem to be solved.

A particular situation is that relating to some paradoxical circumstances which confront us with situations that are not easily decipherable. The most classic example is due to Joseph Louis François Bertrand (1822-1900). The paradox arises from a seemingly simple question: if we take at random a chord of a circle, what is the probability that it is longer than the side of the inscribed equilateral triangle?

The stimulating aspect of the problem is that there are three possible solutions, all correct. This shows how the concept of probability presents ambiguities if the mechanism for defining random variables is not clear. The paradox is known but perhaps it is worth summing up the three solutions:

1. Taking the vertex A of the triangle as a reference point, a chord is drawn at random. If the end falls in the arc BC identified by the base of the triangle, the string is longer than the side. Since the three arcs are of equal length the probability is p = 1/3;
2. A chord parallel to one side of the triangle is drawn at random, if it is less than r/2 from the center the chord is longer than the side. For obvious reasons of symmetry the probability is p = 1/2;
3. A second circle is inscribed in the inscribed equilateral triangle. Only the chords passing through the inner circle are longer than the side of the triangle. Since the area of the small circle is exactly 1/4 of the large circle, the required probability is p = 1/4.

The three cases can be easily represented (Figure 2):

Figure 2. Bertrand Paradox

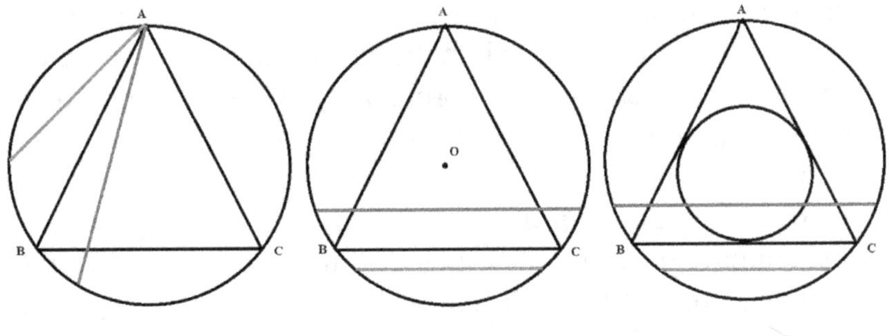

What is the correct solution? All three because they refer to three different random experiments.

Many scholars have studied which methods are most effective in teaching probability. According to Heinz Steinbring (1991), there are two main alternatives: the first is based on intuition through experiments, games of chance and real situations; the second focuses on concepts such as outcomes, events, event space, aimed at providing unambiguous foundations. The only thing that is certain is that, rightly so, research is still open and new ideas and perspectives are welcome, as there are many obstacles "Probability is increasingly taking part in the school mathematics curriculum; yet most teachers have little experience with probability and share with their students a variety of probabilistic misconceptions. Probability is difficult to teach for various reasons, including disparity between intuition and conceptual development even as regards apparently elementary concepts. Since an education that only focuses on technical skills is unlikely to help teachers overcome their erroneous beliefs, it is important to find new ways to teach probability to them, while at the same time bridging their content knowledge and their pedagogical content knowledge" (Batanero et al., 2005).

Returning to the content of the book, at least three key questions can be asked:

1. Which is the most appropriate approach: classic, frequentist or subjective?
2. What role can computer tools play in learning?
3. How can some artifacts help students understand the fundamental concepts of probability?

The first question can be answered in a simple way: it makes no sense to formulate a ranking in terms of didactics but we have to think about which are the best practices (Hulsizer & Woolf, 2009). It should be remembered that all three definitions are "constructivist", i.e. they indicate a way to calculate probabilities, only the axiomatic approach is not, but indicates the axioms to which a coherent system of probability calculation must be subject. For this reason only the first three definitions have found their place in this probability teaching text.

To explain this setting just provide a small exemplary problem, really used in class. A common deck of cards includes, for each suit, a King, Queen, Jack and ranks from one (Ace) to ten. If you draw two cards, what is the probability of having a figure (F) and a card (C) not a figure in your hand?

Presented in this way the problem appears to us as one of the usual examples that can be found in all textbooks, normally treated in combinatorial terms. Let us try to put the question in the opposite direction instead. First, you can ask students to make an estimate of the required value individually, then you can compare these estimates by opening a discussion. In other words, we are at a subjective level where the objective is to arrive at both an evaluation considered credible by most people and a discussion on how to verify its reliability.

The second step is to search, operationally, a possible solution through the repeated extraction of two cards from a deck recording the results. This operation is possible by distributing several decks of cards in order to have the result of a certain number of experiments quickly enough. The procedure is obviously long and quite boring if carried out over a long period, so it is preferable, after a while, to use a spreadsheet in order to obtain results that bring us closer to the empirical law of the case. With 5000000 "draws" the following values were obtained: FF = 5.3%, CC = 59.1%, FC + CF = 35.5%. At this point it is possible to proceed to the construction of the model corresponding to the a priori probability according to the classical approach by counting favorable and total cases. Once the first card has been chosen among the 52 of the deck, the second choice can be made among the remaining 51 but, since it does not matter the order, it must be divided by two; that is, we will have 1326 possible cases. Similarly we can calculate the FF pairs: 12 • 11 /2 = 66; the CC pairs: 40 • 39 /2 = 780; the other cases are given from 1326 - 66 -780 = 480. The theoretical probabilities are given by the favorable cases over the possible ones, so FF = 5%, CC = 59%, FC + CF = 36%. From this simple example we can derive many points of discussion, thinking about the path that led from the comparison of conjectures, through an experiment, to a classical formalization. It should be underlined that students normally estimate the probability of figure and paper between 15 and 20%, therefore with a definite underestimation. Underestimating or amplifying phenomena is a widespread trend (see the Prelec graph in chapter 5) and this shows a co-responsibility of the school.

The example of the cards may seem artificial and therefore we choose a problem more related to everyday life. Let's think about the correct answer to the question: given an individual who claims to have a gluten allergy (G), what is the probability that he really has an allergy (A)?

The answer is not trivial. It is estimated that the perception (subjective probability) of gluten allergy involves 20% of the population while some studies believe (frequentist probability) that the incidence of the phenomenon affects only 1% of the population. At this point we can get Bayes to help us. We can reasonably assume that those who are really allergic claim to be allergic, i.e. p(G|A) = 1. We can calculate the probability that those who think they are allergic are really allergic. Place p(G) = 0.02 and p(A) = 0.01 we can write:

$$p(A|G) = \frac{p(G\mid A)\bullet p(A)}{p(G|A)\bullet p(A) + p(G\mid \bar{A})\bullet p(\bar{A})} = \frac{1\bullet 0.01}{1\bullet 0.01 + 0.2\bullet 0.99} = 0.048$$

The answer to the second question should emerge clearly from the previous example of cards. Today more than ever the question arises: in the didactics of a teacher who wants to introduce probability in the classroom is the use of a computer necessary? It would be too trivial to answer yes. The answer, as often happens in mathematics, is ... it depends. Those who play cards probably read books on the subject, maybe they use dedicated software, but they cannot replace the materiality and physicality of those colored rectangles. This is not to diminish the role of instrumentation; research suggests that the use of automatic calculation tools and specific educational software in learning activity can make possible new ways of giving meaning to concepts and changing the interactions that take place in class between teacher and pupils and between pupils themselves.

In many parts of the book reference was made to simulations reporting the results, except that in one case no specialized software was used, but only a spreadsheet with macros. The choice of simulations was mainly used to support, integrate, and corroborate the concepts presented starting from the objects. When we talk about laboratory didactics we must not refer to a physical place with computers, tablets, calculators, or other devices, but a sort of place of ideas. When we talk about the laboratory, we mean a place of learning where we see it done and we do it. According to Arzarello (2006) .: "the *Mathematics Laboratory* is compared to what happened in the *Bottega d'Arte* of Renaissance artists, where the novices learned through a *cognitive apprenticeship*, namely by doing and watching what was done by experts".

In order to answer the third question, possible benefits should be assessed on the ground. For many years, the objects proposed in the text have been tested with different K6-12 classes with excellent results.

Doing research and developing deductions involves commitment and effort. Trying to identify objects and situations that speak of "different mathematics" offers more opportunities. The use of realia allows, as we have seen, to introduce in a concrete way concepts such as the foundations of probability, combinatorial calculus, Expected value, theoretical distributions of probability, etc..

In the text there was much insistence, not only on the choice of objects, but also on the idea that learning must follow a laboratory strategy. Of course, in a workshop it is not possible to fully bring out the close links between random events and the real world. However, it is possible to face a problem, as we have already had the opportunity to say, starting perhaps from classical probability, enriching the reasoning with the frequentist model, to arrive at subjective probability (or vice versa). The contrasts between these approaches are evident on a conceptual level, but less so on an operational level: when it comes to calculation, in the evaluation of probabilities there is not such a substantial difference and "objectivists" cannot disregard subjective elements, while "subjectivists" must take into account classical or frequentist evaluations in order to establish a reliable probabilistic forecast. As is known, one of the anecdotes circulating around the figure of Euclid says that King Ptolemy asked: " Is there a shorter road in geometry than ELEMENTS?. Euclid answered him, "Sire, there is no royal road direct path to geometry." Even in the case of probability there is no easy way. It is, however, a fairly widespread belief among those involved in teaching that the combination of concrete stimuli and the use of tools is fundamental to make the students' path less impassable. In this sense, the proposal of an approach that has in realia both the inspiration and the incentive for the development of a learning path should be understood.

The path is not without difficulties. As Guy Brousseau (1983) rightly says, in learning mathematics there are numerous obstacles with ontogenic, didactic and epistemological origin. These obstacles, all the more so, are found in the field of probability. The variables to be taken into account are many and often are innate with the gender composition of the class and, in a less evident way, age differences. A relatively recent South African research by Paul & Hlanganipai (2014) found that there was a significant difference in probability cognitive levels of students according to gender. This study found that there was no significant difference in probability cognitive levels of students according to grade". These factors, combined with social background, previous school experience, teachers' expertise, etc., make it difficult to provide general guidance.

The accumulated knowledge itself can be distorted in the sense that the student uses this knowledge to produce the correct answer in a given and known context. However, when this knowledge is used in another context, it can generate errors. An apparently banal example: the student knows that the probability of getting head in the toss of an equitable coin is 1/2 but if we ask him "after you have witnessed the HTTTHHTHTTTTTT sequence in 14 tosses of an equitable coin what would you bet on for the next toss? When flipping a coin, several students will think that a series of coin flips with an approximately equal number of heads and tails is more likely than a series with many more tails than heads and therefore should bet on H.

Four misconceptions are listed in an interesting study by Ang & Shahrill (2014): "Several research studies from different theoretical perspectives seemed to show that students tend to have perceptions which hinder their learning of Probability concepts. Some common ways of thinking which impede the learning of Probability are Representativeness, Equiprobability bias, Beliefs and Human control". Particularly interesting is the last point, to a precise question many answer that the results of objects that can be interpreted as random generators (coins, dice, roulette, Spinner) depend on how one handles these different devices. In fact, a discussion about the role that human control could have and that cannot be left out is almost never taken into consideration by teachers; it can be considered an axiom, that is a principle that can be admitted without discussion, evident in itself. In reality, it is not time wasted reflecting on this aspect. Let's take the case of roulette, according to some players the croupier would be able to direct the ball towards certain sectors; according to experts this is impossible because the lozenges on the edge prevent the ball from having a regular downward spiral. The diamond was inserted in the last century just to avoid that with launches at constant speed the croupier was able to direct the game. Other details are also influential, we have moved from ivory balls to synthetic materials with greater elasticity that makes the ball's stroke more unpredictable.

A second example can serve to clarify this *cognitive bias*, the mentalist Derren Brown showed in a video that he obtained a sequence of ten consecutive heads, event with probability $1/2^{10}$ » 0.001. This result led many to think that it was possible, for an expert, to manipulate the exit of the desired face. Later, Brown himself showed in a video that it took him nine hours to get that sequence!

If we want to formulate a final reflection we can start from a consideration of Kapadia & Borovcnik (2012) "Probability is the theory which organizes the world of chance phenomena. One has to start with phenomena which beg to be organized: intuition about unpredictable events, games of chance, occurrences which seem to happen without regularity, etc. One then has to teach the learner to constitute the mental object for himself/herself. This is the stage preceding concept attainment but too often ignored".

Winning the reluctance to overcome repetitive examples of textbooks to expand the horizon of the arguments is not simple and may seem unnecessarily tiring, but the results pay off the effort. Perhaps it is the best way to combat the innumeracy that Paulos (2001) complained about many years ago. For example, it can be significant, instead of a simple exercise, to propose a cooperative game. What we present here (Table 54) is based on a question that links probability and statistics: determine the age, gender, and geographical origin (North America, Central America, South America, Insular America) of five people on the basis of the information contained in the cards.

Each student is given a card (Table 1) that contains useful information, but not enough to solve the question. The information is different and only by sharing it can the solution be found.

Table 1. The six cards to distribute

Card 1 There are five people, living on the American continent, of whom we have this information: we know their life expectancy: A. 34 B. 43 C. 57 D. 34 E. 47 What is the age, gender, and geographical origin (North America, Central America, South America, Insular America) of the five people?	**Card 2** There are five people, living on the American continent, of whom we have this information: D is a 45-year-old female from Central America What is the age, gender, and geographical origin (North America, Central America, South America, Insular America) of the five people?
Card 3 There are five people, living on the American continent, of whom we have this information: The average age is 34 What is the age, gender, and geographical origin (North America, Central America, South America, Insular America) of the five people?	**Card 4** There are five people, living on the American continent, of whom we have this information: The youngest person, born in North America, is 20 years younger than the oldest person. What is the age, gender, and geographical origin (North America, Central America, South America, Insular America) of the five people?
Card 5 There are five people, living on the American continent, of whom we have this information: The difference in survival probability between males in their thirties shows a four-year advantage for those born in the North over those born in the South What is the age, gender, and geographical origin (North America, Central America, South America, Insular America) of the five people?	**Card 6** There are five people, living on the American continent, of whom we have this information: The sum of the ages of the females, born in the North, Centre, and Insular America, is 110 What is the age, gender, and geographical origin (North America, Central America, South America, Insular America) of the five people?

As you can see, some prerequisites such as "average" and "life expectancy" are necessary to understand the information. Every data is indispensable but taken individually is not enough. The solution is the result of a series of logical steps involving knowledge and understanding of the terms used, which is only possible if students cooperate. If the group has failed in the task, the following solution can be provided.

Card 2 shows the data of D. From cards 5 and 1 we can see that B and E are 30 years old and they are male; B is from the South and E from the North. The information taken from cards 5 and 3 allows to define that the sum of the ages of A, C and D is 110 and therefore, according to card 6, they are females. On the basis of the information of card 2 we deduce that the sum of the ages of A and C is 65. From cards 4 and 1 it results that C is the youngest person (highest life expectancy) and therefore she is 25 years old and lives in the North. Consequently, A is 40 and she was born in Insular America.

In conclusion you will have:

1. 40, female, Insular America
2. 30, male, South America
3. 25, female, North America
4. 45, female Central America
5. 30, male, North America

Working in groups, manipulating objects, exploiting digital resources should be the cornerstones of a constructive didactics able to stimulate curiosity, interest, and participation.

As an example of the importance of concretely stimulating students, I would like to show a fragment of the letter sent to me by an Italian student after a mathematics laboratory based on the use of objects: "Thanks to the meeting my mind has been opened so to say: who would ever know how much Mathematics can be found inside a road sign or inside a toy commonly used every day? Well, yes, now I know that the great Achilles could have run all his life and never reach the tortoise; I know how to calculate the perfect distance to admire a monument or a painting, and I can do all this thanks to just 2 hours that, almost mathematically speaking, are equal to a whole life".

A similar discourse can be made for laboratories more specifically dedicated to probability, if possible even more engaging. A phrase by the aforementioned Ivars Peterson (1998) can condense the meaning of this book: "Randomness, chaos, uncertainty, and chance are all a part of our lives. They reside at the ill-defined boundaries between what we know, what we can know, and what is beyond our knowing. They make life interesting".

In this regard it is useful to try to answer a question asked at the beginning: what are the simple and deep problems?

REFERENCES

Ang, L. H., & Shahrill, M. (2014). Identifying students' specific misconceptions in learning probability. *International Journal of Probability and Statistics*, *3*(2), 23–29.

Arzarello, F., Paola, D., & Robutti, O. (2006, December). Curricular innovation: an example of a learning environment integrated with technology. *Proceedings of 17th ICMI Study Conference*.

Batanero, C., Biehler, R., Maxara, C., Engel, J., & Vogel, M. (2005). Using simulation to bridge teachers content and pedagogical knowledge in probability. *15th ICMI Study Conference: The Professional education and development of teachers of mathematics*.

Batanero, C., Chernoff, E. J., Engel, J., Lee, H. S., & Sánchez, E. (2016). Research on teaching and learning probability. In *Research on teaching and learning probability* (pp. 1–33). Springer. doi:10.1007/978-3-319-31625-3_1

Batanero, C., & Díaz, C. (2007). The meaning and understanding of mathematics. In *Philosophical dimensions in mathematics education* (pp. 107–127). Springer. doi:10.1007/978-0-387-71575-9_6

Brousseau, G. (1983). Les obstacles épistémologiques et les problèmes en mathématiques. *Recherches en Didactique des Mathématiques*, *4*(2), 164–198.

France, A. (2015). *The Garden of Epicurus*. Andesite Press.

Hulsizer, M. R., & Woolf, L. M. (2009). *A guide to teaching statistics: innovations and best practices* (Vol. 10). John Wiley & Sons.

Jones, G. A. (Ed.). (2006). *Exploring probability in school: Challenges for teaching and learning* (Vol. 40). Springer Science & Business Media.

Kapadia, R., & Borovcnik, M. (2012). *Chance encounters: Probability in education* (Vol. 12). Springer Science & Business Media.

Paul, M., & Hlanganipai, N. (2014). The nature of misconceptions and cognitive obstacles faced by secondary school mathematics students in understanding probability: A case study of selected Polokwane secondary schools. *Mediterranean Journal of Social Sciences*, *5*(8), 446.

Paulos, J. A. (2001). *Innumeracy: Mathematical Illiteracy and Its Consequence*. Hill & Wang Pub.

Peterson, I. (1998). *The jungles of randomness: A mathematical safari*. Penguin.

Steinbring, H. (1991). The theoretical nature of probability in the classroom. In *Chance encounters: Probability in education* (pp. 135–167). Springer Netherlands. doi:10.1007/978-94-011-3532-0_5

Zadeh, L. A. (2002). Toward a perception-based theory of probabilistic reasoning with imprecise probabilities. In Soft Methods in Probability, Statistics and Data Analysis (pp. 27-61). Physica. doi:10.1007/978-3-7908-1773-7_2

KEY TERMS AND DEFINITIONS

Bertrand Paradox: Example in which, in order to well define the concept of probability, it is necessary to specify the mechanisms by which random variables are generated.

Cognitive Bias: Pattern of deviation from the norm or from objectivity in the judgment.

About the Author

Alessio Drivet obtained a degree at the University of Trento in 1975. Since 1976 he has been a lecturer in Mathematics, Probability and Statistics at the Commercial Institutes, retired in 2011. He has published books and articles on mathematics and computer science at various publishing houses. Since 1987 he has been engaged in contract training activities for the University of Turin and the Italian Ministry of Education. He is part of the scientific and organizational committee of DI.FI.MA and is a member of the GeoGebra Institute of Turin. He is currently working on the diffusion of mathematics among secondary school students.

Index

www.ingramcontent.com/pod-product-compliance
Lightning Source LLC
LaVergne TN
LVHW080341011026
849161LV00009B/2147